PID CONTROL, IMPLEMENTATION AND TUNING

Edited by **Tamer Mansour**

PID Control, Implementation and Tuning
Edited by Tamer Mansour

Published by InTech
Janeza Trdine 9, 51000 Rijeka, Croatia

Publishing Process Manager Ivana Lorkovic

Technical Editor Teodora Smiljanic

Cover Designer InTech Design Team

Additional hard copies can be obtained from orders@intechopen.com

PID Control, Implementation and Tuning, Edited by Tamer Mansour
p. cm.
ISBN 978-953-307-166-4

For EU product safety concerns:
IN TECH d.o.o., Prolaz Marije Krucifikse Kozulić 3, 51000 Rijeka, Croatia,
info@intechopen.com or visit our website at intechopen.com.

Meet the editor

Dr. Tamer Mansour graduated from Tohoku University at 2008. Since then, he had been involved with the Aerospace Engineering Department at Tohoku University as a visiting researcher. He had published many papers in international journals like "Advanced Robotics" and "Journal of Robotics and Mechatronics." He served as a reviewer for "Journal of Sound and Vibration" and "Robotica." He had the experience in teaching and assisting during the period from 1996-2004. Since October 2004, he started his Ph.D. course and finished in July 2008. During this period, he had the experience as teaching assistant and research assistant in the graduate school of Engineering in Tohoku University. His major fields of interest include control of flexible manipulators and application of intelligent controllers in space robots, PID control, Mechatronics Systems.

Contents

Preface

The Proportional, Integral and Derivative –PID– controller is the most widely used controller in industrial applications. Since its first appearance in the late nineteen century, it had attracted researchers from all over the world because of its simplicity and the ability to provide an excellent control performance. The PID controller now represents more than ninety percent of the controllers used in the market.

This book is a result of contributions and inspirations from many researchers worldwide in the field of control engineering. The book consists of two parts; the first is related to the implementation of PID control in various applications whilst the second part concentrates on the tuning of PID control to get best performance.

Firstly, I wish to thank the authors, who contributed to the production of this book. Also, I wish to convey deepest gratitude to reviewers who devoted their time to review the manuscripts and selected the best of them. We hope that this book can give aid to new research in the field of PID control, in addition to stimulating research in the area of PID control for better utilization in the industrial world.

Tamer Mansour

Part 1

PID Implementation

1

Multivariable PID control of an Activated Sludge Wastewater Treatment Process

[1]Norhaliza Abdul Wahab, [2]Reza Katebi and [2]Jonas Balderud
[1]*Department of Control and Instrumentation Engineering, Faculty of Electrical Engineering, Universiti Teknologi Malaysia, Johor, Malaysia*
[2]*Industrial Control Centre, Dept of Electronic and Electrical Engineering, University of Strathclyde, Glasgow, UK*

1. Introduction

In general, wastewater treatment plant (WWTP) consists of several stages before it is released to a receiving water body. There are, preliminary and primary treatment (mechanical treatment), a secondary treatment (biological treatment) and a tertiary treatment (chemical treatment). In this chapter, since the work involve of identification and control design of activated sludge process to improve the performance of the system, and most of the control priorities are centred on the biological treatment process, only the secondary treatment will be highlighted.

System modelling and identification of the activated sludge process has provided a wider understanding and a powerful tool to predict the behaviour of the system under different conditions. In control design, system modelling and identification are the most important parts which need be taken into account. Often, models developed for controller design have to be as simple as possible. The simplicity means models can be obtained directly from input-output (or experimental) data and used for control design of WWTP. This type of model is called *black box* or data-driven model, see for example Box Jenkins (El-Din et al., 2002) and Artificial Neural Network (ANN) proposed by (Cote et al., 1995). It will be shown that the identified data-driven control model describes the activated sludge wastewater system well, at least around an operating point. One of the popular techniques used in the system identification is the subspace identification algorithm and this algorithm is used for the design of control in WWTP. Another approach to modelling is to use model reductions or simplifications. The reduced order (linear) model can be later used for controller design and/or stochastic simulation, see for example (Robertson and Cameron, 1996).

The biochemical processes involved in the activated sludge wastewater treatment process are complex and their understanding was very limited. However, due to the importance of providing concise and efficient information in describing a complicated set of activated sludge system behaviours, several mathematical models have been developed for gaining a better understanding of a real system. In the late 80s, a more scientific perspective of this biotechnology process was achieved by the first development of International Association for Water Quality (IAWQ) Activated Sludge Model no.1(ASM1) proposed by (Henze et al.,1987), followed by a series of mathematical models known as ASM2 and ASM3. Such advanced models of activated sludge processes, i.e. ASM1, have been developed over the years but have not been used for control design due to their high complexity.

As previously mentioned, models developed for controller design have to be as simple as possible. This work attempts to identify simple data-driven control model of activated sludge system. The multivariable identification is performed into a wastewater system using subspace identification technique that provide multivariable model for designing of multivariable PID controller. PID controller is one of the popular conventional methods used from several decades ago. The implementation of this form of feedback controller have been widely used in any industrial processes. Often, this controller is implemented as a local controller, whereby the PID controller is cascaded with the more advanced control method such as model predictive control (MPC). In that case, there are two different control loops in the system that is outer loop (MPC) and inner loop (PID). The outer loop will decide what is the setpoint to be given to the PID control loop. In such cases, the response time of the control variable in the inner loop must be much faster than that given by the outer loop. In any process control such as wastewater treatment plant, scalar PID controller is extensively used to control the process variables of wastewater system. Unfortunately, this type of controller is of no longer sufficient due to the inherently multivariable nature of wastewater system.

For highly multivariable process of wastewater treatment plant, multivariable control systems are therefore needed to handle the inevitable changes in the plant and its effluent characteristics. In literature, several control strategies of interest have been developed to improve effluent quality control of activated sludge wastewater treatment system given by (Chotkowski, W et. al, 2005), (Y. Ma et al., 2005), (Piotrowski, R. Et al., 2005), (A. Stare et al., 2007) and (E. Mats et al., 2006). (A. Stare et al., 2007) for example, reports that the application of advanced control becomes more cost effective despite the need for possible investment in purchasing additional sensors and actuators. This motivate to the use of data-driven control model for the activated sludge process using MPID controller. In multivariable PID control, the control handles more than one input and output in the systems and hence there are usually a number of interacting control loops in the system. This process interaction is of importance issues need to be taken into consideration to ensure better performance of the closed loop plant as well as to meet the current and future demands on effluent water quality. The work of this chapter highlights the effectiveness of using multivariable PID (MPID) control design with the application to activated sludge wastewater treatment process. The design of MPID controller is performed using data driven models developed from system identification techniques based on subspace approach.

2. Activated sludge wastewater treatment systems

The activated sludge process is a biological process in which an organic matter is oxidised and mineralised by microorganisms. Oxygen is used by microorganisms to oxidise organic matter. The influent of particulate inert matter and the growth of the microorganisms is removed from the plant as excess sludge to maintain a reasonable suspended solids concentration. A simple activated sludge is usually comprised of an aerator and a settler. The bioreactor includes a secondary clarifier (or settler) that serves to retain the biomass in the system while producing a high quality effluent. Part of the settled biomass is recycled to allow the right concentration of microorganisms in the aerated tank. In practice, more than one reactors are commonly applied in the activated sludge process for simultaneous nitrification and denitrification such as one designed in the benchmark COST simulation.

2.1 Benchmark COST simulation

A schematic depicting the COST simulation benchmark model is shown in Fig. 1. There are five series biological reactors (or bioreactor) which contain two anoxic and three aerobic tanks and a 10-layer non-reactive secondary settling tank. A pre-denitrifying plant structure has been applied, whereby anoxic process is located at the beginning of the tank, as seen in Fig. 1.

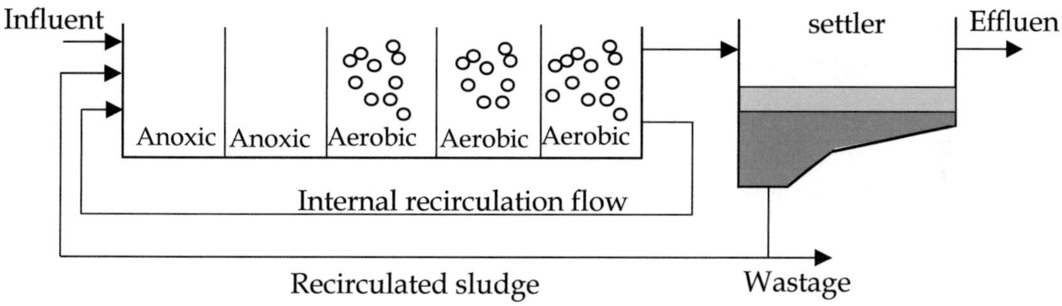

Fig. 1. Activated sludge with pre-denitrification

Each unit of the bioreactor is modelled using IAWQ's ASM1 given by (Henze et al., 1987). The settler is modelled using a double-exponential settling velocity function by (Takács et al., 1991). The bioreactor of ASM1 model describes the removal of organic matter, nitrification and denitrification. To allow for consistent experiment evaluation, the model provides three dynamic data influent flow conditions (or disturbances) and each is meant to be a representative of a different weather condition: dry, rain and storm. For a detailed description of the COST simulation benchmark models, see (Copp, 2002).

2.2 Control structures of activated sludge with pre-denitrification

Two different control structures for the activated sludge process are studied. These structures of multivariable control are developed using subspace identification which later used for MPID controller design.

Case 1

The controller maintains the DO levels in the last three aerobic tanks as seen in Fig. 1, by manipulation of oxygen transfer coefficients (K_{La}). Models are developed at three different operating conditions, i.e. constant influent flow, dry influent flow and rain influent flow conditions.

Case 2

In this case, the simultaneous control of DO level (DO_5) in the last aerobic tank and the control of nitrate (S_{NO2}) level in the second anoxic tank are considered by manipulation of oxygen transfer coefficient (K_{La5}) and internal recirculation rate (Q_{intrn}). Models are developed for two different operating conditions, i.e. constant influent flow and dry influent flow.

3. Subspace method of System Identification

Subspace identification techniques have been (more than 10 years old) developed and have attracted much attention due to their computational simplicity and effectiveness in identifying dynamic state space linear multivariable systems. The subspace identification technique was developed by (De Moor et al., 1988), (Moonen et al., 1989) and (Verhaegen, 1994) and widely known as direct subspace state space system identification (4SID) methods. The advantage of a subspace method is that it is based on reliable numerical algorithms of the QR decomposition and the singular value decomposition (SVD). Moreover, this algorithm can easily be implemented for multi input multi output (MIMO) system identification. The subspace identification uses projection methods and SVD to obtain the model. The identified models in discrete time describe the activated sludge process around an operating point and have been converted to standard continuous linear time invariant state space system:

$$\dot{x}(t) = Ax(t) + B_p u(t) + B_d d(t) \tag{1}$$

$$y(k) = Cx(t) \tag{2}$$

where $x(t)$ is the state vector, $u(t)$ is the input vector, $y(t)$ is the output vector and d(t) is the measurable disturbance vector. A, B_p, B_d and C are matrices of appropriate dimensions. Combining the inputs into a single vector gives the following:

$$\dot{x} = Ax(t) + \begin{pmatrix} B_p & B_d \end{pmatrix} \begin{pmatrix} u(t) \\ d(t) \end{pmatrix} \tag{4}$$

$$y(t) = Cx(t) \tag{5}$$

The system transfer function is defined as:

$$G(s) = C(sI - A)^{-1} \begin{pmatrix} B_p & B_d \end{pmatrix} \tag{6}$$

The COST simulation benchmark is used as a data generator for multivariable identification in the activated sludge process. For a better identification result, the data is pre-processed. In this system which is running at steady state operating point different from zero and hence introducing some DC offsets, subtraction of the sample mean from the data set is done in order to remove these offsets. This is common operation in system identification, as given by (Söderström and Stoica, 1989). In this work, as the data set is generated from a simulation model, no data filtering is necessary. The data set is finally detrended to remove linear trends from input-output data before it can be later applied to the identification algorithm. The results of model identifications for Case 1 for dry and rain scenarios and for Case 2 for dry influent flow are shown in Fig. 2 (a-c). The sampling time were adjusted to 0.001 days for Case 1 and 0.01 for Case 2. The figure shows only the model responses for aerated tank 4 (DO_4).

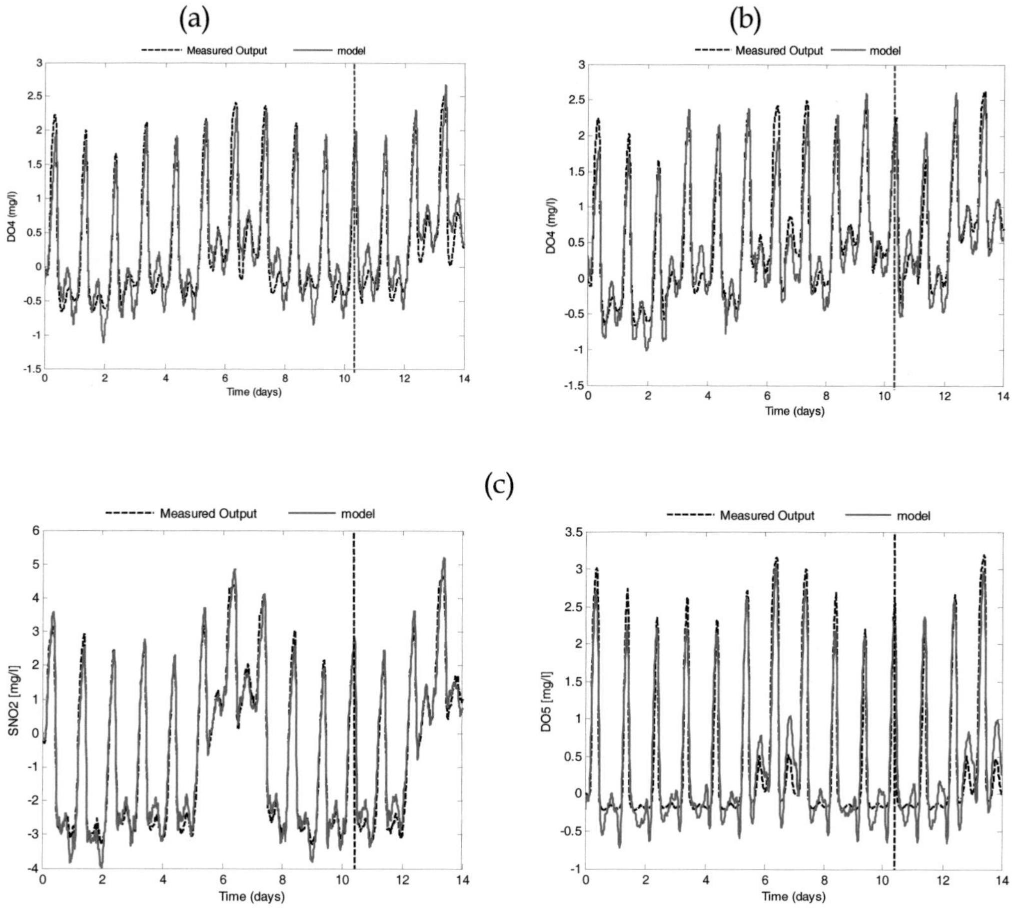

Fig. 2. Response comparison of dynamic influent flows for Cases 1 and 2 - (a) Case 1- dry weather; (b) Case 1- rain weather; (c) Case 2- dry weather

Almost similar results were obtained for the other two outputs (DO_3 and DO_5). In Case 2, the responses are presented for both outputs (S_{NO2} and DO_5). In dry influent flow, the model identification uses 3/4 of the generated data and the other 1/4 are used for validation. As it can be observed, the identified model for a given operating conditions correctly reproduces the main dynamic characteristics of the activated sludge process. In both cases, the simulation started at zero initial conditions. The performance quality of the models are performed by measuring percentage Variance Accounted For (VAF) as follows:

$$VAF(\%) = \left(1 - \frac{\text{var}(y - \hat{y})}{\text{var}(y)}\right) * 100 \tag{7}$$

where y and $\hat{y}$ are the measured outputs and predicted outputs, respectively. The best-identified models are demonstrated by smaller deviations obtained between y and $\hat{y}$ as shown in Tables 1 and 2.

Model	**Order**	**DO3**	**DO4**	**DO5**
Constant	6	96.65	96.05	91.4
Dry	4	87.81	88.85	84.84
Rain	4	87.28	89.41	82.83

Table 1. Multivariable DO model identification (%VAF) validation results (Case 1)

Model	**Order**	**DO5**	**SNO2**
Constant	4	92.23	97.03
Dry	4	88.42	85.63

Table 2. Multivariable DO-Nitrate model identification (%VAF) validation results (Case 2)

On average, good models were obtained from a given percentage of VAF at around 85% and above. The identified models obtained were controllable and observable. In both cases, the best responses were obtained for models of order 4 for dynamic influent (i.e. dry and rain) whilst models of order 6 and 4 for constant influent in Case 1 and Case 2, respectively. The poles (eigenvalues of A) shows that both cases the models are open-loop stable. The interaction measure using the Relative Gain Array (RGA) is studied in the following Section.

3.1 Model analysis

The interaction analysis is of importance when considering multivariable systems. The RGA analysis should not be interpreted as drawing specific conclusions about the control design but rather it is an indication of how inputs and outputs are interacting and hence the most appropriate control structure can be selected. The most widely used interaction measure for multivariable linear systems so far, is the RGA introduced by (Bristol, 1996).

3.1.1 Steady state analysis

The steady state RGA(0) can be calculated as follows:

$$G(0) = -CA^{-1}B_p \tag{8}$$

$$RGA(0) = \left(-CA^{-1}B_p\right) \circ \left(-CA^{-1}B_p\right)^{-T} \tag{9}$$

where *G(0)* is the steady state transfer function matrix and $\circ$ denotes the Schur product (i.e. element-wise multiplication). It can be noted that the calculation for RGA is displayed with three decimal points.

Case 1

The steady state RGA, $\Lambda(0)$ was calculated for different operating points, i.e. constant, dry and rain influent data sets as follows:

$$\Lambda^{const}(0) = \begin{bmatrix} 1.042 & -0.017 & -0.024 \\ -0.009 & 1.059 & -0.049 \\ -0.032 & -0.041 & 1.073 \end{bmatrix} \quad (10)$$

$$\Lambda^{dry}(0) = \begin{bmatrix} 1.508 & -0.104 & -0.401 \\ 0.560 & 2.076 & -1.640 \\ -1.068 & -0.974 & 3.041 \end{bmatrix} \quad (11)$$

$$\Lambda^{rain}(0) = \begin{bmatrix} 1.588 & -0.052 & -0.540 \\ 0.185 & 1.492 & -0.675 \\ -0.777 & -0.439 & 2.217 \end{bmatrix} \quad (12)$$

Clearly, most of the off-diagonal elements in the RGA matrix corresponding to the above operating points are negative. For both dry and rain data sets, large values on the diagonal and some negative values on the off-diagonal means that the system is difficult to control using non-interacting control structure since the process exhibit strong and difficult interactions. Here, the RGA matrix represents a system with various extents of interactions: dry influent indicates the strongest interaction within control loops, following by a moderate interaction for rain condition. The lowest interaction is thereby illustrated by constant influent flow.

Case 2

The analysis of interaction for Case 2 is slightly different from Case 1 so as to allow investigations into the effect of nonlinearities. In this case, the simultaneous controls of nitrate (SNO_2) level in the second anoxic tank and DO (DO_5) level in the last aerobic tank is considered using the manipulation of internal recirculation rate and oxygen transfer coefficient, respectively. Models are developed for two different operating conditions, i.e. constant influent flow and dry influent flow. Under constant influent, three different operating points (refered $\tilde{u}_1, \tilde{u}_2$ and $\tilde{u}_3$) are considered to cover a wider range of operating points, i.e. i.e. $\tilde{u}_1 = \begin{bmatrix} 57552 & 88 \end{bmatrix}^T$; $\tilde{u}_2 = \begin{bmatrix} 58104 & 210 \end{bmatrix}^T$ and $\tilde{u}_3 = \begin{bmatrix} 83007 & 84.84 \end{bmatrix}^T$. $\Lambda(0)$ were obtained as follows:

$$\Lambda_{\tilde{u}_1}^{const}(0) = \begin{bmatrix} 1.031 & -0.031 \\ -0.031 & 1.031 \end{bmatrix} \quad (13)$$

$$\Lambda_{\tilde{u}_2}^{const}(0) = \begin{bmatrix} 0.682 & 0.318 \\ 0.318 & 0.682 \end{bmatrix} \tag{14}$$

$$\Lambda_{\tilde{u}_3}^{const}(0) = \begin{bmatrix} 0.948 & 0.052 \\ 0.052 & 0.948 \end{bmatrix} \tag{15}$$

The off-diagonal elements in the RGA matrix corresponding to the first operating point are negative and the diagonal elements are close to one, the RGA in this case suggests a diagonal controller; that is, Q_{intrn} should control nitrate concentration and, KLa_5 should be used to control DO concentration. For the second operating point, the diagonal elements are quite far from one and a big value in the off-diagonal elements indicates strong interaction between the control loops. This indicates that a full multivariable control structure is required. The diagonal elements in RGA for the third operating point are also close to one with low interaction in control loops. In the following study, the second operating point will be considered for control design. In addition to that, $\Lambda(0)$ for dry influent flow is as follows:

$$\Lambda^{dry}(0) = \begin{bmatrix} 1.558 & -0.558 \\ -0.558 & 1.558 \end{bmatrix} \tag{16}$$

The analysis for the dry influent flow shows almost identical results to constant flow whereby, the anti-diagonal elements in the RGA matrix are negative.

3.1.2 Dynamic RGA analysis

Effective control at nonzero frequencies can be studied using the dynamic RGA. Since the controller design methods investigated in this paper require system decoupling at specific frequencies, it is useful to examine dynamic RGA and use the resulting information to decouple the system at frequency points with highest interactions. In the dynamic RGA, the plant gain, G is allowed to be measured at any frequency, w. This dynamic version is the extension of the RGA and was proposed by (Kinnaert, 1995) (see reference for a more complete discussion). Not surprisingly, the dynamic version of RGA possesses the same properties as the steady state RGA and is defined as:

$$RGA\big(G(iw)\big) = G(iw) \circ \left(G(iw)^{-1}\right)^{-T} \tag{17}$$

In this case, this RGA version is also denoted by $\Lambda(G)$. It is advisory to study this dynamic RGA which can provide useful information about the behaviour of $\Lambda(G)$ in the interesting frequency range. The $\Lambda(G)$ has been evaluated in both cases of 1 and 2.

Case 1

The dynamic study of RGA is evaluated in this case for the three influent flow conditions: constant, dry and rain. Fig. 3 (a-b) shows the behaviour of the real part of $\Lambda(G)$ for dry and rain respectively, over different frequency ranges.

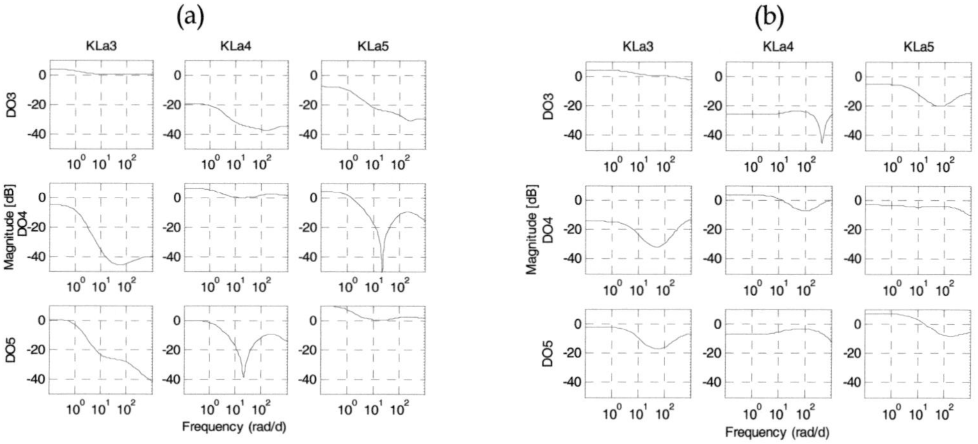

Fig. 3. DRGA gains for dynamic influent flows- a) Case 1- dry weather and b) Case 1 - rain weather

It can be clearly seen that for low and middle frequencies (between 10^{-1}rad/d and 10^{1}rad/d) even higher frequencies the real part is very close to zero for constant influent flow. Hence, the RGA does not suggest a different pairing dynamically than statically. The real part of diagonal elements in both scenarios of dry and rain indicate the process exhibits strong and difficult interactions. For higher frequencies the two dynamic influent conditions (dry and rain) have a real part of $\Lambda(G)$ with a deep valley in some part of the off-diagonals. The curve corresponding to the constant influent flow does not have this property. Overall, dynamic analysis demonstrates that the interactions occur mainly at frequencies about a decade below the open loop bandwidth. Therefore, the low frequency decoupling is most likely to decentralise the control system and minimise the effect of interactions.

Case 2

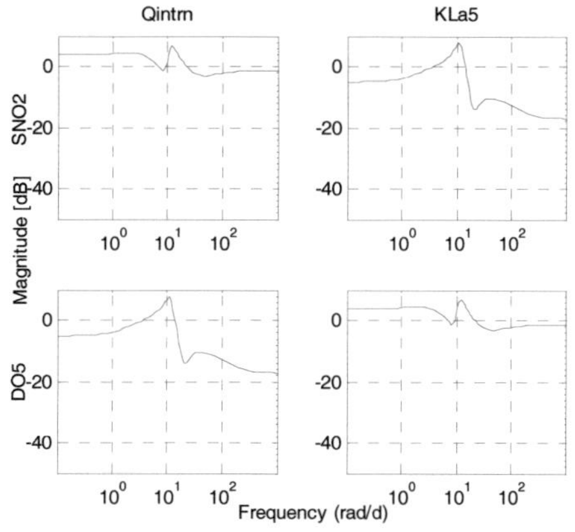

Fig. 4. DRGA gains for dynamic influent flow- a) Case 2- dry weather

The dynamic behaviour of the real part is studied under dry influent flow condition as shown in Fig. 4. Nothing of interest happens for the relevant low and intermediate frequency parts for both conditions in this case and it can be conclude that the plots demonstrate the interactions occur mainly at frequencies about a decade below the open loop bandwidth. The low frequency decoupling is therefore most likely to decentralize the control system and to minimise the effect of interactions.

4. MPID Control Design

In an attempt to improve the industry acceptance of multivariable control techniques, this study investigates three existing multivariable tuning methods and proposes a new one. These methods require only simple data-driven model of step or frequency response type. Most of the existing controller on WWTPs are not designed or tuned effectively. Hence, a systematic control design method is proposed, which reduces the controller commissioning time as well as the tuning efforts. The methods considered are those suggested in (Davison, 1976), (Penttinen and Koivo, 1980) and (Maciejowski, 1989) and these are compared with a new proposed method.

The design of MPID controllers is best carried out using simple linear models which can be derived from step or frequency tests. These models are usually valid for a single operating point and the procedure should be repeated for other points of interest. Linear models can also be derived by linearising the ASM model around a desired operating point but the resulting model requires to be reduced in size and validated using real data. Hence, the use of data-driven model is preferred. The motivation for using data-driven model is to gain additional insight into the dynamic behaviour of the WWTP and to allow for a more precise determination of the best tuning parameters for each control technique investigated, where the latter will subsequently enable a more objective comparison of the control techniques. Disturbances, in the form of variations of the influent flow rate, Q_{in}, influent ammonium concentration, S_{NH} and influent substrate S_S are considered in this study. The loop interactions are taken into account to determine suitable controller structures for a more effective decoupling.

4.1 Tuning methods

This section study tuning of control structures for multivariable systems. For controller tuning, simplicity, as well as optimality, is important. Our intention is to present a framework for multivariable PID controller design which is simple to understand and implement. The control structures and tuning methods investigated in this study are briefly described below.

4.1.1 Davison method

The Davison method uses only integral action. The control law is given by:

$$\underline{u}(s) = K_i \frac{1}{s} \underline{e}(s) \tag{18}$$

where $K_i = \varepsilon G^{-1}(0)$ is the integral feedback gain, *G(0)* is the zero frequency gain of the open loop transfer function matrix, *G(s)*, and *e(s)* denote the output error. The scalar parameter $\boldsymbol{\varepsilon}$ is the tuning parameter. Since the integral gain is proportional to the inverse of the plant dynamics at zero frequency, this method is expected to provide good decoupling characteristics at low frequencies.

4.1.2 Penttinen – Koivo method

The Penttinen- Koivo is slightly more advanced than the Davison method. A proportional term has been added to the control law, giving:

$$\underline{u}(s) = K_c\underline{e}(s) + K_i\frac{1}{s}\underline{e}(s) \tag{19}$$

where, $K_c = \rho\left(CB_p\right)^{-1}$ and $K_i = \varepsilon G^{-1}(0)$. The Davison and Penttinen are similar in the sense that the integral gains of both controllers are linearly related to the inverse of the plant dynamics at zero frequency, and both controllers are therefore expected to provide good control-loop decoupling characteristics at low frequencies. Unlike the Davison, the Penttinen controller also includes proportional control action, where the feedback gain is linearly related to the inverse of the plant dynamics at high frequencies. Therefore, by following the same line of reasoning as above, the latter controller is expected to exhibit good decoupling characteristics at high frequencies. The term CB_p represents the initial slope of the step output response, i.e.:

$$CB_p = \begin{bmatrix} \dot{y}_{1,1} & \cdots & \dot{y}_{1,m} \\ \vdots & \ddots & \vdots \\ \dot{y}_{m,1} & \cdots & \dot{y}_{m,m} \end{bmatrix} \tag{20}$$

where m is the system order and $\dot{y}_{i,j}$ is the initial slope of output, i, in response to a step at input, j. It can be shown that CG_p is the inverse of the plant dynamics at high frequencies by writing the Laurent series expansion of the transfer function G(s) as follows:

$$G(s) = \frac{CG_p}{s} + \frac{CFG_p}{s^2} + \frac{CF^2G_p}{s^3} + \ldots \tag{21}$$

A good approximation of $G(s)$ at high frequencies is $G(s) = CB_p/s$ is given by (21). As K_i/s terms are also negligible at high frequencies compared to K_c, so it can be concluded that $G(s)K_c = I/s$, thus giving the following closed-loop transfer function:

$$(I+GK)^{-1}GK = \begin{bmatrix} H_1(s)\cdots & & 0 \\ \vdots & \ddots & \vdots \\ 0 & \cdots & H_n(s) \end{bmatrix} \quad \text{for large } s \tag{22}$$

The tuning parameters, ρ and ε can be used to tune the proportional and integral gains.

4.1.3 Maciejowski method

M3 extends M2 to non-zero frequencies and hence the controller gains are linearly related to the inverse of the plant dynamics at a particular design frequency, w_b, i.e.

$K_c = \rho G^{-1}(jw_b)$, and $K_i = \varepsilon G^{-1}(jw_b)$. The calculation $G^{-1}(jw_b)$ will typically lead to a complex matrix, and hence a real approximation of $G^{-1}(jw_b)$ is required. This can be achieved by solving the following optimisation problem:

$$J(K,\Theta) = \left[G(jw_b)K - e^{j\Theta}\right]^T \left[G(jw_b)K - e^{j\Theta}\right], \qquad (23)$$
$$\Theta = diag(\theta 1, ..., \theta n)$$

By appropriately selecting the matrix K to minimise J the product of $G(jw_b)$ and K will be close to the identity matrix at the design frequency, and therefore this will provide good control-loop decoupling characteristics around this frequency. This method suffers from a non-trivial frequency analysis.

4.1.4 A proposed new method

Before entering the method description, a short remark on the relevance of the problem is presented. Nowadays many wastewater treatment plants use very simple control technologies such as PID control. To this point, the study presented herein is then an attempt to give a quantitative basis, as rigorously as possible, to a practice that is widely adopted in industrial process. The initial benchmark result indicates that a multivariable PID controller was very effective for the control problem posed by the WWTP benchmark problem. The studied control design strategies presented a reasonable performance of system. Since the main characteristic of the proposed approach is to improve control performance while retaining the simplicity of the multiloop strategy, it will involve enhancements to the PID control calculations; such that, we try to combine some specification of different existing methods to obtain both a good performance of control as well as disturbance rejection, also to minimise the interaction. To devise the proposed method, some quantities useful to characterise an existing tuning method is discussed. The Davison is of no use where integrators are present in the process. Penttinen-Koivo requires the system that have a high frequency motion. The design technique proposed by Maciejowski approximates decoupling at a selected frequency. It has many tractable properties and an intuitive control structure. Initial results also indicated that the controller was effective only for the control problem where all the loops have similar bandwidth frequencies and it also requires a rigorous frequency analysis. This work therefore proposes a new control design technique that retains some of the properties that makes the Maciejowski controller tractable, but eliminates the need for frequency analysis and it is more effective for systems which have control loops of different bandwidths. The proposed control design technique assumes the following control structure:

$$\underline{u}(s) = \underline{e}(s)\left(\rho K + \varepsilon K \frac{1}{s}\right) \qquad (24)$$

where,

$$K = \left[\alpha G(0) + (1-\alpha) C G_p\right]^{-1} \qquad (25)$$

The proportional and integral feedback gain of the proposed controller is a blend between the inverse of the plant dynamics at zero frequency and the inverse of the plant dynamics at high frequency. Thus, provided the plant have low-pass frequency characteristics, a good approximation of $G^{-1}(jw_b)$ can be obtained by appropriately selecting the additional controller tuning parameter, $\alpha \in \begin{bmatrix} 0 & 1 \end{bmatrix}$.

4.2 Optimal tuning of MPID controller

To allow for an objective comparison of the performance achieved by the MPID controllers, the tuning parameters for each controller has been adjusted such that the following penalty function, J is minimised:

$$J = \int_0^{\infty} \tilde{x}(t)^T Q \tilde{x}(t) + u(t)^T R u(t) \tag{26}$$

where (26) minimises the energy corresponds in some sense to keep the state and the control close to zero. $\tilde{x}(k) = \begin{bmatrix} x(k) & v(k) \end{bmatrix}^T$ denotes the controller integrator states. The weighting matrices, Q and R, are non-negative definite symmetric matrices; tuned in such a way that a satisfactory closed loop performance is obtained. In this case, we obtain Q = diag (10^6, 10^6, 10^6) and R = $0.001I$ that produces good performance. It was assumed that the process dynamics and controller states could be described using:

$$\dot{\tilde{x}}(t) = A\tilde{x}(t) + Bu(t) \tag{27}$$

$$y(t) = C\tilde{x}(t) \tag{28}$$

Under these assumptions the MPID control laws could be expressed as:

$$u(t) = -K\tilde{x}(t) \tag{29}$$

$$\dot{u}(t) = -K\dot{\tilde{x}}(t) = -K\left(A\tilde{x}(t) + Bu(t)\right) \tag{30}$$

where $K = [K_c \; K_i]$. The penalty function may be expressed in terms of K as:

$$J = \int_0^{\infty} \tilde{x}(t)^T \left(Q + K^T R K\right) \tilde{x}(t) = \tilde{x}(t)^T P \tilde{x}(t) \tag{31}$$

By assuming that the closed loop system is asymtotically stable so that J becomes:

$$J = x(0)^T P x(0) \tag{32}$$

where P denotes the solution to the following steady state Lyapunov equation:

$$A_c^T P + P A_c + Q + K^T R K = 0 \tag{33}$$

where $A_c = A - BK$. Thus, for each MPID control scheme, the controller parameters, φ is selected such that the matrix norm of P is minimised, i.e.:

$$\min_{\varphi} \|P\|, \tag{34}$$

where φ is given in Table 3 and Table 4 for both Cases 1 and 2, respectively.

	Constant	**Dry**	**Rain**
M1	$K_i = \begin{bmatrix} 68.11 & 0.238 & -6.30 \\ -18.7 & 64.95 & -16.63 \\ -5.09 & -11.7 & 62.33 \end{bmatrix}$ $\varepsilon = 126$	$K_i = \begin{bmatrix} 59.14 & 26.06 & -40.77 \\ -37.30 & 124.33 & -65.83 \\ -9.23 & -23.39 & 28.20 \end{bmatrix}$ $\varepsilon = 195$	$K_i = \begin{bmatrix} 75.294 & 13.37 & -41.365 \\ -18.162 & 80.695 & -39.243 \\ -20.946 & -19.925 & 38.292 \end{bmatrix}$ $\varepsilon = 239$
M2	$K_c = \begin{bmatrix} 0.164 & 0.004 & 0.01 \\ 0.0 & 0.183 & 0.003 \\ 0.003 & 0.002 & 0.144 \end{bmatrix}$ $K_i = \begin{bmatrix} 68.11 & 0.238 & -6.30 \\ -18.7 & 64.95 & -16.63 \\ -5.09 & -11.7 & 62.33 \end{bmatrix}$ $\rho = 283 \quad \varepsilon = 545$	$K_c = \begin{bmatrix} 0.201 & -0.04 & -0.02 \\ 0.009 & 0.127 & -0.232 \\ -0.019 & -0.014 & 0.297 \end{bmatrix}$ $K_i = \begin{bmatrix} 59.14 & 26.06 & -40.77 \\ -37.30 & 124.33 & -65.83 \\ -9.23 & -23.39 & 28.20 \end{bmatrix}$ $\rho = 166 \quad \varepsilon = 510$	$K_c = \begin{bmatrix} 0.159 & -0.088 & 0.185 \\ -0.063 & 0.113 & 0.002 \\ -0.041 & -0.043 & 0.131 \end{bmatrix}$ $K_i = \begin{bmatrix} 75.294 & 13.37 & -41.365 \\ -18.162 & 80.695 & -39.243 \\ -20.946 & -19.925 & 38.292 \end{bmatrix}$ $\rho = 500 \; \varepsilon = 784$
M3	$K = \begin{bmatrix} 0.013 & -0.0 & -0.001 \\ 0.001 & 0.013 & -0.0 \\ 0.0 & 0.001 & 0.011 \end{bmatrix}$ $\rho = 4800 \; \varepsilon = 2581800 \; \omega = 53$	$K = \begin{bmatrix} 0.014 & 0.0 & 0.0 \\ 0.003 & 0.015 & 0.002 \\ 0.001 & 0.002 & 0.014 \end{bmatrix}$ $\rho = 4000 \; \varepsilon = 1326700 \; \omega = 50$	$K = \begin{bmatrix} 0.013 & 0.002 & 0.004 \\ -0.001 & 0.016 & -0.0 \\ 0.003 & 0.011 & 0.031 \end{bmatrix}$ $\rho = 9300 \; \varepsilon = 1733700 \; \omega = 100$
M4	$K = \begin{bmatrix} 37.029 & 0.165 & -0.860 \\ -6.271 & 37.464 & 0.004 \\ -1.665 & -3.497 & 32.878 \end{bmatrix}$ $\rho = 2 \; \varepsilon = 1250 \; \alpha = 0.98$	$K = \begin{bmatrix} 25.530 & -2.595 & -8.371 \\ 0.097 & 21.744 & -17.936 \\ -5.620 & -2.438 & 12.232 \end{bmatrix}$ $\rho = 2 \; \varepsilon = 8669 \; \alpha = 0.95$	$K = \begin{bmatrix} 25.849 & -4.399 & 1.839 \\ -9.249 & 21.571 & -5.454 \\ -6.849 & -5.581 & 14.655 \end{bmatrix}$ $\rho = 13.8 \; \varepsilon = 4914 \; \alpha = 0.96$

Table 3. Parameters for MPID controllers for different Methods (Case 1)

	Constant	**Dry**
M2	$K_c = \begin{bmatrix} 181.673 & 29.368 \\ 1.627 & 0.511 \end{bmatrix}$ $K_i = \begin{bmatrix} 3427 & -5302 \\ 3.5 & 36.9 \end{bmatrix}$ $\rho = 63.266 \; \varepsilon = 170.561$	$K_c = \begin{bmatrix} 283.386 & -494.794 \\ 0.058 & 0.635 \end{bmatrix}$ $K_i = \begin{bmatrix} 5642.8 & 6996 \\ 2.2 & 7.5 \end{bmatrix}$ $\rho = 31.038 \; \varepsilon = 117.231$
M3	$K = \begin{bmatrix} 0.002 & 0.013 \\ -0.0 & 0.008 \end{bmatrix}$ $\rho = 4800 \; \varepsilon = 2581800 \; \omega = 0.027$	$K = \begin{bmatrix} 0.001 & -0.03 \\ 0.0 & 0.025 \end{bmatrix}$ $\rho = 4000 \; \varepsilon = 1326700 \; \omega = 0.002$
M4	$K = \begin{bmatrix} 1694.5 & -564.1 \\ 11.3 & 8.7 \end{bmatrix}$ $\rho = 3.798 \; \varepsilon = 518.408, \alpha = 0.988$	$K = \begin{bmatrix} 4483.1 & 4624.4 \\ 1.7 & 6.2 \end{bmatrix}$ $\rho = 25 \; \varepsilon = 3183, \alpha = 0.985$

Table 4. Parameters for MPID controllers for different Methods (Case 2)

Therefore, the controller parameters φ are optimal in the sense of minimising the cost function J for specific Q and R. For each method, the above problem was solved using the Matlab numerical optimisation function. This approach is justified when the process interaction is strong and the trial-and-error tuning approach would be time consuming. The optimal tuning matrices for all MPID controllers for Cases of 1 and 2 at various operating points are evaluated. The input and output weights in the cost function may be tuned in such a way that satisfactory closed loop performance, as well as effluent quality performance e.g. nitrogen removal improvement could be achieved.

4.3 Evaluation criteria

The MPID control strategies are tested using the nonlinear ASM1 model and the controller performance is evaluated using an index of the Aeration Energy (AE) as described in (Copp, 2002):

$$AE = \frac{24}{T}\int_{t=7d}^{t=14d}\sum_{i=1}^{5}\left[0.4032K_{Lai}(t)^2 + 7.8408K_{Lai}(t)\right]dt \tag{35}$$

where K_{La_i} is the oxygen transfer coefficient (d^{-1}) in each reactor. d is the unit of time (a day). The average AE (kWh/d) is calculated for the last 7 days of the dynamic data (T).

5. Simulation Results

The MPID controller was evaluated in a simulation study where the full ASM1 was used to model the process. The nonlinear ASM1 was used for simulating the process. The constant influent flow has been utilised first to assess the controllers' ability to respond to set point changes, whilst the varying influent flow (dry and rain weather conditions) are used to provide a statistical evaluation of the controllers' performance with respect to disturbance rejection. Note that the time constants for DO and SNO are of the order of minutes (DO) and hours (SNO), respectively. The aim of the controller in Case 1 is to maintain the DO levels in the last three aerobic tanks at DO_3=1.5mg/l, DO_4=3mg/l and DO_5=2mg/l. In Case 2, the set points for DO and the nitrate were set at 2mg/l and 1mg/l, respectively. Notice that, for simplification, each method of tuning is denoted as M1, M2, M3 and M4 methods for Davison, Penttinen-Koivo, Maciejowski and proposed new method, respectively. For the first operating condition (constant influent flow), for both cases of 1 and 2, M4 clearly gives a promising result for compensating the changes in setpoints and better performance for disturbance rejection. This can be first revealed from Table 5 that summarise the results obtained for each control strategy for Case 1. The simulation result for Case 2 is plotted in Fig. 5(a-b).

The result demonstrates that M4 gives a better performance compared to the others for both setpoint tracking and disturbance rejection. M4 exhibits somewhat faster responses than the other controllers. The overshoots to setpoint changes are small and the settling time is about 10-15 minutes as shown in Table 5. The closed loop response for a setpoint change in M2 is satisfactory. The average settling time for DOs given by all control strategies is about 20 minutes which seems reasonable, except for DO5 given by M1 which takes much longer to settle. M3 needs to be fairly tuned in order to obtain a good tracking and disturbance rejection performance. M2 tends to make the system unstable as the controller gain is increased. M3 has better performance than M1 or M2, but it has slightly bigger overshoot than M4. Although the performance of M3 is satisfactory in some outputs, it uses the more time-consuming "sequential" identification procedure for obtaining the tuning constant. The performance of M1 is worst with the slowest response and large overshoot, as seen in Table 5. This method is not applicable in Case 2.

	OS(%)	T_s (min)	SSQ
$DO3_{M1}$	6.7	28	4.18e-5
$DO4_{M1}$	8.3	43.2	3.08e-4
$DO5_{M1}$	25	57.6	9.8e-4
$DO3_{M2}$	0.7	7.2	3.28e-5
$DO4_{M2}$	2	8.64	2.19e-4
$DO5_{M2}$	15	5.76	9.76e-4
$DO3_{M3}$	0.2	8.64	3.28e-5
$DO4_{M3}$	1.8	8.65	1.65e-4
$DO5_{M3}$	10	8.64	9.59e-4
$DO3_{M4}$	0.3	2.85	9.97e-6
$DO4_{M4}$	2	2.88	7.10e-5
$DO5_{M4}$	6	2.80	4.26e-4

Table 5. Dynamic performance comparison of MPID controllers (Case 1)- (OS: Overshoot, T_s: Settling time, SSQ: the residual sum of squares)

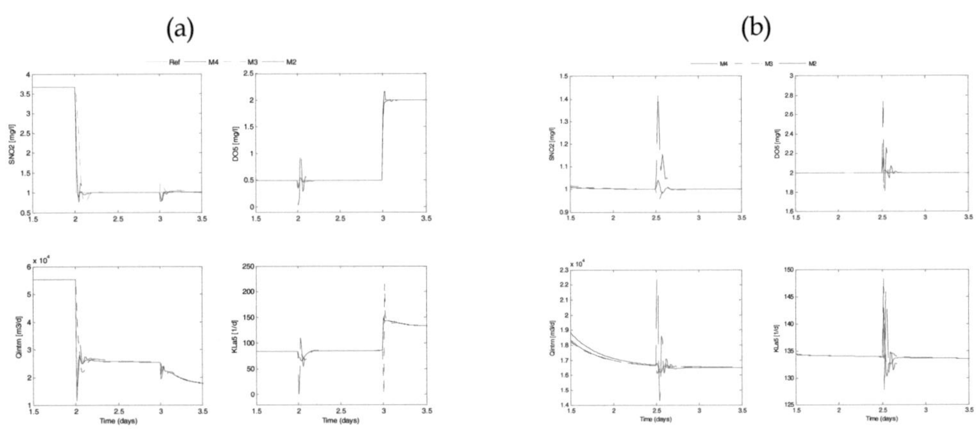

Fig. 5. Dynamic performance comparison of MPID controllers (Case 2)- a) set point tracking; b) disturbance rejection

It is also of great interest to study how the controllers perform under different operating conditions (dynamic influent flows). The statistical evaluation of the performance for Case 1 for each control strategy under dry and rain condition is depicted in Fig. 6, whilst Fig. 7 reveals the performance (Case 2) of disturbance rejection under dry condition.

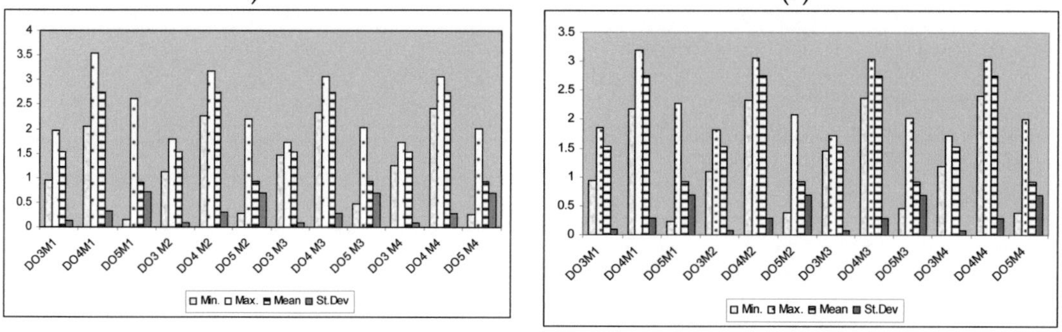

Fig. 6. Dynamic influent statistics (Case 1)- a) Dry weather; b) Rain weather

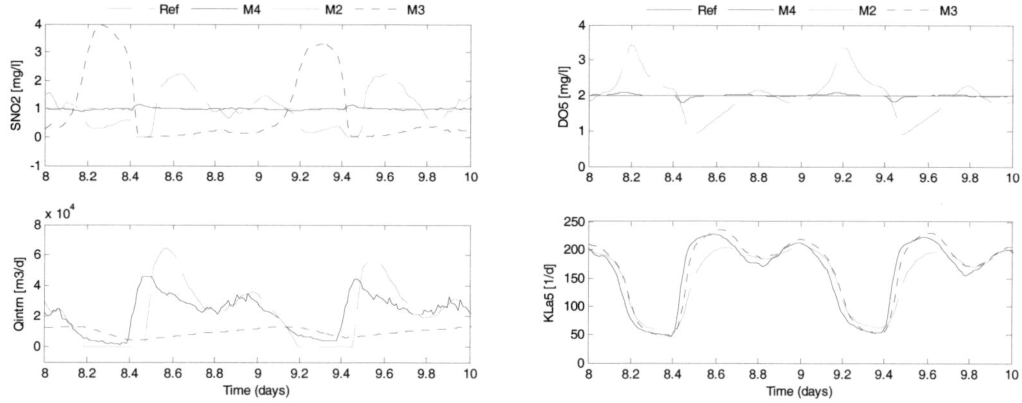

Fig. 7. Disturbance rejection under dry influent flow (Case 2)

Due to high nonlinearities in Case 2, only dry influent flow has been investigated and an adaptive controller is required to design the controller for rain condition. In all cases the result from the statistical evaluation of the performance (Fig. 6) shows lower output error for M4. The result of simulation from the 8th to the 10th day of influent data is shown in Fig. 7. These results will also confirm that M4 has the best performance. M3 shows good tracking properties and compensates the disturbances for DO_5, but it has no control on S_{NO2} as it is evident from the low value of Q_{intrn}. M4 is also more flexible and the tuning parameter, α makes the plant frequency analysis easier to handle. In addition, M2 performs better than M3, but not as good as the M4.

5.1 Robustness performance analysis

The control design strategy is also analysed in term of robustness performance requirement and in this case, constant influent condition is applied. Fig. 8 shows the open loop singular values for Cases 1 and 2

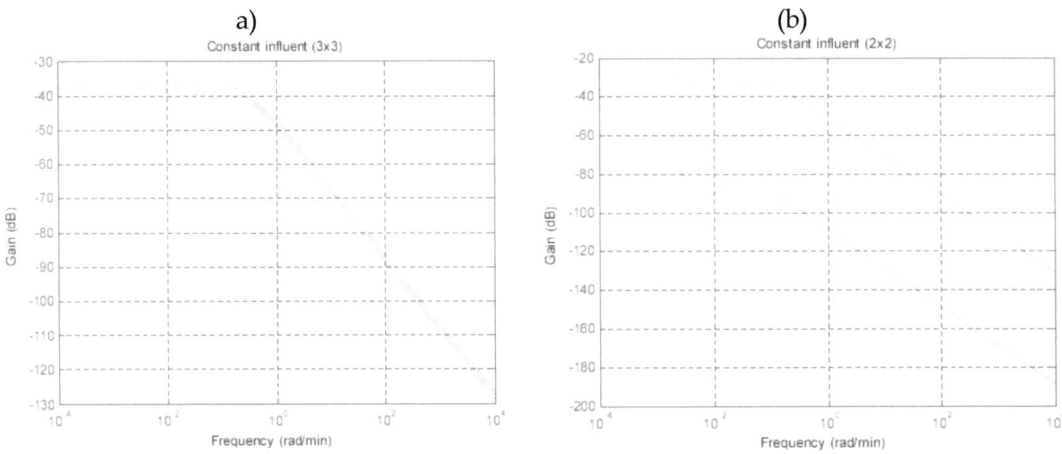

Fig. 8. Open loop singular values - a) Case 1; b) Case 2

The singular values are relatively small at low frequencies in both cases indicating that controlling the variables of interest are not an easy task. Moreover, there is a significant difference in magnitude in each loop for design case 2 indicating that controlling the variables are therefore more difficult. The ability of multivariable PID controller to deal with this difficulty is especially of importance since its closed loop performance is dictacted by low frequency gains of the variable of interest. The open loop bandwidth of 0.02rad/min is given by Case 1 whilst Case 2 shows a significant difference of bandwidth frequency in each control loop.

Fig. 9 compares the results of sensitivity, $(I + GK)^{-1}$ and complementary sensitivity, $GK\,(I + GK)^{-1}$ plots of different control strategies in Case 1. It can be seen that the magnitudes of sensitivity for the three variables (DOs) at low frequency are higher for M1 compared to other control strategies. This implies that performance of M1 in rejecting disturbance is worst. The magnitude of $(I + GK)^{-1}$ for M2 is lowest followed by M4 and M3. This means that M2 is less susceptible to disturbances. Note that although the closed loop sensitivity resulting from M2 is superior to that with the other three control strategies (M1, M3 and M4), the worst-case gain behaviour is much worse as can be seen in Fig. 9. This is also leads to a lower stability margin provided by M2 controller design. For robustness, we also need to keep $GK\,(I + GK)^{-1}$ small. Although M1 gives the best result in terms of noise immunity, it is however the lowest performance in terms of closed loop bandwidth and in rejecting disturbance. The methods of M3 and M4 give satisfactory results, being particularly effective for a given frequency range. However, M4 gives slightly better results compared to M3 especially the closed loop bandwidth and disturbance rejection. Considering the overall performance characteristics given by all different control strategies, the method M4 is the most reliable.

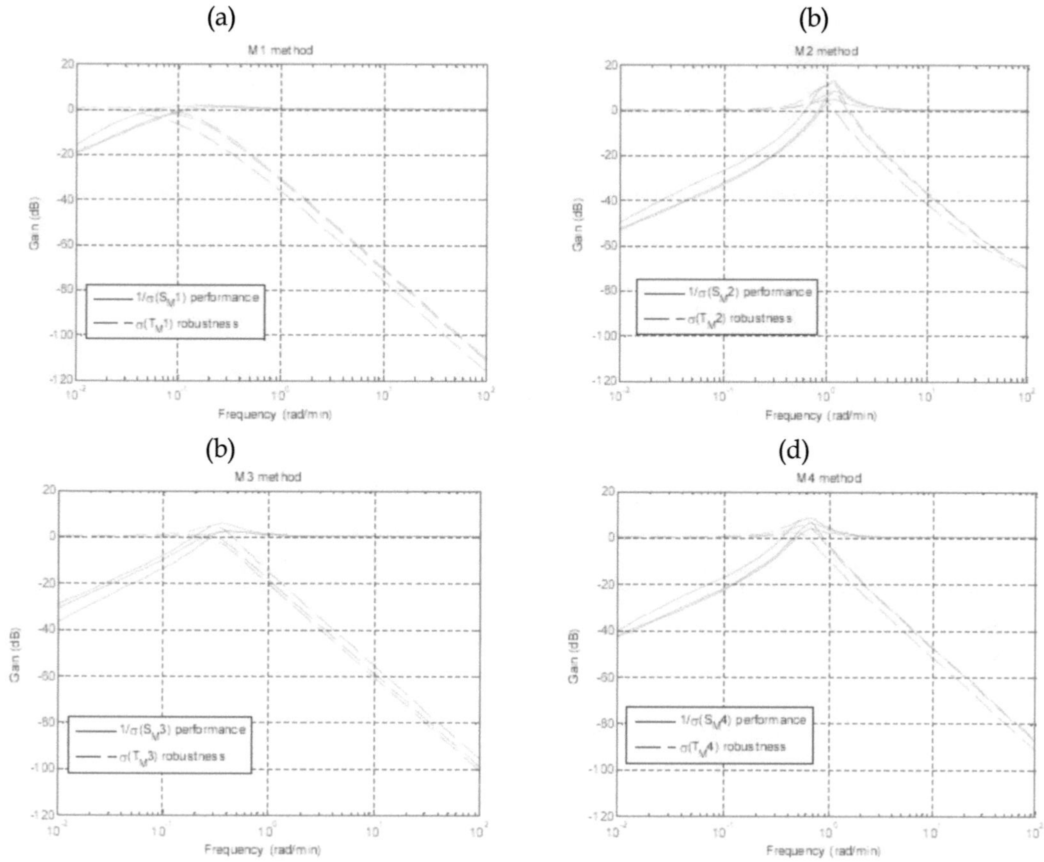

Fig. 9. Performance robustness analysis of Case 1 - sensitivity- a) Davison method; b) Penttinen method; c) Maciejowski method; d) Proposed new method

Fig. 10 compares the results of sensitivity, $(I + GK)^{-1}$ and complementary sensitivity, $GK (I + GK)^{-1}$ plots of different control strategies in Case 2. Method M1 is not applicable, therefore it is not applied in this case. In this case, we have two different frequency bandwidth in the control loops. This leads to challenges in control tuning to obtain simulataneously a good performance in both of loops. It can be seen that the measurement noise is being amplified over a smaller range of frequencies in method M2. However, M2 considers the worst performance in term of disturbance rejection, i.e. highest magnitude of $(I + GK)^{-1}$ at low frequency. As previously discussed in Case 1, M3 and M4 also give better performance in disturbance rejection in Case 2. Fig. 10 shows that, although M3 gives the best result in rejecting disturbance of loop 2 (DO5), i.e. lowest magnitude of $(I + GK)^{-1}$ at low frequency, it is however the worst in noise suppression, i.e. highest magnitude of $GK (I + GK)^{-1}$ for loop 1 (SNO;2) at high frequency. Moreover, M3 has lower stability margin compared to M4 and M2. Overall, M4 provides satisfactory results in the simultaneous multiloop control tuning. It shows good performance in both loops in terms of closed loop bandwidth and can suppress noise better.

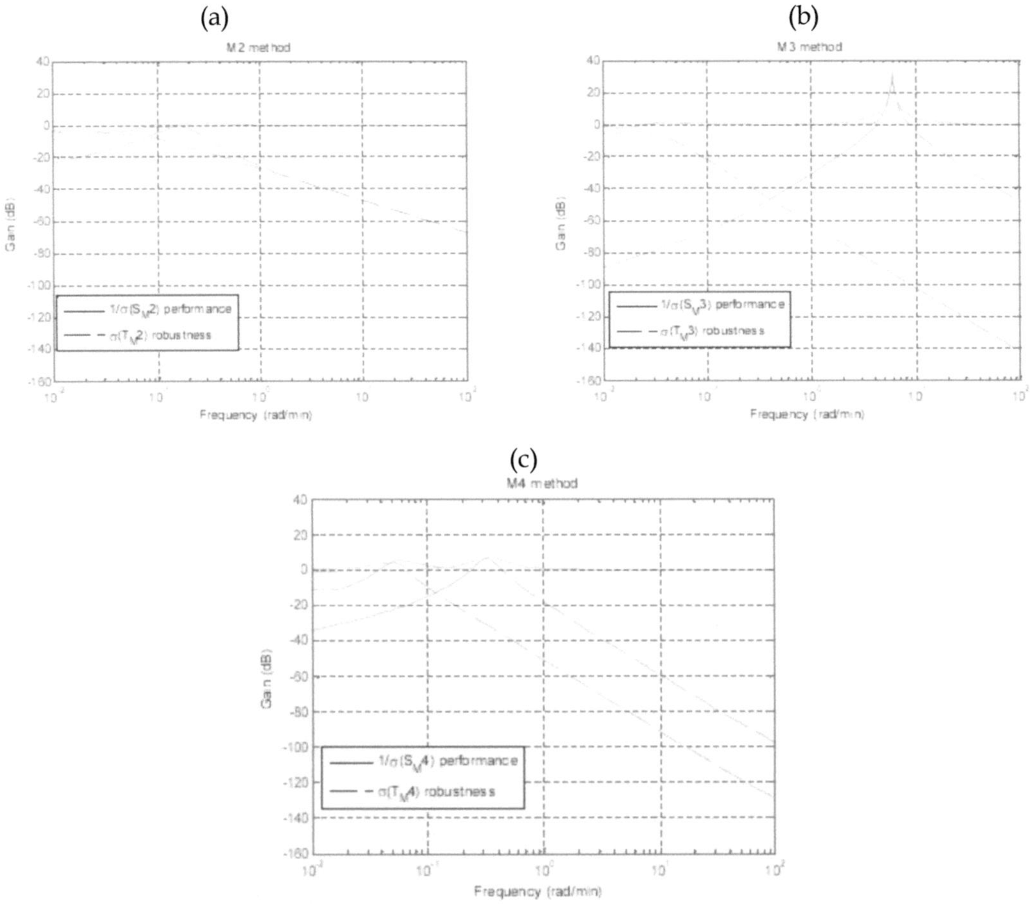

Fig. 10. Performance robustness analysis of Case 2 - sensitivity- a) Penttinen method; b) Maciejowski method; c) Proposed new method

Fig. 11 shows the plots of input disturbance, $\bar{\sigma}(I+GK)^{-1}G$ for both cases of 1 and 2. In this case, the variables of control should prevail in zero steady state errors subject to input disturbances and/or changes in setpoint, i.e. changes in the oxygen transfer coefficients or internal recirculation flow. This can be clearly observed from the positive gradients at low frequency regions of the plots given by all control strategies. It can also be seen from Fig. 11 that the magnitude of $\bar{\sigma}(I+GK)^{-1}G$ is relatively higher for M2 (50-55 dB at 10^{-2} rad/min) compared with M4 (40-45 dB at 10^{-2} rad/min), M3 (30-35 dB at 10^{-2} rad/min rad/min) and M1 (15-20 dB at 10^{-2} rad/min rad/min). Though M2 shows a good performance to input disturbance in Case 1, it appears to be the worst performance due to input disturbance in Case 2. Since the performance measure given by M4 is satisfactory in both cases, the method is proven to be useful for different frequency bandwidth.

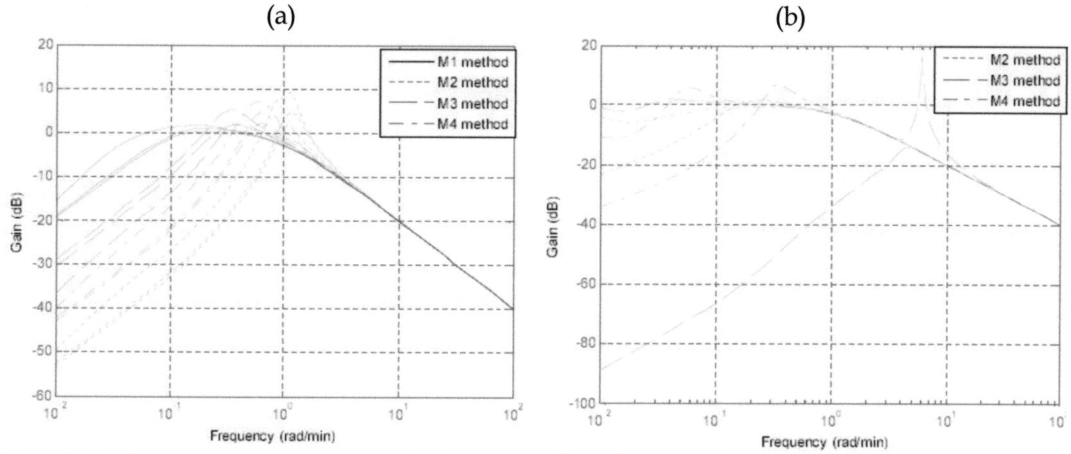

Fig. 11. Performance robustness analysis - input disturbance- a) Case 1; b) Case 2

5.2 Performance evaluation

Here, the performance of the plant is presented for Cases 1 and 2. In Case 1, the effect of controlling three dissolved oxygen in the last three aerated tanks is shown in Fig. 12. As seen in Fig. 12, the DO in reactor 1 and reactor 2 are not controlled. Clearly, the same output of DO, both in the effluent and under flow are demonstrated, as the ones given by the DO in the last aerated tank (reactor), both for dry and rain flow conditions. Control strategies were also evaluated against the criteria described in (35) for Case 2 as shown in Table 6.

	Aeration energy (kWh/d)	Average NH_4-N_{eff}(mg/l)	Average NO_3-N_{eff}(mg/l)
Benchmark	7241.27	2.528	12.439
M2	6532.14 (-9.8%)	3.029 (+19.8%)	12.489 (+0.4%)
M3	6387.12 (-11.7%)	2.267 (-10.3%)	14.53 (+16.8%)
M4	6376.11(-11.9%)	2.411 (-4.6%)	12.045 (-3.2%)

Table 6. Evaluation criteria for different control tuning strategies for dry influent case.

The basic control strategy in benchmark simulation study, proposed by (Copp, 2002) is used as a reference case for comparison. The evaluation criteria considered are aeration energy, effluent ammonia nitrogen and effluent nitrate nitrogen. The MPID control strategies were evaluated for DO-Nitrate dry weather model against single loop PI controllers used in the COST benchmark. A lower aeration cost (AE) is achieved with MPID. These are about 9.8%, 11.7% and 11.9%, for M2, M3 and M4, respectively. The average effluent ammonia (NH_4-N_{eff}) was reduced by 10.3% and 4.6%, for M3 and M4, respectively. M2 gave slightly higher average effluent ammonia but still below the discharge limit (4mg/l). Better total nitrogen removal is achieved using M4 for both ammonia and nitrate in the effluent.

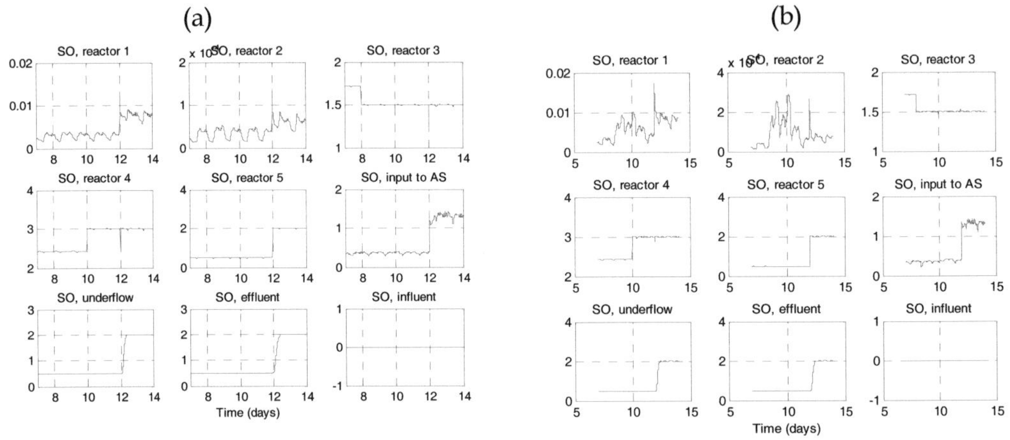

Fig. 12. Plant performance of DO for Case 1, (a) dry influent condition; (b) rain influent condition

6. Conclusion

The objective of the study was to use MPID controllers to improve closed loop performance and reduce loop interactions. Three tuning strategies were compared and a new one was introduced. All methods require information only from simple step or frequency tests. The methods are based on decoupling the system at different frequency points. To identify the most effective control strategy, RGA analysis were performed. It was proposed to use DRGA to find the best frequency point for decoupling. A procedure was also developed to fine-tune the controllers using an optimisation procedure. Extensive simulation studies on a nonlinear ASM1 model demonstrated that the proposed method performed significantly better in setpoint tracking properties and disturbance rejection and gave the best performance with respect to decoupling capabilities. The results suggest considerable improvement can be achieved in terms of energy savings and nitrogen removal with a properly tuned MPID controller. The methods demonstrate that the controller tuning influences multiloop system performance.

7. References

El-Din, A. G. Smith & D. W. (2002). A combined transfer function noise model to predict the dynamic behaviour of a full scale primary sedimentation tank model. *Water Research*, Vol 36, pages 3747-3764.

Cote, P. L., G. A. Ekama & G.v.R. Marais (1995). Dynamic modelling of the activated sludge process: improving prediction using neural networks. *Water Research*, Vol 29, pages 995-1004.

Robertson, G. A. & T. Cameron (1996). Analysis of dynamic process models for structural insight and model reduction -part 1. structural identification measures. *Computer and Chemical Engineering*, Vol 21, No.5, pages 455-473.

M. Henze, C.P.L.Grady Jr., W.Gujer, G.v.R.Marais & T.Matsuo (1987). *Activated sludge model no.1. IAWQ*. Scientific and Technical Report no.1, IAWQ, London.

Chotkowski, W., Brdys, M. A. & Konarczak, K. (2005). Dissolved oxygen control for activated sludge processes, *International Journal of System Science*, Vol 12, pages 727-736.

Y. Ma, Y. Peng & S. Wang, (2005). Feedforward - feedback control of dissolved oxygen concentration in a predenitrification system, Bioprocess. *Biosyst. Eng.* Vol 27, pages 223-228.

Piotrowski, R. & Brdys M. A. (2005) Lower-level controller for hierarchical control of dissolved oxygen concentration in activated sludge processes, *In Proceeding of the 16th IFAC world congress*, Prague, pages 4-8.

A. Stare, D. Vrečko, N. Hvala & S. Strmčnik. (2007) Comparison of control strategies for nitrogen removal in an activated sludge process in terms of operating costs: A simulation study, *Water Res.* Vol 41, pages 2004-2014.

E. Mats, B. Berndt & A. Mikael. (2006). Control of the aeration volume in an activated sludge process using supervisory control strategies, *Water Res.* Vol 40, pages 1668-1676.

I. Takács, G.G.Patry & D.Nolasco. (1991). A dynamic model of the clarification thickening process. *Water Res.* Vol 25, pages 1263-1271.

J.B. Copp. (2002). *COST Action 624, The COST simulation benchmark-Description and simulator manual*, European Communities, Luxembourgh.

De Moor, B. (1988). *Mathematical concepts and technique for modelling of static and dynamic systems*. PhD thesis. Dept. of Electrical Engineering, Katholieke Universiteit Leuven, Belgium.

Moonen, M., B. De Moor, L. Vandenberghe & J. Vandewalle (1989). On and offline identification of linear state space models. *Internal Journal Control*, Vol 49, No. 1, pages 219-232.

Verhaegen, M. (1994). Identification of the deterministic part of mimo state space models given in innovation form from input-output data. *Automatica*, Vol 30, No. 1, pages 61-74.

Söderström, T. and P. Stoica (1989). *System Identification*. Prentice Hall, Inc., Englewood Cliffs, New Jersey, USA.

Bristol, E.H. (1996). On a new measure of interaction for multivariable process control. *IEEE Trans. On Auto Control*, Vol 11, pages 133-134.

Kinnaert, M. (1995). Interaction measures and pairing of controlled and manipulated variables for multiple-input multiple-output systems: A survey. *Journal A*, Vol 36, No.4, pages 15-23.

Davison, E. (1976). Multivariable tuning regulator. *IEEE Transaction on Automatic Control*, Vol 21, pages 35-47.

Penttinen, J. & Koivo, N.H. (1980). Multivariable tuning regulators for unknown systems, *Automatica*, Vol 16, pages 393-398.

Maciejowski, J. M. (1989). *Multivariable feedback design*, Addison Wesley, Wokingham, England.

2

Stable Visual PID Control of Redundant Planar Parallel Robots

Miguel A. Trujano, Rubén Garrido and Alberto Soria
Departamento de Control Automático, CINVESTAV-IPN
México

This chapter presents an image-based Proportional Integral Derivative (PID) controller for a redundant overactuated planar parallel robot; the control objective is to drive the robot end effector to a desired constant reference position. The main feature of the proposed approach is the use of a vision system for obtaining the end effector position. This approach precludes the use of the robot forward kinematics. The Lyapunov method and the LaSalle invariance principle allow assessing asymptotic closed-loop stability. Experiments in a laboratory prototype permit evaluating the performance of the closed-loop system.

1. Introduction

Most of today industrial robots are controlled using joint-level PID controllers (Arimoto & Miyazaki, 1984; Wen & Murphy, 1990; Kelly, 1995; Spong, et al., 2005). In the case of parallel robots, their forward kinematics allows computing the end-effector position and orientation (Kock & Schumacher, 1998; Cheng et al., 2003); using the forward kinematics in real time may be computational demanding for some robot designs and sometimes it does not have an analytical solution; besides, a prior calibration procedure estimate the forward kinematics parameters. Any error in this estimation procedure would translate into positioning errors. An approach explored in this chapter is to use a vision system for measuring the end-effector coordinates; this methodology avoids solving in real time the forward kinematics and any calibration procedure. The chapter focuses on redundant planar parallel robots of the RRR-type studied in (Cheng et al., 2003) and shown in Fig. 1. This type of robot is well suited for laser and water cutting systems and in tasks requiring positioning in a plane. It is also worth remarking that over actuation reduces or even eliminates some kinds of singularities and improves Cartesian stiffness in the robot workspace.

Visual Servoing represents an attractive solution to position and motion control problems of autonomous robot manipulators evolving in unstructured environments (Corke, 1996; Hutchinson et al., 1996; Kelly, 1996; Papanikolopoulos & Khosla, 1993; Weiss et al., 1987; Wilson et al., 1996; Chaumette & Hutchinson, 2006 & 2007; Kragic & Christensen, 2005). There exist two approaches for this robot control strategy: camera-in-hand and fixed-camera. In the camera-in-hand configuration, the robot end-effector carries on the camera; the objective of this approach is to move the manipulator in such a way that the projection of a moving or static object is always at a desired location in the image given by the camera.

In contrast, in fixed-camera robotic systems, one or several cameras, fixed with respect to a global coordinate frame, capture images of the robot and its environment; the objective is to move the robot in such a way that its end-effector reaches a desired target. The proposed control law uses this later approach.

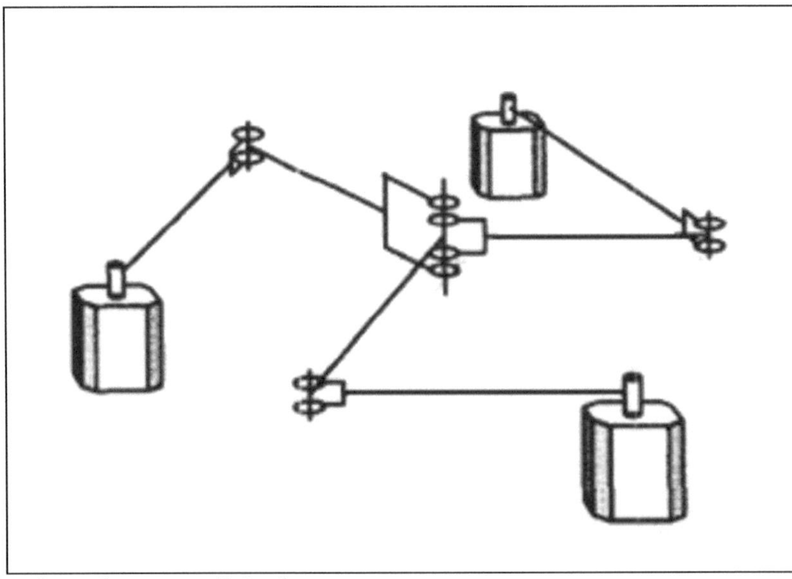

Fig. 1. Redundant planar parallel robot

Visual Servoing of parallel robots is an emerging field and until recently, some papers report interesting research in this area. Using a vision system in parallel robots allows calibrating their Forward Kinematics; moreover, in some instances it permits obtaining the position and orientation of some part of the robot mechanical structure thus dispensing the use of the Forward Kinematic for closed-loop control. Visual information of the robot legs allows controlling a Gough-Stewart parallel robot in (Andreff & Martinet, 2006) and (Andreff et al., 2007). Another interesting approach in (Dallej et al., 2007) shows how to control an I4R parallel robot using only visual feedback. Simulation results using a realistic robot model show satisfactory closed-loop performance. A visual control scheme, applied to the delta robot RoboTenis, is a key feature in (Angel et al., 2008) and (Sebastian et al., 2007). This approach uses the robot native joint controller as an inner loop, and a camera, which rests on the robot end-effector and closes an outer control loop; moreover, the authors show uniform ultimate boundedness of the tracking error. Experiments validate the proposed approach.

This Chapter proposes a control law that solves the position control problem for a redundant overactuated planar parallel robot by using direct vision feedback into the control loop; in this way, the proposed approach does not stem on solving the robot Forward Kinematics. The proposed algorithm exploits a PID-like control structure, similar to those proposed previously for open-loop kinematic chain robot manipulators (Kelly, 1998; Santibañez & Kelly, 1998). Moreover, compared with previous approaches on visual control of parallel robots, the stability analysis presented here, based on the Lyapunov method and the LaSalle principle, takes into account the robot dynamics. Experiments in a laboratory prototype permit assesing the performance of the closed-loop system.

2. Parallel robot modeling

A parallel manipulator is a closed-loop kinematic chain mechanism whose end-effector is linked to its base by several independent kinematic chains. References (Merlet, 2000; Tsai, 1999) describe a rather exhaustive enumeration of parallel robots mechanical architectures and their diverse applications are described in Singularities, which also appear in open chain robots, are abundant in parallel robots; when a manipulator is in a singular configuration, it loses stiffness and becomes uncontrollable. Singularities make the limited workspace of parallel manipulators even smaller. Redundant actuation is a method for removing singularities over the workspace; in this case, the number of actuators is greater than the number of end-effector coordinates. Besides removing singularities over the workspace, redundant actuation also has the advantages of making the robot structure lighter and faster, optimizing force distribution and improving Cartesian stiffness. The following paragraphs describe the modeling issues concerning the kinematics and dynamics of redundant planar parallel robots of the RRR-type.

2.1 Kinematics of parallel manipulators

The kinematic analysis of parallel robots comprises two parts: The Inverse Kinematics and the Forward Kinematics. In the Inverse Kinematics, given an end-effector position and orientation, the problem is to find the robot active joint values leading to these position and orientation. In the case of the Forward Kinematics, the robot active joint values are given and the problem is to find the position and orientation of the end-effector. As a rule, as the number of closed cinematic chains in the mechanism increases, the difficulty of the Forward Kinematics solution also increases, whereas the difficulty for the Inverse Kinematics solution diminishes.

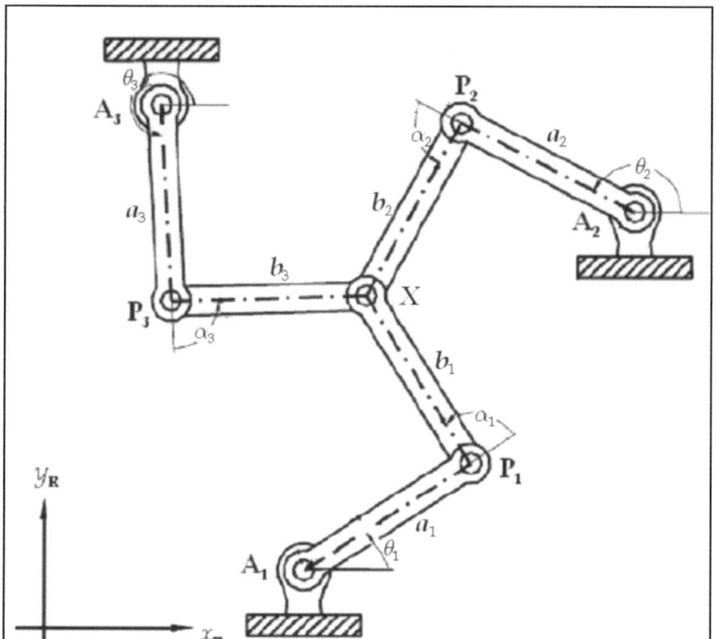

Fig. 2. Parallel Robot coordinate frame.

Figure 2 depicts a sketch of the redundant planar parallel robot. The robot kinematics assumes that all chain links have equal lengths, i.e. $L=a_i$ and $L=b_i$, $i=1,2,3$. Typically, a parallel robot has both active and passive joints; the robot actuators drive only the active joints. Symbol $\mathbf{A}_i$ represents the ith active joint with coordinates $\mathbf{X}_{\mathbf{A}_i}=[x_{\mathbf{A}_i}\ \ y_{\mathbf{A}_i}]^T$ with respect to the global Cartesian reference frame. Symbol $\mathbf{P}_i$ stands for the ith passive joint with coordinates $\mathbf{X}_{\mathbf{P}_i}=[x_{\mathbf{P}_i}\ \ y_{\mathbf{P}_i}]^T$. Variable $\mathbf{X}=[x\ \ y]^T$ defines the end-effector position, variable θ_i denotes the angle of the ith active joint, and variable α_i is the angle of the ith passive joint. These angles permits defining the active and passive joint position vectors

$$\mathbf{q}_\mathbf{a}=[\theta_1\ \ \theta_2\ \ \theta_3]^T, \tag{1}$$

$$\mathbf{q}_\mathbf{p}=[\alpha_1\ \ \alpha_2\ \ \alpha_3]^T. \tag{2}$$

Concatenating the above vectors produce a vector corresponding to all the robot joints

$$\mathbf{q}=[\mathbf{q}_\mathbf{a}^T\ \ \mathbf{q}_\mathbf{p}^T]^T. \tag{3}$$

Forward Kinematics

The following relationship describes the robot Forward Kinematics (Cheng et al., 2003)

$$x=\frac{\|\mathbf{X}_{\mathbf{P}1}\|^2(y_{\mathbf{P}2}-y_{\mathbf{P}3})+\|\mathbf{X}_{\mathbf{P}2}\|^2(y_{\mathbf{P}3}-y_{\mathbf{P}1})+\|\mathbf{X}_{\mathbf{P}3}\|^2(y_{\mathbf{P}1}-y_{\mathbf{P}2})}{2[x_{\mathbf{P}1}(y_{\mathbf{P}2}-y_{\mathbf{P}3})+x_{\mathbf{P}2}(y_{\mathbf{P}3}-y_{\mathbf{P}1})+x_{\mathbf{P}3}(y_{\mathbf{P}1}-y_{\mathbf{P}2})]}, \tag{4}$$

$$y=\frac{\|\mathbf{X}_{\mathbf{P}1}\|^2(x_{\mathbf{P}3}-x_{\mathbf{P}2})+\|\mathbf{X}_{\mathbf{P}2}\|^2(x_{\mathbf{P}1}-x_{\mathbf{P}3})+\|\mathbf{X}_{\mathbf{P}3}\|^2(x_{\mathbf{P}2}-x_{\mathbf{P}1})}{2[x_{\mathbf{P}1}(y_{\mathbf{P}2}-y_{\mathbf{P}3})+x_{\mathbf{P}2}(y_{\mathbf{P}3}-y_{\mathbf{P}1})+x_{\mathbf{P}3}(y_{\mathbf{P}1}-y_{\mathbf{P}2})]}, \tag{5}$$

$$\mathbf{X}_{\mathbf{P}i}=\mathbf{X}_{\mathbf{A}i}+L\begin{bmatrix}\cos\theta_i\\ \sin\theta_i\end{bmatrix},\ i=1,2,3. \tag{6}$$

It is worth remarking that the end-effector position $\mathbf{X}=[x\ \ y]^T$ does not depend on all the robot joint angles but only on the active joints angles θ_i. Therefore, it is possible to write down the robot Forward Kinematics as

$$\mathbf{X}=\varphi(\mathbf{q}_\mathbf{a}). \tag{7}$$

Workspace

The set Ω defines the robot workspace; therefore, the end effector position must belong to this set, i.e. $\mathbf{X}\in\Omega\subset\mathbb{R}^2$. Fig. 3 shows workspace plots for $a_i=b_i=L$ and the general case $a_i\neq b_i$; variable d corresponds to the distance between the centers of two consecutive active joints. The robot under control has the configuration $a_i=b_i=L$, and $a_i+b_i<d$.

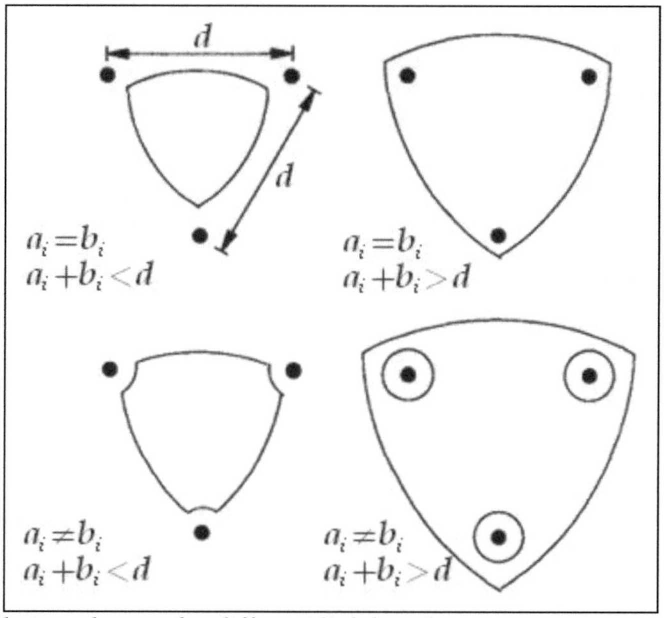

Fig. 3. Parallel Robot workspace for different link lengths.

Inverse kinematics

In this case the active joint angles depends only on the robot end-effector coordinates $\mathbf{X}$, i.e.

$$\theta_i = \arctan 2\left(\frac{\gamma_i}{\phi_i}\right) + \arctan 2\left(\frac{\pm\sqrt{\phi_i^2 + \gamma_i^2 - \xi_i^2}}{\xi_i}\right), \quad i = 1,2,3. \tag{8}$$

$$\begin{aligned} \phi_i &= 2L(x - x_{\mathbf{A}i}), \\ \gamma_i &= 2L(y - y_{\mathbf{A}i}), \\ \xi_i &= \|\mathbf{X} - \mathbf{X}_{\mathbf{A}i}\|^2. \end{aligned} \tag{9}$$

Subsequently, the active joint angles allows computing the passive joint angles as follows

$$\alpha_i = \operatorname{atan}\left(\frac{y - y_{\mathbf{A}_i} - l_1 \sin\theta_i}{x - x_{\mathbf{A}_i} - l_1 \cos\theta_i}\right) - \theta_i; \quad i = 1,2,3. \tag{10}$$

These solutions represent two different configurations for each leg that produce to $2^3 = 8$ solutions for the manipulator, as depicted in Fig. 4.

Configurations *a,* and *e* are preferable because they have shown more symmetric and isotropic force transmission throughout the workspace.

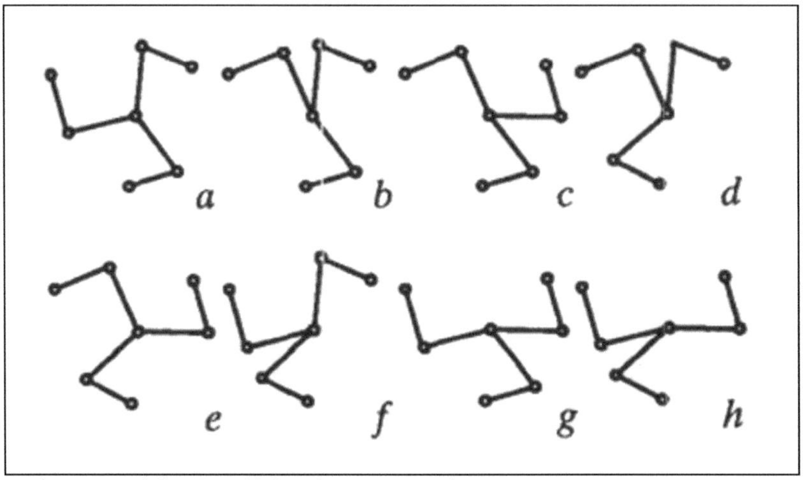

Fig. 4. All the solutions of the Parallel Robot inverse kinematics.

Differential kinematics

The following equations describe the relationship between the velocities at the joints and at the end effector

$$\dot{\mathbf{q}}_a = \begin{bmatrix} \dot{\theta}_1 \\ \dot{\theta}_2 \\ \dot{\theta}_3 \end{bmatrix} = \begin{bmatrix} \dfrac{\cos(\theta_1+\alpha_1)}{L\sin\alpha_1} & \dfrac{\sin(\theta_1+\alpha_1)}{L\sin\alpha_1} \\ \dfrac{\cos(\theta_2+\alpha_2)}{L\sin\alpha_2} & \dfrac{\sin(\theta_2+\alpha_2)}{L\sin\alpha_2} \\ \dfrac{\cos(\theta_3+\alpha_3)}{L\sin\alpha_3} & \dfrac{\sin(\theta_3+\alpha_3)}{L\sin\alpha_3} \end{bmatrix} \begin{bmatrix} \dot{x} \\ \dot{y} \end{bmatrix} = \mathbf{S}\dot{\mathbf{X}}, \tag{11}$$

$$\dot{\mathbf{q}}_p = \begin{bmatrix} \dot{\alpha}_1 \\ \dot{\alpha}_2 \\ \dot{\alpha}_3 \end{bmatrix} = \begin{bmatrix} -\dfrac{d_{1_x}}{L^2\sin\alpha_1} & -\dfrac{d_{1_y}}{L^2\sin\alpha_1} \\ -\dfrac{d_{2_x}}{L^2\sin\alpha_2} & -\dfrac{d_{2_y}}{L^2\sin\alpha_2} \\ -\dfrac{d_{3_x}}{L^2\sin\alpha_3} & -\dfrac{d_{3_y}}{L^2\sin\alpha_3} \end{bmatrix} \begin{bmatrix} \dot{x} \\ \dot{y} \end{bmatrix} = \mathbf{H}\dot{\mathbf{X}}. \tag{12}$$

$$\begin{aligned} d_{i_x} &= L\left[\cos\theta_i + \cos(\theta_i+\alpha_i)\right], \quad i=1,2,3. \\ d_{i_y} &= L\left[\sin\theta_i + \sin(\theta_i+\alpha_i)\right], \quad i=1,2,3. \end{aligned} \tag{13}$$

Concatenating (11) and (12) yields

$$\dot{\mathbf{q}} = \begin{bmatrix} \dot{\mathbf{q}}_a \\ \dot{\mathbf{q}}_p \end{bmatrix} = \begin{bmatrix} \mathbf{S} \\ \mathbf{H} \end{bmatrix} \dot{\mathbf{X}} = \mathbf{W}\dot{\mathbf{X}} \tag{14}$$

2.2 Dynamics of redundant planar parallel robot

In accordance with (Cheng et al., 2003), the Lagrange-D'Alembert formulation yields a simple scheme for computing the dynamics of redundantly actuated parallel manipulators; this approach uses the equivalent open-chain mechanism of the robot shown in Fig. 5. In order to apply this scheme, the first step is to obtain a relationship between the joint torques associated to all the robot joints and the robot active joint torques. The following Proposition gives a method for obtaining this relationship

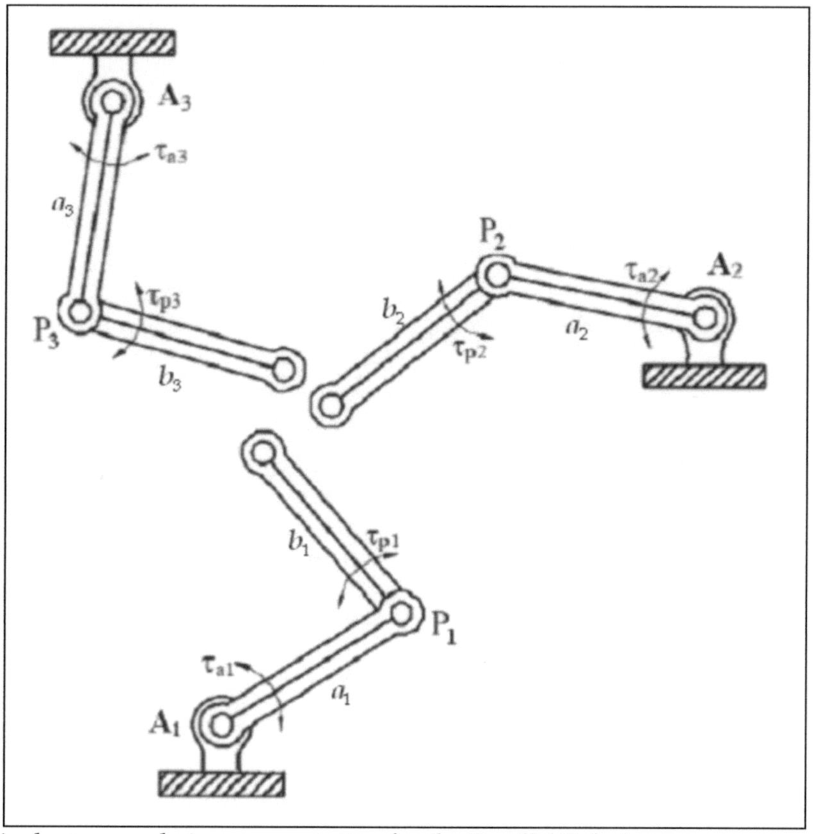

Fig. 5. Equivalent open-chain representation for the Parallel Robot.

***Proposition 1*:** Let the joint torque $\boldsymbol{\tau} \in \mathbb{R}^n$ of the equivalent open-chain system and the joint torque $\boldsymbol{\tau}_a$ of the redundantly actuated closed-chain system required to generate the same motion; then, both torques are related as follows

$$\mathbf{S}^T \boldsymbol{\tau}_a = \mathbf{W}^T \boldsymbol{\tau}. \tag{15}$$

***Proof of Proposition 1*:** We denote by $\mathbf{q}_e$ the vector of independent generalized coordinates of the mechanism. In the case of redundant actuation, the virtual displacement $\partial \mathbf{q}_a$ of the actuated joints is constrained. Using the kinematic constrains allows expressing $\mathbf{q}_a$ and $\mathbf{q}_p$ as

$$\mathbf{q}_a = \mathbf{q}_a(\mathbf{q}_e) \quad \text{and} \quad \mathbf{q}_p = \mathbf{q}_p(\mathbf{q}_e). \tag{16}$$

Differentiating the above equations gives

$$\delta\mathbf{q}_a = \frac{\partial \mathbf{q}_a}{\partial \mathbf{q}_e}\delta\mathbf{q}_e \text{ and } \delta\mathbf{q}_p = \frac{\partial \mathbf{q}_p}{\partial \mathbf{q}_e}\delta\mathbf{q}_e. \tag{17}$$

Applying the above results to the Lagrange-D'Alembert equations yields

$$\begin{aligned}\left(\frac{d}{dt}\left(\frac{\partial L}{\partial \dot{\mathbf{q}}}\right) - \frac{\partial L}{\partial \mathbf{q}} - \boldsymbol{\tau}\right)^T \delta\mathbf{q} &= \left(\frac{d}{dt}\left(\frac{\partial L}{\partial \dot{\mathbf{q}}_a}\right) - \frac{\partial L}{\partial \mathbf{q}_a} - \boldsymbol{\tau}_a\right)^T \delta\mathbf{q}_a + \left(\frac{d}{dt}\left(\frac{\partial L}{\partial \dot{\mathbf{q}}_p}\right) - \frac{\partial L}{\partial \mathbf{q}_p} - \boldsymbol{\tau}_p\right)^T \delta\mathbf{q}_p \\ &= \left[\left(\frac{d}{dt}\left(\frac{\partial L}{\partial \dot{\mathbf{q}}_a}\right) - \frac{\partial L}{\partial \mathbf{q}_a} - \boldsymbol{\tau}_a\right)^T \frac{\partial \mathbf{q}_a}{\partial \mathbf{q}_e} + \left(\frac{d}{dt}\left(\frac{\partial L}{\partial \dot{\mathbf{q}}_p}\right) - \frac{\partial L}{\partial \mathbf{q}_p} - \boldsymbol{\tau}_p\right)^T \frac{\partial \mathbf{q}_p}{\partial \mathbf{q}_e}\right]\delta\mathbf{q}_e = 0.\end{aligned} \tag{18}$$

Variable $\boldsymbol{\tau}_p$ is the actuating torque on the passive joints. Since $\delta\mathbf{q}_e$ is now free to vary, the following expression follows from (18)

$$\left[\left(\frac{d}{dt}\left(\frac{\partial L}{\partial \dot{\mathbf{q}}_a}\right) - \frac{\partial L}{\partial \mathbf{q}_a}\right)^T, \left(\frac{d}{dt}\left(\frac{\partial L}{\partial \dot{\mathbf{q}}_p}\right) - \frac{\partial L}{\partial \mathbf{q}_p}\right)^T\right]\begin{bmatrix}\frac{\partial \mathbf{q}_a}{\partial \mathbf{q}_e} \\ \frac{\partial \mathbf{q}_p}{\partial \mathbf{q}_e}\end{bmatrix} - \left[\boldsymbol{\tau}_a{}^T \frac{\partial \mathbf{q}_a}{\partial \mathbf{q}_e} + \boldsymbol{\tau}_p{}^T \frac{\partial \mathbf{q}_p}{\partial \mathbf{q}_e}\right] = 0. \tag{19}$$

Or equivalently

$$\left[\frac{d}{dt}\left(\frac{\partial L}{\partial \dot{\mathbf{q}}}\right) - \frac{\partial L}{\partial \mathbf{q}}\right]^T \begin{bmatrix}\frac{\partial \mathbf{q}_a}{\partial \mathbf{q}_e} \\ \frac{\partial \mathbf{q}_p}{\partial \mathbf{q}_e}\end{bmatrix} = \boldsymbol{\tau}_a{}^T \frac{\partial \mathbf{q}_a}{\partial \mathbf{q}_e} + \boldsymbol{\tau}_p{}^T \frac{\partial \mathbf{q}_p}{\partial \mathbf{q}_e}. \tag{20}$$

Ignoring friction at the passive joints allows setting $\boldsymbol{\tau}_p = 0$. Note also that $\frac{d}{dt}\left(\frac{\partial L}{\partial \dot{\mathbf{q}}}\right) - \frac{\partial L}{\partial \mathbf{q}} = \boldsymbol{\tau}$.

These facts allow writing (20) as

$$\mathbf{W}^T\boldsymbol{\tau} = \mathbf{S}^T\boldsymbol{\tau}_a \tag{21}$$

$$\mathbf{W} = \begin{bmatrix}\frac{\partial \mathbf{q}_a}{\partial \mathbf{q}_e} \\ \frac{\partial \mathbf{q}_p}{\partial \mathbf{q}_e}\end{bmatrix} = \frac{\partial \mathbf{q}}{\partial \mathbf{q}_e} \tag{22}$$

$$\mathbf{S} = \frac{\partial \mathbf{q}_a}{\partial \mathbf{q}_e}. \tag{23}$$

The Euler-Lagrange's well-known formalism (Spong et al., 2005) allows modeling each of the legs of the open-chain mechanism in Fig. 5. Assuming that the robot moves in the horizontal plane, the following equations model the equivalent open chain mechanism

$$\bar{\mathbf{M}}_i \begin{bmatrix} \ddot{\theta}_i \\ \ddot{\alpha}_i \end{bmatrix} + \bar{\mathbf{C}}_i \begin{bmatrix} \dot{\theta}_i \\ \dot{\alpha}_i \end{bmatrix} + \bar{\mathbf{N}}_i = \begin{bmatrix} \tau_{\mathbf{a}_i} \\ \tau_{\mathbf{p}_i} \end{bmatrix}, \ i=1,2,3 \tag{24}$$

where

$$\bar{\mathbf{M}}_i = \begin{bmatrix} \bar{M}_{i_{11}} & \bar{M}_{i_{12}} \\ \bar{M}_{i_{21}} & \bar{M}_{i_{22}} \end{bmatrix} = \begin{bmatrix} \lambda_i + 2\beta_i \cos\alpha_i & \gamma_i + \beta_i \cos\alpha_i \\ \gamma_i + \beta_i \cos\alpha_i & \gamma_i \end{bmatrix}, \quad \bar{\mathbf{C}}_i = \begin{bmatrix} \bar{C}_{i_{11}} & \bar{C}_{i_{12}} \\ \bar{C}_{i_{21}} & \bar{C}_{i_{22}} \end{bmatrix} = \begin{bmatrix} -\beta_i \dot{\alpha}_i \sin\alpha_i & -\beta_i(\dot{\alpha}_i + \dot{\theta}_i)\sin\alpha_i \\ \beta_i \dot{\alpha}_i \sin\alpha_i & 0 \end{bmatrix}$$

$$\lambda_i = m_{i1} r_{i1}{}^2 + I_{i1} + m_{i2}(a_i^2 + r_{i2}^2) + I_{i2}, \ \beta_i = m_{i2} a_i r_{i2}, \ \gamma_i = m_{i2} r_{i2}{}^2 + I_{i2}$$

Parameters I_{ij}, m_{ij} and r_{ij}, $i,j:1,2,3$, correspond to the inertia, mass, and center of mass of each link. Combining the equations described above gives the dynamics of the open-chain system in the form

$$\bar{\mathbf{M}}\ddot{\mathbf{q}} + \bar{\mathbf{C}}\dot{\mathbf{q}} + \bar{\mathbf{N}} = \boldsymbol{\tau} \tag{25}$$

$$\bar{\mathbf{M}} = \begin{bmatrix} \bar{M}_{1_{11}} & 0 & 0 & \bar{M}_{1_{12}} & 0 & 0 \\ 0 & \bar{M}_{2_{11}} & 0 & 0 & \bar{M}_{2_{12}} & 0 \\ 0 & 0 & \bar{M}_{3_{11}} & 0 & 0 & \bar{M}_{3_{12}} \\ \bar{M}_{1_{12}} & 0 & 0 & \bar{M}_{1_{22}} & 0 & 0 \\ 0 & \bar{M}_{2_{12}} & 0 & 0 & \bar{M}_{2_{22}} & 0 \\ 0 & 0 & \bar{M}_{3_{12}} & 0 & 0 & \bar{M}_{3_{22}} \end{bmatrix}, \quad \bar{\mathbf{C}} = \begin{bmatrix} \bar{C}_{1_{11}} & 0 & 0 & \bar{C}_{1_{12}} & 0 & 0 \\ 0 & \bar{C}_{2_{11}} & 0 & 0 & \bar{C}_{2_{12}} & 0 \\ 0 & 0 & \bar{C}_{3_{11}} & 0 & 0 & \bar{C}_{3_{12}} \\ \bar{C}_{1_{21}} & 0 & 0 & 0 & 0 & 0 \\ 0 & \bar{C}_{2_{21}} & 0 & 0 & 0 & 0 \\ 0 & 0 & \bar{C}_{3_{21}} & 0 & 0 & 0 \end{bmatrix}, \quad \bar{\mathbf{N}} = \begin{bmatrix} \bar{\mathbf{N}}_1 \\ \bar{\mathbf{N}}_2 \\ \bar{\mathbf{N}}_3 \end{bmatrix},$$

$$\tau = \begin{bmatrix} \tau_{\mathbf{a}_1} & \tau_{\mathbf{p}_1} & \tau_{\mathbf{a}_2} & \tau_{\mathbf{p}_2} & \tau_{\mathbf{a}_3} & \tau_{\mathbf{p}_3} \end{bmatrix}^T$$

The term $\bar{\mathbf{M}}$ is the inertial matrix, $\bar{\mathbf{C}}$ the Coriolis and centrifugal force terms, and $\bar{\mathbf{N}}$ is a constant disturbance vector. The number of active and passive joints is n, $\mathbf{q_a} = [\theta_1 \ \theta_2 \ \theta_3]^T \in \mathbb{R}^m$ stands for the active joints and $\mathbf{q_p} = [\alpha_1 \ \alpha_2 \ \alpha_3]^T \in \mathbb{R}^{n-m}$ for the angles of the passive joints. In the same way, vectors $\boldsymbol{\tau}_\mathbf{a} = [\tau_{\mathbf{a}_1} \ \tau_{\mathbf{a}_2} \ \tau_{\mathbf{a}_3}]^T \in \mathbb{R}^m$, $\boldsymbol{\tau}_\mathbf{p} = [\tau_{\mathbf{p}_1} \ \tau_{\mathbf{p}_2} \ \tau_{\mathbf{p}_3}]^T \in \mathbb{R}^{n-m}$ correspond to the torques in the active and passive joints respectively. It is worth noting that in most parallel robots the angles of the active joints cannot play the role of generalized coordinates because their Forward Kinematics do not have a closed form solution., Therefore, it is not possible to write down the dynamic equations in terms of the active joints. For that reason, the development of the parallel robot dynamic model will consider the robot end-effector coordinates as a set of generalized coordinates, i.e. $\mathbf{q_e} = \mathbf{X}$. Substituting $\boldsymbol{\tau}$ in (25) into(21), we have

$$\mathbf{W}^T(\bar{\mathbf{M}}\ddot{\mathbf{q}} + \bar{\mathbf{C}}\dot{\mathbf{q}} + \bar{\mathbf{N}}) = \mathbf{S}^T \boldsymbol{\tau}_\mathbf{a}. \tag{26}$$

Taking the time derivative of (14) leads to

$$\ddot{\mathbf{q}} = \dot{\mathbf{W}}\dot{\mathbf{X}} + \mathbf{W}\ddot{\mathbf{X}} \tag{27}$$

Substituting $\dot{\mathbf{q}}$ and $\ddot{\mathbf{q}}$ given in(14) and (27) into (26) produces the following dynamic model

$$\mathbf{M}\ddot{\mathbf{X}}+\mathbf{C}\dot{\mathbf{X}}+\mathbf{N}=\mathbf{S}^T\boldsymbol{\tau}_{\mathrm{a}}, \tag{28}$$

where

$$\mathbf{M}=\mathbf{W}^T\bar{\mathbf{M}}\mathbf{W},$$
$$\mathbf{C}=\mathbf{W}^T\bar{\mathbf{M}}\dot{\mathbf{W}}+\mathbf{W}^T\bar{\mathbf{C}}\mathbf{W},$$
$$\mathbf{N}=\mathbf{W}^T\bar{\mathbf{N}}.$$

□

Note that the above model relates the active joint torques $\boldsymbol{\tau}_{\mathrm{a}}$ and the end-effector coordinates $\mathbf{X}$. The inertia matrix $\mathbf{M}$ and the Coriolis matrix $\mathbf{C}$ satisfy the following structural properties as long as matrix $\mathbf{W}$ has full rank

Property 1. Matrix $\mathbf{M}$ is a symmetric and positive definite.

Property 2. Matrix $\dot{\mathbf{M}}-2\mathbf{C}$ is skew-symmetric.

Property 3. There exists a positive constant k_{C1} such that

$$\|\mathbf{C}\|\leq k_{C1}\|\dot{\mathbf{X}}\|. \tag{29}$$

3. Model of the vision system

Consider the redundant planar parallel robot described previously together with its Cartesian coordinate frame $x_{\mathbf{R}}-y_{\mathbf{R}}$ (see Fig. 6). This coordinate frame defines a plane where the motion of the robot end-effector takes place. A camera providing an image of the whole robot workspace, including the robot end-effector, is perpendicular to the plane where the robot evolves. The optical center is located at a distance z with respect to the $x_{\mathbf{R}}-y_{\mathbf{R}}$ plane, and the intersection $\mathbf{O}=[O_x \;\; O_y]^T$ between the optical axis and the robot workspace is located anywhere in the robot workspace. Variable β denotes the orientation of the camera around the optical axis with respect to the negative side of axis $x_{\mathbf{R}}$ of the robot coordinate frame, measured clockwise.

The camera sensor has associated a coordinate frame called the image coordinate frame with axes x_i and y_i; they are parallel to the robot coordinate frame. The camera sensor captures the image that is later stored in the computer frame buffer and displayed in the computer screen. The visual feature of interest is the robot end-effector position $\mathbf{X}_{\mathrm{i}}=[x_i \;\; y_i]^T$ defined in the image coordinate frame; the units for $\mathbf{X}_{\mathrm{i}}$ are pixels. Image-processing algorithms, allow the estimation of the coordinate $\mathbf{X}_{\mathrm{i}}$. Thus, this estimate feeds the control algorithm without further processing. This later feature is common to all image-based Visual Servoing algorithms and permits avoiding camera calibration procedures.

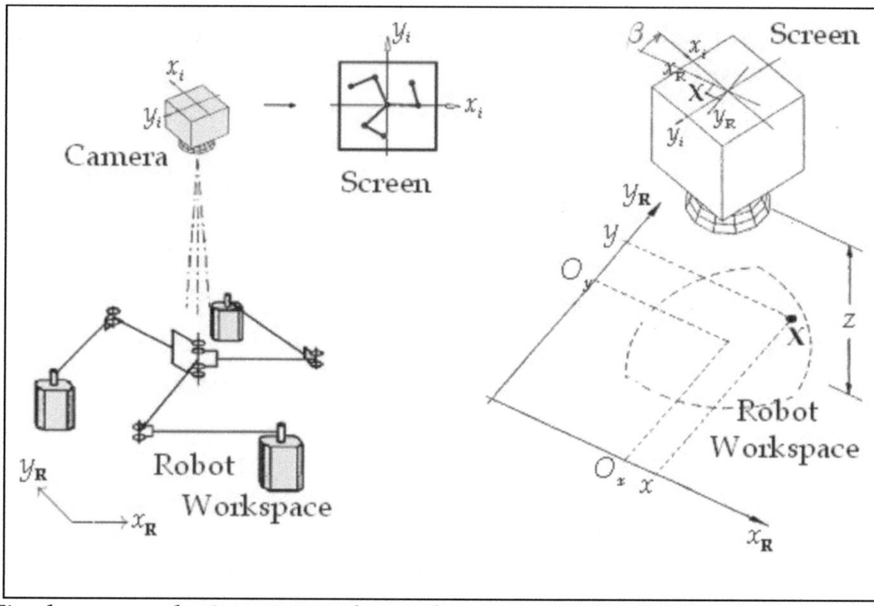

Fig. 6. Fixed-camera robotic system, robot and camera coordinate frames.

Let assume a perspective transformation as an ideal pinhole camera model (Kelly, 1996), the next relationship describes the position of the end-effector given in the image coordinate frame in terms of its position in the robot workspace

$$\mathbf{X}_\mathrm{i}=\eta h\mathbf{R}(\beta)(\mathbf{X}-\mathbf{O})+\mathbf{C}_\mathrm{i} \tag{30}$$

Parameter $\mathbf{C}_\mathrm{i}=\begin{bmatrix}C_{\mathrm{i}x} & C_{\mathrm{i}y}\end{bmatrix}^T$ is the image center, η is a scale factor in *pixels/m*, which is assumed negative, h is the magnification factor defined as

$$h=\frac{\lambda}{\lambda-z}<0 \tag{31}$$

where λ is the camera focal distance. $\mathbf{R}(\beta)\in SO(2)$ is the rotation matrix generated by clockwise rotating the camera about its optical axis by β radians

$$\mathbf{R}(\beta)=\begin{bmatrix}\cos\beta & -\sin\beta \\ \sin\beta & \cos\beta\end{bmatrix}. \tag{32}$$

The time derivative of (30) gives the end-effector linear velocity in terms of the image coordinate frame

$$\dot{\mathbf{X}}_\mathrm{i}=\eta h\mathbf{R}(\beta)\dot{\mathbf{X}}. \tag{33}$$

The following equation gives the desired end-effector position $\mathbf{X}^* = [x^* \quad y^*]^T$ expressed in terms of the image coordinate frame

$$\mathbf{X}_i^* = \eta h \mathbf{R}(\beta)(\mathbf{X}^* - \mathbf{O}) + \mathbf{C}_i \tag{34}$$

where $\mathbf{X}^* = [x^* \quad y^*]^T$ denotes the desired end-effector position expressed in the robot coordinate frame and located strictly inside the robot workspace, so there exists at least one (unknown) constant joint position vector, say $\mathbf{q_d} \in \mathbb{R}^6$ for which the robot end-effector reaches the desired position, in other words, there exists a nonempty set $Q \subset \mathbb{R}^n$ such that $\mathbf{X}^* = f(\mathbf{q_{da}}) \in \Omega$ for $\mathbf{q_{da}} \in Q$. At this point, it is convenient to introduce the definition of the image position error $\tilde{\mathbf{X}}_i$ as the visual distance between the measured and desired end-effector positions, see Fig. 7, i.e.

$$\tilde{\mathbf{X}}_i = \mathbf{X}_i^* - \mathbf{X}_i = \begin{bmatrix} x_i^* \\ y_i^* \end{bmatrix} - \begin{bmatrix} x_i \\ y_i \end{bmatrix}. \tag{35}$$

Therefore, expressions (30),(34), and (35) allow defining the image error vector $\tilde{\mathbf{X}}_i$ as

$$\tilde{\mathbf{X}}_i = \eta h \mathbf{R}(\beta)[\varphi(\mathbf{q_{da}}) - \varphi(\mathbf{q_a})]. \tag{36}$$

Assuming a fixed desired position, taking the time derivative of the image position error yields

$$\frac{d\tilde{\mathbf{X}}_i}{dt} = -\dot{\mathbf{X}}_i = -\eta h \mathbf{R}(\beta)\dot{\mathbf{X}}. \tag{37}$$

4. Visual PID control algorithm

4.1 Preliminaries

A standard linear PID control law has the following form

$$u = K_P e + K_I \int_0^t e(\sigma) d\sigma + K_D \dot{e} \tag{38}$$

Here, variable $e = r - y$ defines the error with r the set point and y the output variable; therefore, the error e as well as its time-integral and time-derivative feed this algorithm. In some cases, the time derivative $-\dot{y}$ replaces $\dot{e}$ leading to the controller

$$u = K_P e + K_I \int_0^t e(\sigma) d\sigma - K_D \dot{y} \tag{39}$$

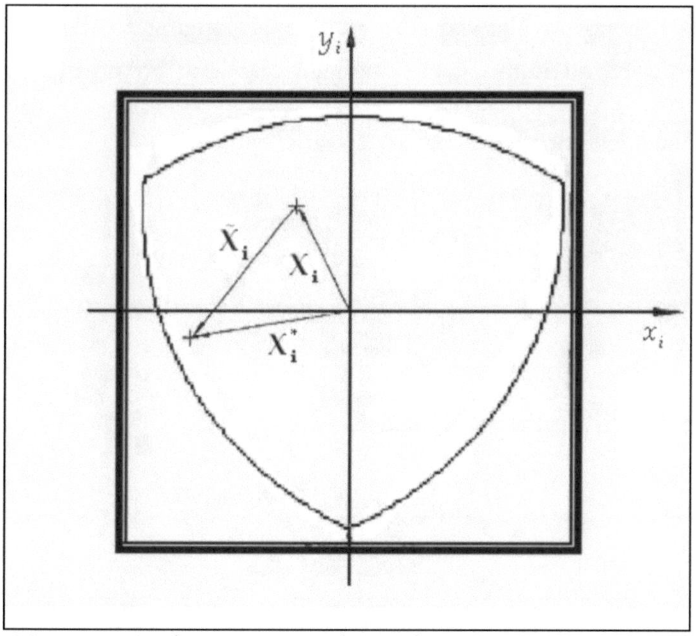

Fig. 7. Image position error in the image coordinate frame.

This last controller attenuates overshoots in face of abrupt changes in the set point value. When applied to joint control of robot manipulators, the linear PID controller leads to local stability or semi-global stability results. Applying a saturating function to the error, the Authors in references (Kelly, 1998) and (Santibañez & Kelly, 1998) were able to obtain global stability results. The next expression is an example of a PID controller using saturating functions

$$u = K_P e + K_I \int_0^t f(e(\sigma))d\sigma - K_D \dot{y}. \tag{40}$$

In this case, the term $f(\cdot)$ corresponds to a saturation function applied to the error. The proposed method for the control the redundant parallel robot will resort on a similar approach. The following definition states some key properties of the saturating functions used in the control law described in subsequent paragraphs.

***Definition** 1.* $\mathcal{F}(m,\varepsilon,\mathbf{x})$ with $1 \geq m > 0$, $\varepsilon > 0$ and $\mathbf{x} \in \mathbb{R}^n$ denotes the set of all continuous differentiable increasing functions $f(\mathbf{x}) = [f(x_1) \quad f(x_2) \quad \cdots \quad f(x_n)]^T$ such that

- $|x| \geq |f(x)| \geq m|x|, \quad \forall\, x \in \mathbb{R} : |x| < \varepsilon;$
- $\varepsilon \geq |f(x)| \geq m\varepsilon, \quad \forall\, x \in \mathbb{R} : |x| \geq \varepsilon;$
- $1 \geq (d/dx) f(x) \geq 0;$

where $|\cdot|$ stands for the absolute value.

Figure 8 depicts the region allowed for functions belonging to the set $\mathcal{F}(m,\varepsilon,\mathbf{x})$. Two important properties of functions $f(\mathbf{x})$ belonging to $\mathcal{F}(m,\varepsilon,\mathbf{x})$ are now stated

***Property* 4.** The Euclidean norm of $f(\mathbf{x}), \mathbf{x}\in\mathbb{R}^n$ satisfies

$$\|f(\mathbf{x})\| \geq \begin{cases} m\|\mathbf{x}\|, & if \|\mathbf{x}\| < \varepsilon \\ m\varepsilon, & if \|\mathbf{x}\| \geq \varepsilon \end{cases},$$

$$\|f(\mathbf{x})\| \leq \begin{cases} \|\mathbf{x}\|, & if \|\mathbf{x}\| < \varepsilon \\ \sqrt{n}\varepsilon, & if \|\mathbf{x}\| \geq \varepsilon. \end{cases}$$

***Property* 5.** The function $f(\mathbf{x})^T\mathbf{x}, \mathbf{x}\in\mathbb{R}^n$ satisfies

$$f(\mathbf{x})^T\mathbf{x} \geq \begin{cases} m\|\mathbf{x}\|^2, & if \ \|\mathbf{x}\| < \varepsilon \\ m\varepsilon\|\mathbf{x}\|, & if \ \|\mathbf{x}\| \geq \varepsilon. \end{cases}$$

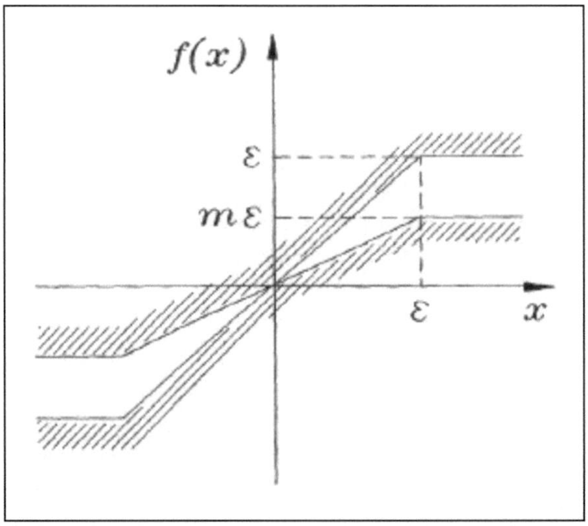

Fig. 8. Saturating functions $\mathcal{F}(m,\varepsilon,\mathbf{x})$.

4.2 Control problem formulation

Consider the robotic system described in Fig.6. Assume that the camera together with the vision system provide the position $\mathbf{X}_{\mathrm{i}}=[x_{\mathrm{i}}\ y_{\mathrm{i}}]^T$ of the robot end-effector expressed in the image coordinate frame. Suppose that measurements of joint position $\mathbf{q}$ and velocity $\dot{\mathbf{q}}$ are available. However, the magnification factor h and the position of the intersection of the camera axis with the robot workspace $\mathbf{O}=[O_x\ O_y]^T$ expressed in terms of the robot coordinate frame are assumed unknown. The control problem can be stated as that of designing a control law for the active joint actuator torques $\boldsymbol{\tau}_{\mathbf{a}}$ such that the robot end-

effector reaches, in the image supplied on the screen, the desired position defined in the robot workspace, i.e., the control law must ensure that $\lim_{t\to\infty}(\mathbf{X}_i^* - \mathbf{X}_i) = \mathbf{0}$ for $\mathbf{X}_i^* \in \Omega \subset \mathbb{R}^2$.
In order to solve the problem stated previously, assume that

$$\mathbf{S}^T \boldsymbol{\tau}_a = \mathbf{u}. \tag{41}$$

Variable $\mathbf{u}$ defines a control signal in terms of the end-effector coordinates, and drives the robot dynamics (28). Hence, torques $\boldsymbol{\tau}_a$ are the solutions of the following equation

$$\boldsymbol{\tau}_a = (\mathbf{S}^T)^{\dagger} \mathbf{u}. \tag{42}$$

The symbol $(\mathbf{S}^T)^{\dagger} = \mathbf{S}(\mathbf{S}^T\mathbf{S})^{-1}$ stands for the Moore-Penrose pseudo-inverse of $\mathbf{S}^T$, satisfying $\mathbf{S}^T(\mathbf{S}^T)^{\dagger} = I$, and $[\mathbf{S}^T(\mathbf{S}^T)^{\dagger}]^T = \mathbf{S}^{\dagger}\mathbf{S} = I$. Solution (42) makes sense only if the pseudo-inverse $(\mathbf{S}^T)^{\dagger}$ is well defined, i.e., if matrix $\mathbf{S}$ has full rank. Matrix $\mathbf{S}$ loose rank if the parallel robot reaches a singular configuration; in the sequel, matrix $\mathbf{S}$ is assumed full rank. Let us propose the following PID control law

$$\mathbf{u} = \mathbf{K}_P \mathbf{Y} + \mathbf{K}_I \int_0^t f(\mathbf{Y}(\sigma))d\sigma - \mathbf{K}_D \dot{\mathbf{X}} \tag{43}$$

Using (41) and (42) allows writing the control law (39) as follows

$$\boldsymbol{\tau}_a = (\mathbf{S}^T)^{\dagger}\left[\mathbf{K}_P \mathbf{Y} + \mathbf{K}_I \int_0^t f(\mathbf{Y}(\sigma))d\sigma - \mathbf{K}_D \dot{\mathbf{X}}\right] \tag{44}$$

The term $\mathbf{Y} = \mathbf{R}(\beta)^T \tilde{\mathbf{X}}_i$ corresponds to the rotated position error, variables $\mathbf{K}_P$, $\mathbf{K}_I$ and $\mathbf{K}_D$ are diagonal positive definite matrices and correspond to the proportional, integral and derivative actions. The above control law is composed of a linear Proportional Derivative (PD) term plus an integral action of the nonlinear function of the position error $f(\mathbf{Y})$. Note that the position error $\tilde{\mathbf{X}}_i$ feeds the proportional and the integral actions, whereas the active joint velocities $\dot{\mathbf{q}}_a$ feed the derivative action using the relationship $\dot{\mathbf{X}} = \mathbf{S}^{\dagger}\dot{\mathbf{q}}_a$. Note also that in order to implement control law (44) it is not necessary to know the parameters η and h; hence, camera calibration is not necessary. The Fig. 9 depicts the corresponding block diagram.

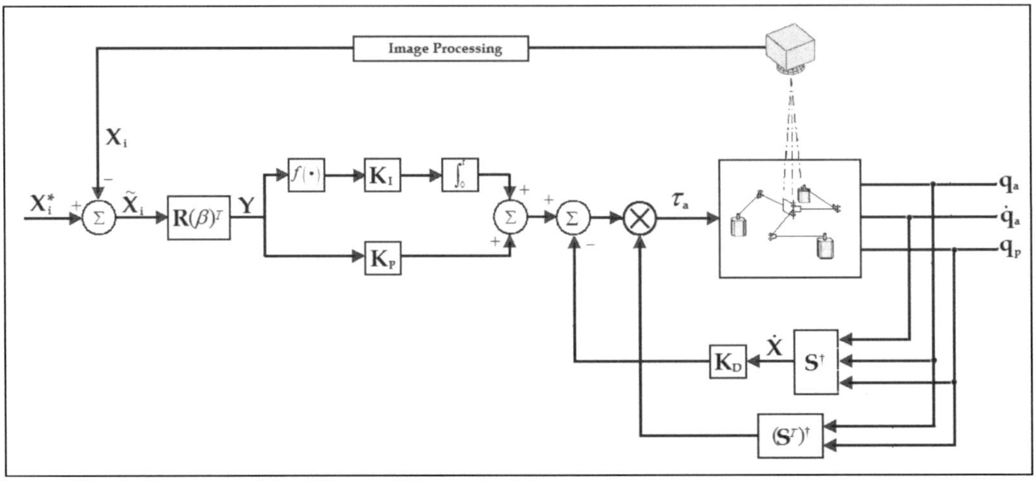

Fig. 9. Block diagram of the Visual PID control law.

Substituting control law (44) into the robot dynamics (28) and defining an auxiliary variable $\mathbf{Z}$ as

$$\mathbf{Z} = \int_0^t f(\mathbf{Y}(\sigma))d\sigma - \mathbf{K}_I^{-1}\mathbf{N} \quad (45)$$

yield the closed-loop dynamics

$$\frac{d}{dt}\begin{bmatrix}\mathbf{Y}\\ \dot{\mathbf{X}}\\ \mathbf{Z}\end{bmatrix} = \begin{bmatrix}-\eta h\dot{\mathbf{X}}\\ \mathbf{M}^{-1}\{\mathbf{K}_P\mathbf{Y}+\mathbf{K}_I\mathbf{Z}-\mathbf{K}_D\dot{\mathbf{X}}-\mathbf{C}\dot{\mathbf{X}}\}\\ f(\mathbf{Y})\end{bmatrix} \quad (46)$$

The following proposition provides conditions on the controller gains $\mathbf{K}_P, \mathbf{K}_D$, and $\mathbf{K}_I$ guaranteeing the asymptotic stability of the equilibrium of the closed-loop dynamics.

***Proposition* 2.** Consider the robot dynamics (28) together with control law (44) where $f(\mathbf{Y}) \in \mathcal{F}(m,\varepsilon,x)$. Assume that the PID controller gains fulfill

$$\lambda_{\min}\{\mathbf{K}_D\} > \lambda_{\max}\{\mathbf{M}\} + k_{C2}, k_{C2} > 0 \quad (47)$$

$$\lambda_{\min}\{\mathbf{K}_P\} > \lambda_{\max}\{\mathbf{K}_I\} + \frac{2}{\eta h}\lambda_{\max}\{\mathbf{M}\} \quad (48)$$

Then, the equilibrium $\begin{bmatrix}\mathbf{Y} & \dot{\mathbf{X}} & \mathbf{Z}\end{bmatrix}^T = \begin{bmatrix}0 & 0 & 0\end{bmatrix}^T$ of (46) is asymptotically stable.

***Proof of Proposition* 2:** The stability analysis employs the following Lyapunov Function Candidate

$$V(\mathbf{Y},\dot{\mathbf{X}},\mathbf{Z})=\frac{1}{2}\left[\dot{\mathbf{X}}-\frac{1}{\eta h}f(\mathbf{Y})\right]^T\mathbf{M}\left[\dot{\mathbf{X}}-\frac{1}{\eta h}f(\mathbf{Y})\right]+\frac{1}{2\eta h}[\mathbf{Z}+\mathbf{Y}]^T\mathbf{K_I}[\mathbf{Z}+\mathbf{Y}]+\frac{1}{\eta^2h^2}\int_0^{\mathbf{Y}}f(w)^T\mathbf{K_D}dw$$
$$+\frac{1}{2\eta h}\mathbf{Y}^T[\mathbf{K_P}-\mathbf{K_I}]\mathbf{Y}-\frac{1}{2\eta^2h^2}f(\mathbf{Y})^T\mathbf{M}f(\mathbf{Y}). \tag{49}$$

The first term is a nonnegative function of $\mathbf{Y}$ and $\dot{\mathbf{X}}$, while the second is a nonnegative function of variables $\mathbf{Y}$ and $\mathbf{Z}$. Using the fact that $\mathbf{K_D}$ is a diagonal positive definite matrix, $f(\mathbf{0})=\mathbf{0}$, and the entries of $f(\mathbf{Y})$ are increasing functions, it is not difficult to show that the third term satisfies

$$\frac{1}{\eta^2h^2}\int_0^{\mathbf{Y}}f(w)^T\mathbf{K_D}dw>0,\ \forall\ \mathbf{Y}\neq 0 \tag{50}$$

Therefore, this term is positive definite with respect to $\mathbf{Y}$. For the remaining terms, notice that using the Rayleigh-Ritz inequality leads to

$$\frac{1}{2\eta h}\mathbf{Y}^T[\mathbf{K_P}-\mathbf{K_I}]\mathbf{Y}-\frac{1}{2\eta^2h^2}f(\mathbf{Y})^T\mathbf{M}f(\mathbf{Y})\geq\frac{1}{2\eta h}\left[\lambda_{\min}\{\mathbf{K_P}\}-\lambda_{\max}\{\mathbf{K_I}\}\right]\|\mathbf{Y}\|^2-\frac{1}{2\eta^2h^2}\lambda_{\max}\{\mathbf{M}\}\|f(\mathbf{Y})\|^2.$$

The above result and Property 4 yields

$$\frac{1}{2\eta h}\left[\lambda_{\min}\{\mathbf{K_P}\}-\lambda_{\max}\{\mathbf{K_I}\}\right]\|\mathbf{Y}\|^2-\frac{1}{2\eta^2h^2}\lambda_{\max}\{\mathbf{M}\}\|f(\mathbf{Y})\|^2\geq\begin{cases}\frac{1}{2\eta h}\left[\lambda_{\min}\{\mathbf{K_P}\}-\lambda_{\max}\{\mathbf{K_I}\}-\frac{1}{\eta h}\lambda_{\max}\{\mathbf{M}\}\right]\|\mathbf{Y}\|^2, & if\ \|\mathbf{Y}\|<\varepsilon\\ \frac{1}{2\eta h}\left[\lambda_{\min}\{\mathbf{K_P}\}-\lambda_{\max}\{\mathbf{K_I}\}-\frac{2}{\eta h}\lambda_{\max}\{\mathbf{M}\}\right]\varepsilon^2, & if\ \|\mathbf{Y}\|\geq\varepsilon.\end{cases} \tag{51}$$

The right-hand side of (51) is a positive definite function with respect to $\mathbf{Y}$ because of inequality (48); therefore, the Lyapunov function candidate (49) is a positive definite function. The following equation gives the time derivative of Lyapunov Function Candidate (49)

$$\frac{d}{dt}V(\mathbf{Y},\dot{\mathbf{X}},\mathbf{Z})=\dot{\mathbf{X}}^T\mathbf{M}\ddot{\mathbf{X}}-\frac{1}{\eta h}\dot{\mathbf{X}}^T\mathbf{M}\dot{f}(\mathbf{Y})-\frac{1}{\eta h}f(\mathbf{Y})^T\mathbf{M}\ddot{\mathbf{X}}+\frac{1}{\eta^2h^2}f(\mathbf{Y})^T\mathbf{M}\dot{f}(\mathbf{Y})+\frac{1}{2}\dot{\mathbf{X}}^T\dot{\mathbf{M}}\dot{\mathbf{X}}-\frac{1}{\eta h}f(\mathbf{Y})^T\dot{\mathbf{M}}\dot{\mathbf{X}}+\frac{1}{2\eta^2h^2}f(\mathbf{Y})^T\dot{\mathbf{M}}f(\mathbf{Y})$$
$$+\frac{1}{\eta h}\mathbf{Z}^T\mathbf{K_I}\dot{\mathbf{Z}}+\frac{1}{\eta h}\mathbf{Z}^T\mathbf{K_I}\dot{\mathbf{Y}}+\frac{1}{\eta h}\mathbf{Y}^T\mathbf{K_I}\dot{\mathbf{Z}}+\frac{1}{\eta h}\mathbf{Y}^T\mathbf{K_I}\dot{\mathbf{Y}}+\frac{d}{dt}\left[\frac{1}{\eta^2h^2}\int_0^{\mathbf{Y}}f(w)^T\mathbf{K_D}dw\right]+\frac{1}{\eta h}\mathbf{Y}^T\mathbf{K_P}\dot{\mathbf{Y}}-\frac{1}{\eta h}\mathbf{Y}^T\mathbf{K_I}\dot{\mathbf{Y}}$$
$$-\frac{1}{\eta^2h^2}f(\mathbf{Y})^T\mathbf{M}\dot{f}(\mathbf{Y})-\frac{1}{2\eta^2h^2}f(\mathbf{Y})^T\dot{\mathbf{M}}f(\mathbf{Y}).$$

Applying the Leibnitz rule to the time derivative of the integral term produces

$$\frac{d}{dt}\left[\frac{1}{\eta^2h^2}\int_0^{\mathbf{Y}}f(w)^T\mathbf{K_D}dw\right]=\frac{1}{\eta^2h^2}f(\mathbf{Y})^T\mathbf{K_D}\dot{\mathbf{Y}}.$$

From the above, the Lyapunov Functions Candidate time derivative becomes

$$\frac{d}{dt}V(\mathbf{Y},\dot{\mathbf{X}},\mathbf{Z})=\dot{\mathbf{X}}^T\mathbf{M}\ddot{\mathbf{X}}-\frac{1}{\eta h}\dot{\mathbf{X}}^T\mathbf{M}\dot{f}(\mathbf{Y})-\frac{1}{\eta h}f(\mathbf{Y})^T\mathbf{M}\ddot{\mathbf{X}}+\frac{1}{2}\dot{\mathbf{X}}^T\dot{\mathbf{M}}\dot{\mathbf{X}}-\frac{1}{\eta h}f(\mathbf{Y})^T\dot{\mathbf{M}}\dot{\mathbf{X}}$$
$$+\frac{1}{\eta h}\mathbf{Z}^T\mathbf{K}_I\dot{\mathbf{Z}}+\frac{1}{\eta h}\mathbf{Z}^T\mathbf{K}_I\dot{\mathbf{Y}}+\frac{1}{\eta h}\mathbf{Y}^T\mathbf{K}_I\dot{\mathbf{Z}}+\frac{1}{\eta h}\mathbf{Y}^T\mathbf{K}_I\dot{\mathbf{Y}}+\frac{1}{\eta^2h^2}f(\mathbf{Y})^T\mathbf{K}_D\dot{\mathbf{Y}}+\frac{1}{\eta h}\mathbf{Y}^T\mathbf{K}_P\dot{\mathbf{Y}}-\frac{1}{\eta h}\mathbf{Y}^T\mathbf{K}_I\dot{\mathbf{Y}} \tag{52}$$

Note that the time derivative of the saturating function $f(\mathbf{Y})$ fulfills $\dot{f}(\mathbf{Y})=-\eta hF(\mathbf{Y})\dot{\mathbf{X}}$. The term $F(\mathbf{Y})$ is a diagonal matrix, and its entries $\partial f(\mathbf{Y}_j)/\partial \mathbf{Y}_j;\ j=1,2$ are nonnegative and smaller than or equal to one. Substituting the closed-loop dynamics (46) into (52) yields

$$\frac{d}{dt}V(\mathbf{Y},\dot{\mathbf{X}},\mathbf{Z})=\dot{\mathbf{X}}^T\mathbf{K}_P\mathbf{Y}+\dot{\mathbf{X}}^T\mathbf{K}_I\mathbf{Z}-\dot{\mathbf{X}}^T\mathbf{K}_D\dot{\mathbf{X}}-\dot{\mathbf{X}}^T\mathbf{C}\dot{\mathbf{X}}+\dot{\mathbf{X}}^T\mathbf{M}F(\mathbf{Y})\dot{\mathbf{X}}$$
$$-\frac{1}{\eta h}f(\mathbf{Y})^T\mathbf{K}_P\mathbf{Y}-\frac{1}{\eta h}f(\mathbf{Y})^T\mathbf{K}_I\mathbf{Z}+\frac{1}{\eta h}f(\mathbf{Y})^T\mathbf{K}_D\dot{\mathbf{X}}+\frac{1}{\eta h}f(\mathbf{Y})^T\mathbf{C}\dot{\mathbf{X}}+\frac{1}{2}\dot{\mathbf{X}}^T\dot{\mathbf{M}}\dot{\mathbf{X}}-\frac{1}{\eta h}f(\mathbf{Y})^T\dot{\mathbf{M}}\dot{\mathbf{X}}$$
$$+\frac{1}{\eta h}\mathbf{Z}^T\mathbf{K}_If(\mathbf{Y})-\mathbf{Z}^T\mathbf{K}_I\dot{\mathbf{X}}+\frac{1}{\eta h}\mathbf{Y}^T\mathbf{K}_If(\mathbf{Y})-\mathbf{Y}^T\mathbf{K}_I\dot{\mathbf{X}}-\frac{1}{\eta h}f(\mathbf{Y})^T\mathbf{K}_D\dot{\mathbf{X}}-\mathbf{Y}^T\mathbf{K}_P\dot{\mathbf{X}}+\mathbf{Y}^T\mathbf{K}_I\dot{\mathbf{X}}.$$

Some simplifications and the use of *Property* 2 lead to the following expression for the time derivative of the Lyapunov Function Candidate (49) along the trajectories of the closed-loop system (46)

$$\dot{V}(\mathbf{Y},\dot{\mathbf{X}},\mathbf{Z})=-\dot{\mathbf{X}}^T[\mathbf{K}_D-\mathbf{M}F(\mathbf{Y})]\dot{\mathbf{X}}-\frac{1}{\eta h}f(\mathbf{Y})^T[\mathbf{K}_P-\mathbf{K}_I]\mathbf{Y}-\frac{1}{\eta h}f(\mathbf{Y})^T\mathbf{C}^T\dot{\mathbf{X}} \tag{53}$$

By using *Properties* 3 and 4 we have

$$-\frac{1}{\eta h}f(\mathbf{Y})^T\mathbf{C}^T\dot{\mathbf{X}}\le\frac{1}{\eta h}k_{C1}\sqrt{2}\varepsilon\|\dot{\mathbf{X}}\|^2=k_{C2}\|\dot{\mathbf{X}}\|^2.$$

On the other hand, note that

$$\dot{\mathbf{X}}^T\mathbf{M}F(\mathbf{Y})\dot{\mathbf{X}}\le\lambda_{\max}\{\mathbf{M}\}\|\dot{\mathbf{X}}\|^2.$$

Therefore, the time derivative of the Lyapunov Function Candidate (53) satisfies

$$\dot{V}(\mathbf{Y},\dot{\mathbf{X}},\mathbf{Z})\le-\gamma\|\dot{\mathbf{X}}\|^2-\frac{1}{\eta h}f(\mathbf{Y})^T[\mathbf{K}_P-\mathbf{K}_I]\mathbf{Y} \tag{54}$$

The parameter $\gamma=\lambda_{\min}\{\mathbf{K}_D\}-\lambda_{\max}\{\mathbf{M}\}-k_{C2}$ is positive because of the selection of $\mathbf{K}_D$ in (47). The fact that $\mathbf{K}_P$ and $\mathbf{K}_I$ are diagonal positive definite matrices and $\mathbf{Y}_if(\mathbf{Y}_i)\ge0$ allows establishing the following upper bound for the second term of (54)

$$-\frac{1}{\eta h}f(\mathbf{Y})^T[\mathbf{K}_P-\mathbf{K}_I]\mathbf{Y}\le-\frac{1}{\eta h}[\lambda_{\min}\{\mathbf{K}_P\}-\lambda_{\max}\{\mathbf{K}_I\}]f(\mathbf{Y})^T\mathbf{Y}.$$

Taking into account *Property 5* leads to

$$-\frac{1}{\eta h} f(\mathbf{Y})^T [\mathbf{K_P} - \mathbf{K_I}] \mathbf{Y} \leq \begin{cases} -\mu \|\mathbf{Y}\|^2, & \text{if } \|\mathbf{Y}\| < \varepsilon \\ -\mu\varepsilon \|\mathbf{Y}\|, & \text{if } \|\mathbf{Y}\| \geq \varepsilon \end{cases} \tag{55}$$

The choice of $\mathbf{K_P}$ in (48) ensures $\mu = \frac{m}{\eta h}[\lambda_{min}\{\mathbf{K_P}\} - \lambda_{max}\{\mathbf{K_I}\}] > 0$.

Therefore, incorporating (55) into(54) produces the following a negative semi-definite function

$$\dot{V}(\mathbf{Y}, \dot{\mathbf{X}}, \mathbf{Z}) \leq \begin{cases} -\gamma \|\dot{\mathbf{X}}\|^2 - \mu \|\mathbf{Y}\|^2, & \text{if } \|\mathbf{Y}\| < \varepsilon \\ -\gamma \|\dot{\mathbf{X}}\|^2 - \mu\varepsilon \|\mathbf{Y}\|, & \text{if } \|\mathbf{Y}\| \geq \varepsilon \end{cases} \tag{56}$$

Using the fact that the Lyapunov Function Candidate (49) is a positive definite function and its time derivative is a negative semi-definite function, allows concluding that the equilibrium of the closed-loop system (46) is stable. Finally, by invoking the LaSalle's invariance principle permits establishing asymptotical stability as follows. Since $\dot{V}(\mathbf{Y}, \dot{\mathbf{X}}, \mathbf{Z}) \equiv 0$ if and only if $\dot{\mathbf{X}}$ and $\mathbf{Y}$ are zero. This implies that $\ddot{\mathbf{X}}$, $\dot{\mathbf{Y}}$, and $f(\mathbf{Y})$ are also zero; then, from the closed-loop system (46) , it follows that $\mathbf{M}^{-1}\{\mathbf{K_P Y} + \mathbf{K_I Z} - \mathbf{K_D}\dot{\mathbf{X}} - \mathbf{C}\dot{\mathbf{X}}\} = \mathbf{0}$. This result allows concluding $\mathbf{K_I Z} = 0$. Therefore, $\mathbf{Z} = 0$ because $\mathbf{K_I}$ is a diagonal positive definite matrix. Thus, $\dot{V}(\mathbf{Y}, \dot{\mathbf{X}}, \mathbf{Z}) \equiv 0$ in the invariant set $\{\mathbf{Y} = \mathbf{0}, \dot{\mathbf{X}} = \mathbf{0}, \mathbf{Z} = \mathbf{0}\}$ and asymptotic stability follows.

□

Some comments regarding the proposed control law are worth making. Firstly, note that the measurements provided by the vision system feed the integral and proportional actions. The Derivative action employs velocity measurements from the active joints; then, using the relationship $\dot{\mathbf{X}} = \mathbf{S}^{\dagger}\dot{\mathbf{q}}_\mathbf{a}$ allows obtaining velocity estimates of the robot end effector. In practice, since in most cases, the robot active joints are endowed only with position sensors, high-pass filter or backward differences approaches would permit estimating $\dot{\mathbf{q}}_\mathbf{a}$ from position measurements. An advantage of using $\dot{\mathbf{q}}_\mathbf{a}$ for generating the Derivative action is that position measurements at the active joints, supplied in most cases by optical encoders, are obtained at higher sampling rates compared with the measurements provided by a vision system. The reader will note in the next section that the sampling rate for the incremental optical encoders associated to the active joints is five times faster than that corresponding to the measurements obtained though the vision system.

5. Experimental Results

Experiments conducted on a laboratory prototype (Fig. 10) display the performance of the proposed controller. The nominal link lengths of the prototype are $L = 15\,cm$. Brushed servomotors from Moog, model C34L80W40 drive the active joints. Incremental optical encoders attached to the motors provide position measurements corresponding to the vector $\mathbf{q_a}$. These motors steer the active joints through timing belts with a 3.6:1 ratio. Pulse width modulation digital amplifiers from Copley

Controls, model Junus 90 and working in current mode, drive the motors. Absolute optical encoders from US Digital, model A2, with 4096 pulses per turn, supply measurements of the robot active and passive joints angles θ_i and α_i that allows computing the Jacobians $\mathbf{S}^{\dagger}$ and $(\mathbf{S}^{T})^{\dagger}$.

Fig. 10. Laboratory prototype.

Two computers compose the control architecture; which is an update of the architecture presented in (Soria et al. 2006). The first computer, called the vision computer and endowed with an Intel Core2 processor running at 2.4 GHz, executes image acquisition; a Dalsa Camera, model CA-1D-128A is connected to this computer by means of a National Instruments card, model NI-1422. Image processing is performed using Visual C++ and the DIAS software[1]. The second computer, called the control computer and endowed with an Intel 4 processor running at 3.0 GHz, executes the control algorithm and performs data logging. This computer receives data from the vision computer through an RS-232 port at 115 *Kbaud*. Data acquisition is carried out through a data card from Quanser consulting, model MultiQ 3. This card reads signals from the optical incremental encoders attached to the motors and supplies control voltages to the power amplifiers. Optical absolute encoders connect to the control computer through an RS-232 using an AD2-B adapter from US Digital.

Algorithms are coded using the Matlab/Simulink 5.2 software under the Wincon 3.02 real-time environment. A counter in the MultiQ 3 card sets a sampling period of

$T_{ie} = 0.5ms$, which corresponds to the master clock of the closed-loop system; this sampling period also sets the sampling time for reading the active joint incremental optical encoders. The image sampling period is $T_{im} = 5ms$; during this time interval, the vision computer executes data acquisition and processing; it also includes the time required to send the robot end-effector coordinates to the control computer through the RS-232 link. It is worth mentioning that T_{im} corresponds to the time delay introduced in the visual measurements. The absolute encoder measurements are sampled every $T_{ab} = 15ms$. The sampling time for the visual and absolute encoder, measurements are synchronized with the master clock. The choice for the numerical method in Simulink was the ODE 45 Dormand-Price algorithm. Gains for the proposed controller were set to $\mathbf{K_P} = diag\{0.22 \quad 0.22\}$, $\mathbf{K_D} = diag\{0.004 \quad 0.004\}$, and $\mathbf{K_I} = diag\{0.176 \quad 0.156\}$. The reference x_i^* is square wave of 16 pixels of amplitude, with a frequency of 0.2 Hz, while the reference y_i^* is a square wave with a frequency of 0.4 Hz. Fig 12 depicts the experimental position control results without and with integral action. The upper part in Fig. 12 corresponds to the x_i coordinate whereas the bottom part corresponds to the y_i coordinate. Fig.13 depicts the image position errors; note that when the reference changes, the position error settles around 0.5 pixels using the integral action. These results indicate that the integral action removes the steady state error without greatly affecting the transient response.

Fig. 11. Camera with image coordinate frame parallel to the robot coordinate frame.

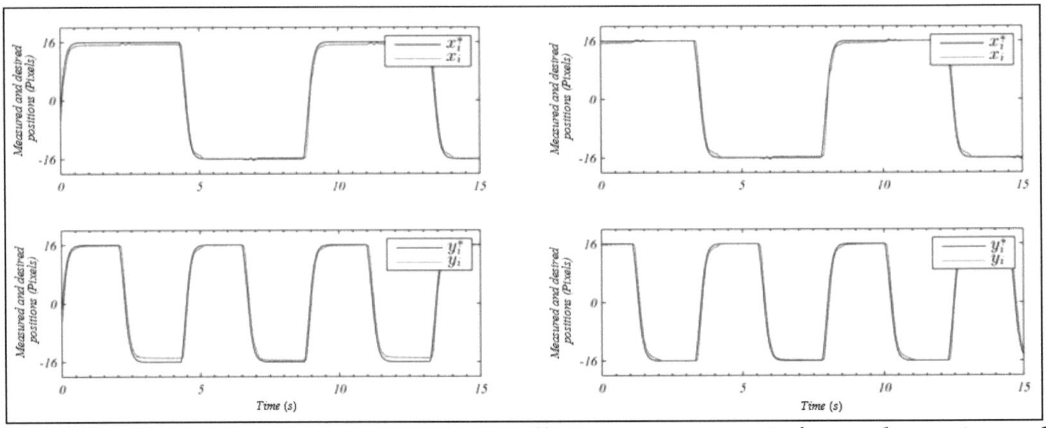

Fig. 12. Desired and measured end effector positions: Left, without integral action $\mathbf{K}_\mathrm{I} = diag\{0 \quad 0\}$; right, with integral action $\mathbf{K}_\mathrm{I} = diag\{0.176 \quad 0.156\}$.

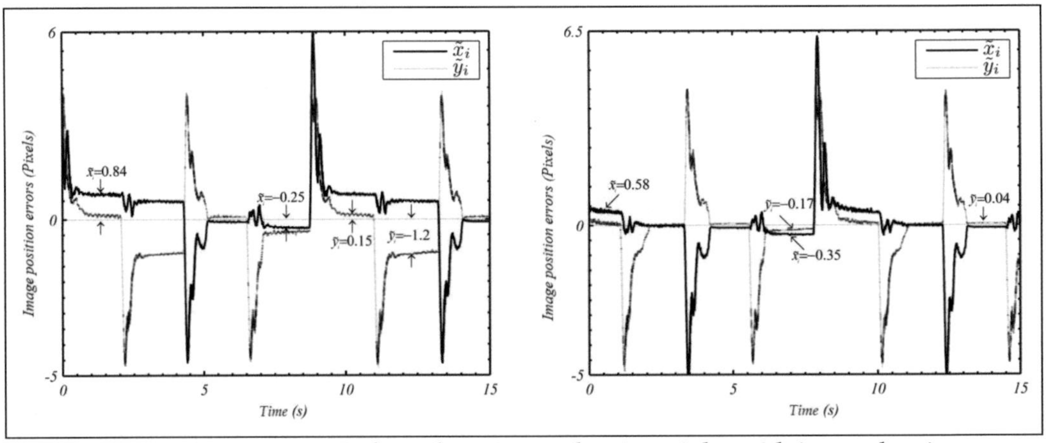

Fig. 13. Image position errors: Left, without integral action; right, with integral action.

6. Conclusion

This chapter has presented some modeling and control issues related to a class of overactuated planar parallel robots. After reviewing the kinematic and dynamic modeling of this kind of robots, the Authors propose a novel imaged-based Proportional-Integral-Derivative regulator. A key element in this control law is the measurement of the end-effector position using a vision system. This feature avoids using the robot Forward Kinematics employed traditionally for controlling planar parallel robots, and which requires an off-line calibration. Moreover, the proposed control law does not rely on camera calibration. A theoretical study provides conditions on the PID gains for obtaining asymptotic closed-loop stability. A practical implementation of the proposed method using a laboratory prototype shows a good performance of the closed-loop system. The experiments indicate that, as expected in a PID controller, the integral action removes the steady state error without a noticeable degradation in the transient response.

7. References

Angel, L.; Traslosheros, A.; Sebastian, J.M.; Pari, L.; Carelli, R. & Roberti, F. (2008). Vision-based control of the robotenis system, *Recent Progress in Robotics*, Springer Verlag. LNCIS 370, pp. 229-240.

Arimoto, S. & Miyazaki, F. (1984). Stability and robustness of PID feedback control for robot manipulators of sensory capability, In: *Robotics Research: First International Symposium*, Brady, M. & Paul, R. P., Ed. Cambridge, 783-799, MIT Press, MA.

Andreff, N. & Martinet, Ph. (2006). Unifying kinematic modeling, identification, and control of a Gough-Stewart parallel robot into a vision-based framework, *IEEE Trans. on Robotics*, Vol. 22, No. 6.

Andreff, N.; Dallej, T. & Martinet, Ph. (2007). Image-based Visual Servoing of a Gough-Stewart Parallel Manipulator using Leg Observations, *The International Journal of Robotic Research*, Vol. 26, pp. 667-687.

Chaumette, F. & Hutchinson, S. (2006). *Visual servo control part I: Basic approaches*. IEEE Robotics & Automation Magazine, December 2006.

Chaumette, F. & Hutchinson, S. (2007). *Visual servo control part II: Advanced approaches*. IEEE Robotics & Automation Magazine, March 2007.

Cheng, H.; Yiu, Y.K. & Li, Z. (2003). Dynamics and Control of Redundantly Actuated Parallel Manipulators, *IEEE/ ASME Transactions on Mechatronics*, Vol. 8, No. 4, December 2003, 483-491.

Corke, P. (1996). *Visual Control of Robots: High performance Visual Servoing*. Taunton, Somerset, England: Research Studies Press.

Dallej, T.; Andreff, N. & Martinet, Ph. (2007). Image-based visual servoing of the I4R parallel robot without propioceptive sensors. *Proceedings of the Int. Conf. on Robotics and Automation*, Roma, Italy.

Hutchinson, S.; Hager, G.D. & Corke P.I. (1996) A tutorial on visual servo control. *IEEE Transactions on Robotics and Automation*, Vol. 12, No. 5, October 1996, 651-670.

Kelly, R. (1995). A tuning procedure for stable PID control of robot manipulators, *Robotica*, pt. 2, Vol. 13, Mar. /Apr. 1995, pp. 141-148.

Kelly, R. (1996). Robust Asymptotically Stable Visual Servoing of Planar Robots, *IEEE Transactions on Robotics and Automation*, Vol. 12, No. 5, October 1996, 759-766.

Kelly, R. (1998). Global Positioning of Robot Manipulators via PD Control plus a Class of Nonlinear Integral Actions, *IEEE Transactions on Automatic Control*, Vol. 43, No. 7, July 1998, pp. 934-938.

Kock, S. & Schumacher, W. (1998). A Parallel x-y Manipulator with Actuation Redundancy for High-Speed and Active-Stiffness Applications, *Proceeding of the 1998 IEEE International Conference on Robotics & Automation*, pp. 2295-2300, Belgium, May 1998, Leuven.

Kragic, D. & Christensen, H.I. (2005). *Survey on Visual Servoing for Manipulation*. Technical report ISNR KTH/NA/P-02/01-SE, Department of Numerical Analysis and Computing Science, University of Stockolm, Sweden.

Merlet, J.P. (2000). *Parallel Robots*, Klwuer Academic Publishers.

Papanikolopoulos, N. & Khosla, P. (1993). Adaptive Robotic Visual Tracking: Theory and Experiments, *IEEE Transactions on Automatic Control*, Vol. 38, No. 3, March 1993, 429-444.

Santibañez, V. & Kelly, R. (1998). A Class of Nonlinear PID Global Regulators for Robot Manipulators", *Proc. IEEE Int. Conf. on Robotics and Automation*, pp. 3601-3606.

Sebastian, J.M.; Traslosheros, A.; Angel, L.; Roberti, F. & Carelli, R. (2007). *Parallel robot high speed object tracking*. M. Kamel and A. Campilho (Eds.): ICIAR 2007, LNCS 4633, pp. 295.306.

Soria, A.; Garrido, R.; Vásquez, I. & Vázquez ,R. (2006). Architecture for rapid prototyping of visual controllers. *Robotics and Autonomous Systems*, Vol. 54, pp. 486-495.

Spong, M.W.; Hutchinson, S. & Vidyasagar, M. M.W. (2005) *Robot Modeling and Control*, Wiley.

Tsai, L.W. (1999). *Robot Analysis*. John Wiley and Sons Inc.

Weiss, L.; Sanderson, A. & Newman, C. (1987). Dynamic Sensor-Based Control of Robots with Visual Feedback, *IEEE Journal of Robotics and Automation*, Vol. RA-3, No. 5, October 1987, 404-417.

Wen, J. T. & Murphy, S. (1990). *PID control for robot manipulators*, CIRSSE Document # 54, Rensselaer Polytechnic Institute.

Wilson, W.; Williams Hulls, C. & Bell, G. (1996). Relative End-Effector Control Using Cartesian Position Based Visual Servoing, *IEEE Transactions on Robotics and Automation*, Vol. 12, No.5, October 1996, pp. 684-696.

[1] Voss, K.; Ortmann,W. & Suesse, H. (1998). DIAS-Interactive Image Processing System, V 5.0, Friedrich-Schiller-University Jena, Germany.

3

Pid Controller with Roll Moment Rejection for Pneumatically Actuated Active Roll Control (Arc) Suspension System

Khisbullah Hudha[1)], Fauzi Ahmad[1)], Zulkiffli Abd. Kadir[2)] and Hishamuddin Jamaluddin[3)]

[1)]*Vehicle Control and Biomechanics Research Group for Vehicle Research and Development (CeVReD) Universiti Teknikal Malaysia Melaka (UTeM) Hang Tuah Jaya, 76100 Durian Tunggal Melaka.*

[2)]*Dept. of Mech. Eng., Faculty of Engineering Universiti Pertahanan Nasional Malaysia (UPNM) Kem Sungai Besi, 57000 Kuala Lumpur, Malaysia*

[3)]*Faculty of Mechanical Engineering Universiti Teknologi Malaysia (UTM) 81310 UTM Skudai, Johor, Malaysia*

Abstract

This chapter presents a successful implementation of PID controller for a pneumatically actuated active roll control suspension system in both simulation and experimental studies. For the simulation model, a full vehicle model which consists of ride, handling and tire subsystems to study vehicle dynamics behavior in lateral direction is derived. The full vehicle model is then validated experimentally using an instrumented experimental vehicle based on the driver input from the steering wheel. Two types of vehicle dynamics test are performed for the purpose of model validation namely step steer test and double lane change test. The results of model validation show that the behaviors of the model closely follow the behavior of a real vehicle with acceptable error. An active roll control (ARC) suspension system is then developed on the validated full vehicle model to reduce unwanted vehicle motions during cornering maneuvers such as body roll angle, body roll rate, vertical acceleration of the body and body heave. The proposed controller structure for the ARC system is PID control with roll moment rejection loop. The ARC system is then implemented on an instrumented experimental vehicle in which four units of pneumatic actuators are installed in parallel arrangement with the passive suspension system. The results of the study shows that the proposed control structure is able to significantly improve the dynamics performance of the vehicle during step steer and double lane change maneuvers compared to a passive vehicle system. It can also be noted that the additional roll moment rejection loop is able to further improve the performance of the PID controller for the ARC system.

1. Introduction

PID controller is the most popular feedback controller used in the process industries. The algorithm is simple but it can provide excellent control performance despite variation in the dynamic characteristics of a process plant. PID controller is a controller that includes three elements namely proportional, integral and derivative actions. The PID controller was first placed on the market in 1939 and has remained the most widely used controller in process control until today (Araki, 2006). A survey performed in 1989 in Japan indicated that more than 90% of the controllers used in process industries are PID controllers and advanced versions of the PID controller (Takatsu *et al.*, 1998).

The use of electronic control systems in modern vehicles has increased rapidly and in recent years, electronic ontrol systems can be easily found inside vehicles, where they are responsible for smooth ride, cruise control, traction control, anti-lock braking, fuel delivery and ignition timing. The successful implementation of PID controller for automotive systems have been widely reported in the literatures such as for engine control (Ying *et al.*, 1999; Yuanyuan *et al.*, 2008; Bustamante *et al.*, 2000), vehicle air conditioning control (Zhang *et al.*, 2010), clutch control (Wu *et al.*, 2008; Wang *et al.*, 2001), brake control (Sugisaka *et al.*, 2006; Hashemi-Dehkordi *et al.*, 2009; Zhang *et al.*, 1999), active steering control (Marino *et al.*, 2009; Yan *et al.*, 2008), power steering control (Morita *et al.*, 2008), drive train control (Mingzhu *et al.*, 2008; Wei *et al.*, 2010; Xu, *et al.*, 2007), throttle control (Shoubo *et al.*, 2009; Tan *et al.*, 1999; Corno *et al.*, 2008) and suspension control (Ahmad *et al.*, 2008; Ahmad *et al.*, 2009a; Ahmad *et al.*, 2009b; Hanafi, 2010; Ayat *et al.*, 2002a).

Over the last two decades, various active chassis control systems for automotive vehicles have been developed and put to commercial utilization. In particular, Vehicle Dynamics Control (VDC) and Electronic Stability Program (ESP) systems have become very active and attracting research efforts from both academic community and automotive industries (Mammar and Koenig, 2002; McCann, 2000; Mokhiamar and Abe, 2002; Wang and Longoria, 2006). The main goals of active chassis control include improvement in vehicle stability, maneuverability and passenger comfort especially in adverse driving conditions.

Ignited by advanced electronic technology, many different active chassis control systems have been developed, such as traction control system (Borrelli *et al.*, 2006), active steering control (Falcone *et al.*, 2007), antilock braking system (Cabrera *et al.*, 2005), active roll control suspension system and others. This study is part of the continuous efforts in the prototype development of a pneumatically actuated active roll control suspension system for passenger vehicles. The proposed ARC system is used to minimize the effects of unwanted roll and vertical body motions of the vehicle in the presence of steering wheel input from the driver.

ARC system is a class of electronically controlled active suspension system. Although active suspension has been widely studied for decades, most of the research are focused on vehicle ride comfort, with only few papers (Williams and Haddad, 1995; Ayat *et al.*, 2002a; Wang *et al.*, 2005, Ayat *et al.*, 2002b) studying how an active suspension system can improve vehicle handling. It is well-known that a vehicle tends to roll on its longitudinal axis if the vehicle is subjected to steering wheel input due to the weight transfer from the inside to the outside wheels. Some control strategies for ARC systems have been proposed to cancel out lateral weight transfer using active force control strategy (Hudha *et al.* 2003), hybrid fuzzy-PID (Xinpeng and Duan, 2007), speed dependent gain scheduling control (Darling and Ross-Martin, 1997), roll angle and roll moment control (Miege and Cebon, 2002), state feedback controller optimized with genetic algorithm (Du and Dong, 2007) and the combination of yaw rate and side slip angle feedback control (Sorniotti and D'Alfio, 2007).

In this study, ARC system is developed using four units of pneumatic system installed between lower arms and vehicle body. The proposed control strategy for the ARC system is the combination of PID based feedback control and roll moment rejection based feed forward control. Feedback control is used to minimize unwanted body heave and body roll motions, while the feed forward control is intended to reduce the unwanted weight transfer during steering input maneuvers. The forces produced by the proposed control structure are used as the target forces by the four unit of pneumatic system.

The use of pneumatic actuator for an active roll control suspension system is a relatively new concept and has not been thoroughly explored. The use of pneumatic system is rare in active suspension application although they have several advantages compared with other actuation systems such as hydraulic system. The main advantage of pneumatic system is their power-to-weight ratio which is better than hydraulic system. They are also clean, simple system and comparatively low cost (Smaoui *et al.*, 2006). The disadvantage of pneumatic system is the unwanted nonlinearity because of the compressibility and springing effects of air (Situm *et al.*, 2005; Richer and Hurmuzlu, 2000). Due to these difficulties, early use of pneumatic actuators was limited to simple applications that required only positioning at the two ends of the stroke. But, during the past decade, many researchers have proposed various approaches to continuously control the pneumatic actuators (Ben-Dov and Salcudean, 1995; Wang *et al.*, 1999; Messina *et al.*, 2005). It is shown that the comparative advantages and difficulties of pneumatic system are still interesting and also a challenging problems in controller design in order to achieve reasonable performance in terms of position and force controls.

The proposed control strategy is optimized for a 14 degrees of freedom (DOF) full vehicle model. The full vehicle model consists of 7-DOF vehicle ride model and 7-DOF vehicle handling model coupled with Calspan tyre model. The full vehicle model can be used to study the behavior of vehicle in lateral, longitudinal and vertical directions due to both road and driver inputs. Calspan tire model is employed due to its capability to predict the behavior of a real tire better than Dugoff and Magic formula tire model (Kadir *et al.*, 2008).

Beside the proposed control structure, another consideration of this chapter is that the proposed control structure for the ARC system is implemented on a validated full vehicle model as well as on a real vehicle. It is common that the controllers, developed on the validated model, are ready to be implemented in practice with high level of confidence and need less fine tuning works. For the purpose of vehicle model validation, an instrumented experimental vehicle has been developed using a Malaysia National Car. Two types of road test namely step steer and double lane change test were performed using the instrumented experimental vehicle. The data obtained from the road tests are used as the validation benchmarks of the 14-DOF full vehicle model.

This chapter is organized as follows: The first section contains introduction and the review of some related works, followed by mathematical derivations of the 14-DOF full vehicle model with Calspan tyre model in the second section. The third section introduces the proposed controller structure for the ARC system. The fourth section presents the results of validation of the full vehicle model. Furthermore, improvements of vehicle dynamics performance on simulation studies and experimental tests using the proposed ARC system are presented in the fifth and the sixth section, respectively. The last section contains some conclusions.

2. Full Vehicle Modeling with Calspan Tire Model

The full-vehicle model of the passenger vehicle considered in this study consists of a single sprung mass (vehicle body) connected to four unsprung masses and is represented as a 14-DOF system as shown in Figure 1. The sprung mass is represented as a plane and is allowed to pitch, roll and yaw as well as to displace in vertical, lateral and longitudinal directions. The unsprung masses are allowed to bounce vertically with respect to the sprung mass. Each wheel is also allowed to rotate along its axis and only the two front wheels are free to steer.

2.1 Modeling Assumptions

Some of the modeling assumptions considered in this study are as follows: the vehicle body is lumped into a single mass which is referred to as the sprung mass, aerodynamic drag force is ignored, and the roll centre is coincident with the pitch centre and located just below the body center of gravity. The suspensions between the sprung mass and unsprung masses are modeled as passive viscous dampers and spring elements. Rolling resistance due to passive stabilizer bar and body flexibility are neglected. The vehicle remains grounded at all times and the four tires never lost contact with the ground during maneuvering. A 4 degrees tilt angle of the suspension system toward vertical axis is neglected ($\cos 4 = 0.998 \approx 1$). Tire vertical behavior is represented as a linear spring without damping, while the lateral and longitudinal behaviors are represented with Calspan model. Steering system is modeled as a constant ratio and the effect of steering inertia is neglected.

2.2 Vehicle Ride Model

The vehicle ride model is represented as a 7-DOF system. It consists of a single sprung mass (car body) connected to four unsprung masses (front-left, front-right, rear-left and rear-right wheels) at each corner of the vehicle body. The sprung mass is free to heave, pitch and roll while the unsprung masses are free to bounce vertically with respect to the sprung mass. The suspensions between the sprung mass and unsprung masses are modeled as passive viscous dampers and spring elements. While, the tires are modeled as simple linear springs without damping. For simplicity, all pitch and roll angles are assumed to be small. A similar model was used by Ikenaga (2000).

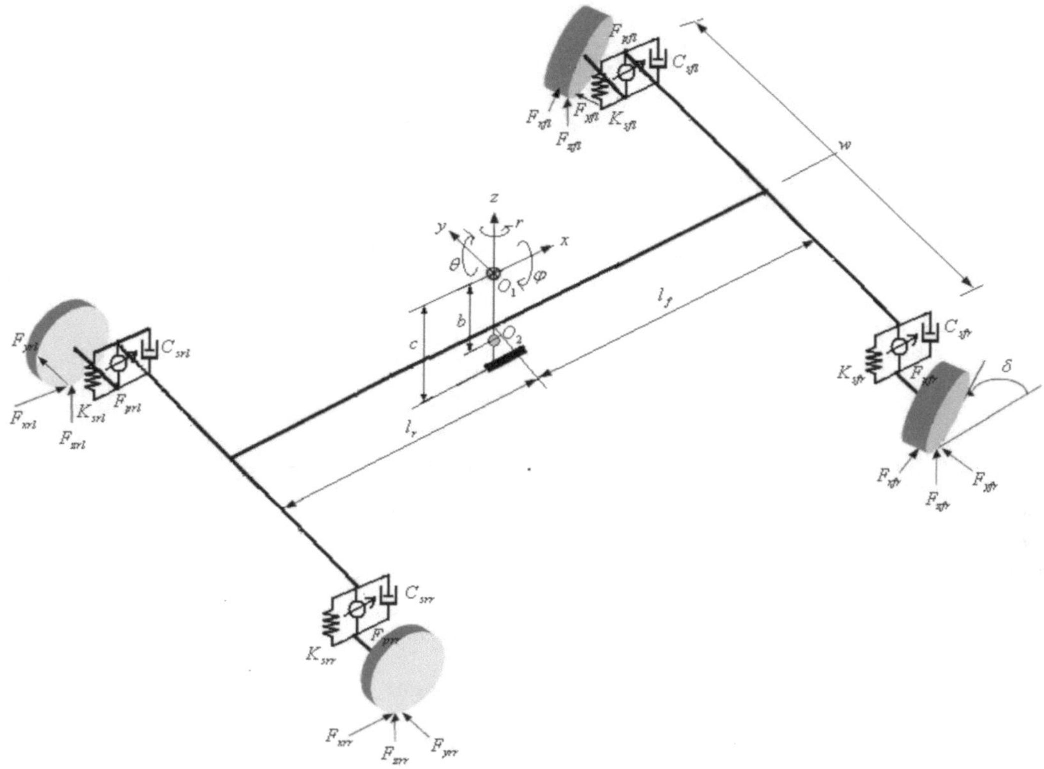

Fig. 1. A 14-DOF full vehicle ride and handling model

Referring to Figure 1, the force balance on sprung mass is given as

$$F_{fl} + F_{fr} + F_{rl} + F_{rr} + F_{pfl} + F_{pfr} + F_{prl} + F_{prr} = m_s \ddot{Z}_s \tag{1}$$

where,

F_{fl} = suspension force at front left corner
F_{fr} = suspension force at front right corner
F_{rl} = suspension force at rear left corner
F_{rr} = suspension force at rear right corner
m_s = sprung mass weight
$\ddot{Z}_s$ = sprung mass acceleration at body centre of gravity
$F_{pfl}; F_{pfr}; F_{prl}; F_{prr}$ = pneumatic actuator forces at front left, front right, rear left and rear right corners, respectively.

The suspension force at each corner of the vehicle is defined as the sum of the forces produced by suspension components namely spring force and damper force as the followings

$$
\begin{aligned}
F_{fl} &= K_{s,fl}\left(Z_{u,fl} - Z_{s,fl}\right) + C_{s,fl}\left(\dot{Z}_{u,fl} - \dot{Z}_{s,fl}\right)\\
F_{fr} &= K_{s,fr}\left(Z_{u,fr} - Z_{s,fr}\right) + C_{s,fr}\left(\dot{Z}_{u,fr} - \dot{Z}_{s,fr}\right)\\
F_{rl} &= K_{s,rl}\left(Z_{u,rl} - Z_{s,rl}\right) + C_{s,rl}\left(\dot{Z}_{u,rl} - \dot{Z}_{s,rl}\right)\\
F_{rr} &= K_{s,rr}\left(Z_{u,rr} - Z_{s,rr}\right) + C_{s,rr}\left(\dot{Z}_{u,rr} - \dot{Z}_{s,rr}\right)
\end{aligned}
\tag{2}
$$

where,

$K_{s,fl}$ = front left suspension spring stiffness
$K_{s,fr}$ = front right suspension spring stiffness
$K_{s,rr}$ = rear right suspension spring stiffness
$K_{s,rl}$ = rear left suspension spring stiffness
$C_{s,fr}$ = front right suspension damping
$C_{s,fl}$ = front left suspension damping
$C_{s,rr}$ = rear right suspension damping
$C_{s,rl}$ = rear left suspension damping
$Z_{u,fr}$ = front right unsprung mass displacement
$Z_{u,fl}$ = front left unsprung mass displacement
$Z_{u,rr}$ = rear right unsprung mass displacement
$Z_{u,rl}$ = rear left unsprung mass displacement
$\dot{Z}_{u,fr}$ = front right unsprung mass velocity
$\dot{Z}_{u,fl}$ = front left unsprung mass velocity
$\dot{Z}_{u,rr}$ = rear right unsprung mass velocity
$\dot{Z}_{u,rl}$ = rear left unsprung mass velocity

The sprung mass position at each corner can be expressed in terms of bounce, pitch and roll given by

$$
\begin{aligned}
Z_{s,fl} &= Z_s - l_f \sin\theta + 0.5w\sin\varphi\\
Z_{s,fr} &= Z_s - l_f \sin\theta - 0.5w\sin\varphi\\
Z_{s,rl} &= Z_s + l_r \sin\theta + 0.5w\sin\varphi\\
Z_{s,rr} &= Z_s + l_r \sin\theta - 0.5w\sin\varphi
\end{aligned}
\tag{3}
$$

It is assumed that all angles are small, therefore Eq. (3) becomes

$$\begin{aligned} Z_{s,fl} &= Z_s - l_f\theta + 0.5w\varphi \\ Z_{s,fr} &= Z_s - l_f\theta - 0.5w\varphi \\ Z_{s,rl} &= Z_s + l_r\theta + 0.5w\varphi \\ Z_{s,rr} &= Z_s + l_r\theta - 0.5w\varphi \end{aligned} \tag{4}$$

where,

l_f	= distance between front of vehicle and center of gravity of sprung mass
l_r	= distance between rear of vehicle and center of gravity of sprung mass
w	= track width
θ	= pitch angle at body centre of gravity
φ	= roll angle at body centre of gravity
$Z_{s,fl}$	= front left sprung mass displacement
$Z_{s,fr}$	= front right sprung mass displacement
$Z_{s,rl}$	= rear left sprung mass displacement
$Z_{s,rr}$	= rear right sprung mass displacement

By substituting Eq. (4) and its derivative (sprung mass velocity at each corner) into Eq. (2) and the resulting equations are then substituted into Eq. (1), the following equation is obtained

$$\begin{aligned} m_S\ddot{Z}_S = &-2(K_{S,f} + K_{S,r})Z_S - 2(C_{S,f} + C_{S,r})\dot{Z}_S + 2(l_f K_{S,f} - l_r C_{S,r})\theta \\ &+ 2(l_f C_{S,f} - l_r C_{S,r})\dot{\theta} + K_{sf} Z_{u,fl} + C_{S,f}\dot{Z}_{u,fl} + K_{sf} Z_{u,fr} + \\ &+ C_{S,f}\dot{Z}_{u,fr} K_{sr} Z_{u,rl} + C_{S,r}\dot{Z}_{u,rl} + K_{sr} Z_{u,rr} + C_{S,r}\dot{Z}_{u,rr} \\ &+ F_{pfl} + F_{pfr} + F_{prl} + F_{prr} \end{aligned} \tag{5}$$

where,

$\dot{\theta}$	= pitch rate at body centre of gravity
Z_s	= sprung mass displacement at body centre of gravity
$\dot{Z}_s$	= sprung mass velocity at body centre of gravity
$K_{s,f}$	= spring stiffness of front suspension ($K_{s,fl} = K_{s,fr}$)
$K_{s,r}$	= spring stiffness of rear suspension ($K_{s,rl} = K_{s,rr}$)
$C_{s,f}$	= $C_{s,fl} = C_{s,fr}$ = damping constant of front suspension
$C_{s,r}$	= $C_{s,rl} = C_{s,rr}$ = damping constant of rear suspension

Similarly, moment balance equations are derived for pitch θ and roll φ, and are given as

$$\begin{aligned}I_{yy}\ddot{\theta} &= 2\left(l_f K_{s,f} - l_r K_{s,r}\right)Z_s + 2\left(l_f C_{s,f} - l_r C_{s,r}\right)\dot{Z}_s - 2\left(l_f^2 K_{s,f} + l_r^2 K_{s,r}\right)\theta + \\ &- 2\left(l_f^2 C_{s,f} + l_r^2 C_{s,r}\right)\dot{\theta} - l_f K_{s,f} Z_{u,fl} - l_f C_{s,f} \dot{Z}_{u,fl} - l_f K_{s,f} Z_{u,fr} + \\ &- l_f C_{s,f} \dot{Z}_{u,fr} + l_r K_{s,r} Z_{u,rl} + l_r C_{s,r} \dot{Z}_{u,rl} + l_r K_{s,r} Z_{u,rr} + l_r C_{s,r} \dot{Z}_{u,rr} \\ &- (F_{pfl} + F_{pfr}) l_f + (F_{prl} + F_{prr}) l_r\end{aligned} \qquad (6)$$

$$\begin{aligned}I_{xx}\ddot{\varphi} &= -0.5w^2\left(K_{s,f} + K_{s,r}\right)\varphi - 0.5w^2\left(C_{s,f} + C_{s,r}\right)\dot{\varphi} + 0.5wK_{s,f} Z_{u,fl} + \\ &+ 0.5wC_{s,f}\dot{Z}_{u,fl} - 0.5wK_{s,f} Z_{u,fr} - 0.5wC_{s,f}\dot{Z}_{u,fr} + \\ &+ 0.5wK_{s,r} Z_{u,rl} + 0.5wC_{s,r}\dot{Z}_{u,rl} - 0.5wK_{s,r} Z_{u,rr} - 0.5wC_{s,r}\dot{Z}_{u,rr} \\ &+ (F_{pfl} + F_{prl})\frac{w}{2} - (F_{pfr} + F_{prr})\frac{w}{2}\end{aligned} \qquad (7)$$

where,

$\ddot{\theta}$	= pitch acceleration at body centre of gravity
$\ddot{\varphi}$	= roll acceleration at body centre of gravity
I_{xx}	= roll axis moment of inertia
I_{yy}	= pitch axis moment of inertia

By performing force balance analysis at the four wheels, the following equations are obtained

$$\begin{aligned}m_u\ddot{Z}_{u,fl} &= K_{s,f}Z_s + C_{s,f}\dot{Z}_s - l_f K_{s,f}\theta - l_f C_{s,f}\dot{\theta} + 0.5wK_{s,f}\varphi + \\ &+ 0.5wC_{s,f}\dot{\varphi} - \left(K_{s,f} + K_t\right)Z_{u,fl} - C_{s,f}\dot{Z}_{u,fl} + K_t Z_{r,fl} - F_{pfl}\end{aligned} \qquad (8)$$

$$\begin{aligned}m_u\ddot{Z}_{u,fr} &= K_{s,f}Z_s + C_{s,f}\dot{Z}_s - l_f K_{s,f}\theta - l_f C_{s,f}\dot{\theta} - 0.5wK_{s,f}\varphi + \\ &- 0.5wC_{s,f}\dot{\varphi} - \left(K_{s,f} + K_t\right)Z_{u,fr} - C_{s,f}\dot{Z}_{u,fr} + K_t Z_{r,fr} - F_{pfr}\end{aligned} \qquad (9)$$

$$\begin{aligned}m_u\ddot{Z}_{u,rl} &= K_{s,r}Z_s + C_{s,r}\dot{Z}_s + l_r K_{s,r}\theta + l_r C_{s,r}\dot{\theta} + 0.5wK_{s,r}\varphi + \\ &+ 0.5wC_{s.r}\dot{\varphi} - \left(K_{s,r} + K_t\right)Z_{u,rl} - C_{s,r}\dot{Z}_{u,rl} + K_t Z_{r,rl} - F_{prl}\end{aligned} \qquad (10)$$

$$\begin{aligned}m_u\ddot{Z}_{u,rr} &= K_{s,r}Z_s + C_{s,r}\dot{Z}_s + l_r K_{s,r}\theta + l_r C_{s,r}\dot{\theta} - 0.5wK_{s,r}\varphi + \\ &- 0.5wC_{s,r}\dot{\varphi} - \left(K_{s,r} + K_t\right)Z_{u,rr} - C_{s,r}\dot{Z}_{u,rr} + K_t Z_{r,rr} - F_{prr}\end{aligned} \qquad (11)$$

where,

$\ddot{Z}_{u,fr}$ = front right unsprung mass acceleration

$\ddot{Z}_{u,fl}$ = front left unsprung mass acceleration

$\ddot{Z}_{u,rr}$ = rear right unsprung mass acceleration

$\ddot{Z}_{u,rl}$ = rear left unsprung mass acceleration

$Z_{r,fr} = Z_{r,fl} = Z_{r,rr} = Z_{r,rl}$ = road profiles at front left, front right, rear right and rear left tyres respectively

2.3 Vehicle Handling Model

The handling model employed in this paper is a 7-DOF system as shown in Figure 2. It takes into account three degrees of freedom for the vehicle body in lateral and longitudinal motions as well as yaw motion (r) and one degree of freedom due to the rotational motion of each tire. In vehicle handling model, it is assumed that the vehicle is moving on a flat road. The vehicle experiences motion along the longitudinal x-axis and the lateral y-axis, and the angular motions of yaw around the vertical z-axis. The motion in the horizontal plane can be characterized by the longitudinal and lateral accelerations, denoted by a_x and a_y respectively, and the velocities in longitudinal and lateral direction, denoted by v_x and v_y, respectively.

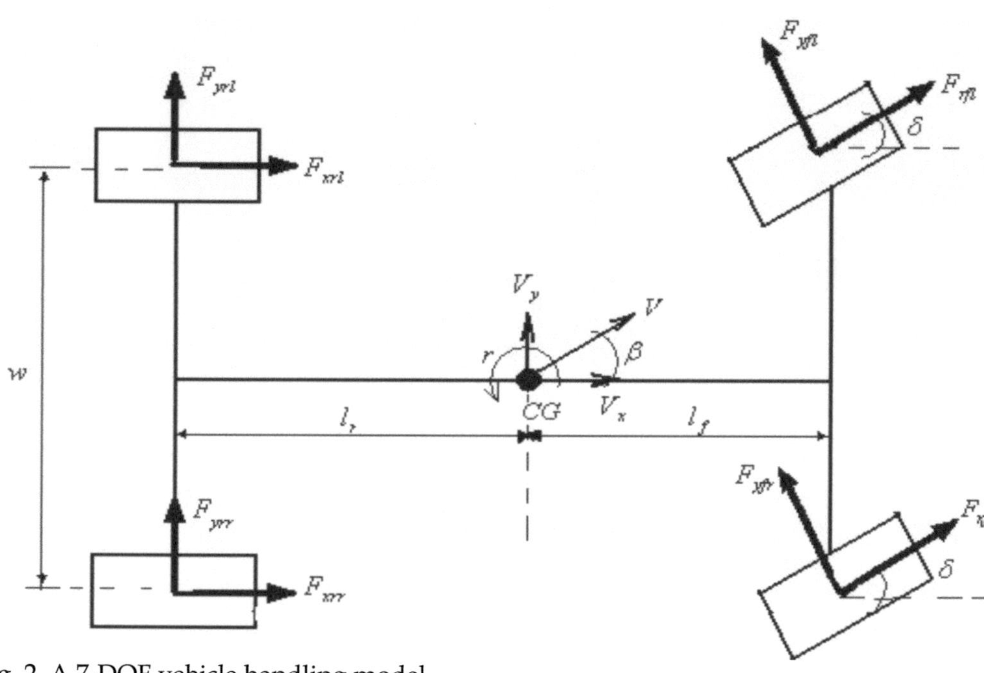

Fig. 2. A 7-DOF vehicle handling model

Acceleration in longitudinal x-axis is defined as

$$\dot{v}_x = a_x + v_y \dot{r} \tag{12}$$

By summing all the forces in x-axis, longitudinal acceleration can be defined as

$$a_x = \frac{F_{xfl}\cos\delta + F_{yfl}\sin\delta + F_{xfr}\cos\delta + F_{yfr}\sin\delta + F_{xrl} + F_{xrr}}{m_t} \tag{13}$$

Similarly, acceleration in lateral y-axis is defined as

$$\dot{v}_y = a_y - v_x \dot{r} \tag{14}$$

By summing all the forces in lateral direction, lateral acceleration can be defined as

$$a_y = \frac{F_{yfl}\cos\delta - F_{xfl}\sin\delta + F_{yfr}\cos\delta - F_{xfr}\sin\delta + F_{yrl} + F_{yrr}}{m_t} \tag{15}$$

where F_{xij} and F_{yij} denote the tire forces in the longitudinal and lateral directions, respectively, with the index (i) indicating front (f) or rear (r) tires and index (j) indicating left (l) or right (r) tires. The steering angle is denoted by δ, the yaw rate by $\dot{r}$ and m_t denotes the total vehicle mass. The longitudinal and lateral vehicle velocities v_x and v_y can be obtained by integrating of $\dot{v}_y$ and $\dot{v}_x$. They can be used to obtain the side slip angle, denoted by α. Thus, the slip angle of front and rear tires are found as

$$\alpha_f = \tan^{-1}\left(\frac{v_y + l_f r}{v_x}\right) - \delta_f \quad ; \tag{16}$$

and

$$\alpha_r = \tan^{-1}\left(\frac{v_y - l_r r}{v_x}\right) \tag{17}$$

where, α_f and α_r are the side slip angles at front and rear tires respectively. While l_f and l_r are the distance between front and rear tire to the body center of gravity respectively.

To calculate the longitudinal slip, longitudinal component of the tire velocity should be derived. The front and rear longitudinal velocity component is given by:

$$v_{wxf} = V_{tf} \cos \alpha_f \tag{18}$$

where, the speed of the front tire is,

$$V_{tf} = \sqrt{(v_y + l_f r)^2 + v_x^2} \tag{19}$$

The rear longitudinal velocity component is,

$$v_{wxr} = V_{tr} \cos \alpha_r \tag{20}$$

where, the speed of the rear tire is,

$$V_{tr} = \sqrt{(v_y + l_r r)^2 + v_x^2} \tag{21}$$

Then, the longitudinal slip ratio of front tire,

$$S_{af} = \frac{v_{wxf} - \omega_f R_w}{v_{wxf}}, \text{ under braking conditions} \tag{22}$$

The longitudinal slip ratio of rear tire is,

$$S_{ar} = \frac{v_{wxr} - \omega_r R_w}{v_{wxr}}, \text{ under braking conditions} \tag{23}$$

where, ω_r and ω_f are angular velocities of rear and front tires, respectively and R_w, is the wheel radius. The yaw motion is also dependent on the tire forces F_{xij} and F_{yij} as well as on the self-aligning moments, denoted by M_{zij} acting on each tire:

$$\begin{aligned}\ddot{r} = \frac{1}{J_z}\Big[& \frac{w}{2}F_{xfl}\cos\delta - \frac{w}{2}F_{xfr}\cos\delta + \frac{w}{2}F_{xrl} - \frac{w}{2}F_{xrr} + \frac{w}{2}F_{yfl}\sin\delta - \\ & \frac{w}{2}F_{yfr}\sin\delta - l_r F_{yrl} - l_r F_{yrr} + l_f F_{yfl}\cos\delta + l_f F_{yfr}\cos\delta - l_f F_{xfl}\sin\delta - \\ & l_f F_{xfr}\sin\delta + M_{zfl} + M_{zfr} + M_{zrl} + M_{zrr} \Big]\end{aligned} \tag{24}$$

where, J_z is the moment of inertia around the z-axis. The roll and pitch motion depend very much on the longitudinal and lateral accelerations. Since only the vehicle body undergoes roll and pitch, the sprung mass, denoted by m_s has to be considered in determining the effects of handling on pitch and roll motions as the following:

$$\ddot{\varphi} = \frac{-m_s c a_y + \varphi(m_s g c - k_\phi) + \dot{\varphi}(-\beta_\varphi)}{J_{sx}} \tag{25}$$

$$\ddot{\theta} = \frac{-m_s c a_y + \theta(m_s g c - k_\theta) + \dot{\theta}(-\beta_\theta)}{J_{sy}} \tag{26}$$

where, c is the height of the sprung mass center of gravity from the ground, g is the gravitational acceleration and k_φ, β_φ, k_θ and β_θ are the damping and stiffness constant for roll and pitch, respectively. The moments of inertia of the sprung mass around x-axis and y-axis are denoted by J_{sx} and J_{sy} respectively.

2.4 Simplified Calspan Tire Model

Tire model considered in this study is Calspan model as described in Szostak *et al.* (1988). Calspan model is able to describe the behavior of a vehicle in any driving scenario including inclement driving conditions which may require severe steering, braking, acceleration, and other driving related operations (Kadir *et al.*, 2008). The longitudinal and lateral forces generated by a tire are a function of the slip angle and longitudinal slip of the tire relative to the road. The previous theoretical developments in Szostak *et al.* (1988) lead to a complex, highly non-linear composite force as a function of composite slip. It is convenient to define a saturation function, $f(\sigma)$, to obtain a composite force with any normal load and coefficient of friction values (Singh *et al.*, 2000). The polynomial expression of the saturation function is presented by:

$$f(\sigma) = \frac{F_c}{\mu F_z} = \frac{C_1\sigma^3 + C_2\sigma^2 + (4/\pi)\sigma}{C_1\sigma^3 + C_2\sigma^2 + C_4\sigma + 1} \tag{27}$$

where, C_1, C_2, C_3 and C_4 are constant parameters fixed to the specific tires. The tire contact patch lengths are calculated using the following two equations:

$$ap_0 = \frac{0.0768\sqrt{F_z F_{ZT}}}{T_w(T_p + 5)} \tag{28}$$

$$ap = \left(1 - \frac{K_a F_x}{F_z}\right) \tag{29}$$

where ap is the tire contact patch, F_z is a normal force, T_w is a tread width, and T_p is a tire pressure. While F_{ZT} and K_a are tire contact patch constants. The lateral and longitudinal stiffness coefficients (K_s and K_c, respectively) are a function of tire contact patch length and normal load of the tire as expressed as follows:

$$K_s = \frac{2}{ap_0^{\;2}}\left(A_0 + A_1 F_z - \frac{A_1 F_z^{\;2}}{A_2}\right) \tag{30}$$

$$K_c = \frac{2}{ap_0^{\;2}} F_z (CS / FZ) \tag{31}$$

where the values of A_0, A_1, A_2 and CS/FZ are stiffness constants. Then, the composite slip calculation becomes:

$$\sigma = \frac{\pi a p^2}{8\mu_0 F_z}\sqrt{K_s^{\;2}\tan^2\alpha + K_c^{\;2}\left(\frac{s}{1-s}\right)^2} \tag{32}$$

Where S is a tire longitudinal slip, α is a tire slip angle, and μ_o is a nominal coefficient of friction and has a value of 0.85 for normal road conditions, 0.3 for wet road conditions, and 0.1 for icy road conditions. Given the polynomial saturation function, lateral and longitudinal stiffness, the normalized lateral and longitudinal forces are derived by resolving the composite force into the side slip angle and longitudinal slip ratio components:

$$\frac{F_y}{\mu F_z} = \frac{f(\sigma) K_s \tan\alpha}{\sqrt{K_s^{\;2}\tan^2\alpha + K_c'^{\;2} S^2}} + Y_\gamma \gamma \tag{33}$$

$$\frac{F_x}{\mu F_z} = \frac{f(\sigma) K_c' S}{\sqrt{K_s^{\;2}\tan^2\alpha + K_c'^{\;2} S^2}} \tag{34}$$

Lateral force has an additional component due to the tire camber angle, γ, which is modeled as a linear effect. Under significant maneuvering conditions with large lateral and longitudinal slip, the force converges to a common sliding friction value. In order to meet this criterion, the longitudinal stiffness coefficient is modified at high slips to transition to lateral stiffness coefficient as well as the coefficient of friction defined by the parameter K_μ.

$$K_c' = K_c + (K_s - K_c)\sqrt{\sin^2 \alpha + S^2 \cos^2 \alpha} \tag{35}$$

$$\mu = \mu_0 (1 - K_\mu)\sqrt{\sin^2 \alpha + S^2 \cos^2 \alpha} \tag{36}$$

3. Controller Structure of Pneumatically Actuated Active Roll Control Suspension System

The proposed controller structure consists of inner loop controller to reject the unwanted weight transfer and outer loop controller to stabilize heave and roll responses due to steering wheel input from the driver. An input decoupling transformation is placed between inner and outer loop controllers that blend the inner loop and outer loop controller. The outer loop controller provides the ride control that isolates the vehicle body from vertical and rotational vibrations induced by steering wheel input and the inner loop controller provides the weight transfer rejection control that maintains load-leveling and load distribution during vehicle maneuvers. The proposed control structure is shown in Figure 3.

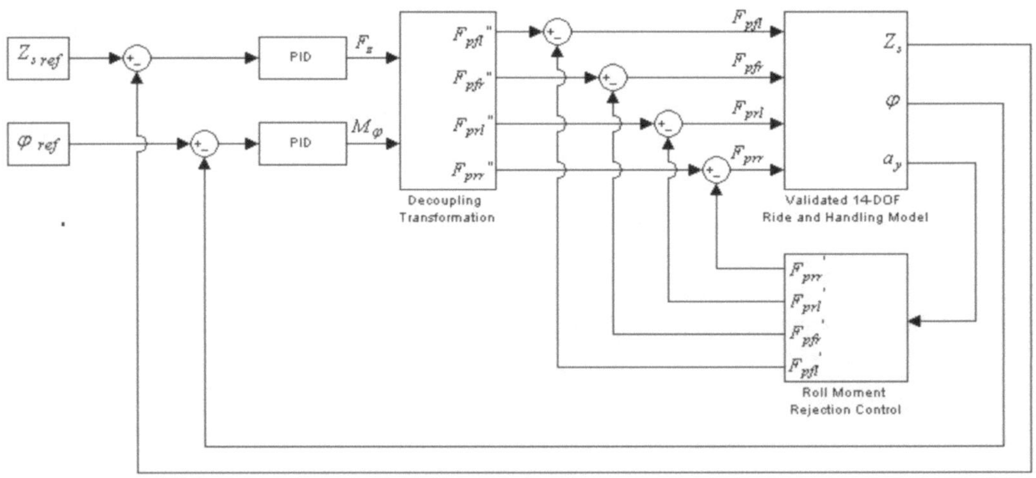

Fig. 3. The proposed control structure for arc system

The outputs of the outer loop controller are vertical forces to stabilize body bounce (M_z) and moment to stabilize roll (M_θ). These forces and moments are then distributed into target forces of the four pneumatic actuators produced by the outer loop controller. Distribution of the forces and moments into target forces of the four pneumatic actuators is performed using decoupling transformation subsystem. The outputs of the decoupling transformation subsystem namely the target forces of the four pneumatic actuators are then subtracted with the relevant outputs from the inner loop controller to produce the ideal target forces of the four pneumatic actuators. Decoupling transformation subsystem requires an understanding of the system dynamics in the previous section. The equivalent force and moment for heave, pitch and roll can be defined by

$$F_z = F''_{pfl} + F''_{pfr} + F''_{prl} + F''_{prr} \tag{37}$$

$$M_\theta = -F''_{pfl} l_f - F''_{pfr} l_f + F''_{prl} l_r + F''_{prr} l_r \tag{38}$$

$$M_\varphi = F''_{pfl}\left(\frac{w}{2}\right) - F''_{pfr}\left(\frac{w}{2}\right) + F''_{prl}\left(\frac{w}{2}\right) - F''_{prr}\left(\frac{w}{2}\right) \tag{39}$$

where $F''_{pfl}, F''_{pfr}, F''_{prl}\ F''_{prr}$ are the pneumatic forces produced by outer loop controller in front left, front right, rear left and rear right corners, respectively. In the case of the vehicle input comes from steering wheel, the pitch moment can be neglected. Equations (37), (38) and (39) can be rearranged in matrix format as the following

$$\begin{bmatrix} F_z(t) \\ M_\theta(t) \\ M_\varphi(t) \end{bmatrix} = \begin{bmatrix} 1 & 1 & 1 & 1 \\ -l_f & -l_f & l_r & l_r \\ \frac{w}{2} & -\frac{w}{2} & \frac{w}{2} & -\frac{w}{2} \end{bmatrix} \begin{bmatrix} F''_{pfl} \\ F''_{pfr} \\ F''_{prl} \\ F''_{prr} \end{bmatrix} \tag{40}$$

For a linear system of equations **y=Cx**, if **C** $\in \Re^{mxn}$ has full row rank, then there exists a right inverse **C**$^{-1}$ such that **C-1 C= 1**$^{\mathbf{mxm}}$. The right inverse can be computed using **C**$^{-1}$**=C**^{T}**(CC**^{T}**)**$^{-1}$. Thus, the inverse relationship of equation (40) can be expressed as:

$$\begin{bmatrix} F''_{pfl} \\ F''_{pfr} \\ F''_{prl} \\ F''_{prr} \end{bmatrix} = \begin{bmatrix} \frac{l_r}{2(l_f+l_r)} & -\frac{1}{2(l_f+l_r)} & \frac{1}{2w} \\ \frac{l_r}{2(l_f+l_r)} & -\frac{1}{2(l_f+l_r)} & -\frac{1}{2w} \\ \frac{l_f}{2(l_f+l_r)} & \frac{1}{2(l_f+l_r)} & \frac{1}{2w} \\ \frac{l_f}{2(l_f+l_r)} & \frac{1}{2(l_f+l_r)} & -\frac{1}{2w} \end{bmatrix} \begin{bmatrix} F_z \\ M_\theta \\ M_\varphi \end{bmatrix} \tag{41}$$

In the outer loop controller, PID control is applied for suppressing both body vertical displacement and body roll angle. The inner loop controller of roll moment rejection control is described as follows: during cornering, a vehicle will produce a sideway force namely cornering force at the body center of gravity. The cornering force generates roll moment to the roll center causing the body center of gravity to shift outward as shown in Figure 4. Shifting the body center of gravity causes a weight transfer from the inside toward the outside wheels. By defining b as the distance between body center of gravity and the roll center, roll moment is defined by

$$M_r = M_s a_y b \tag{42}$$

The two pneumatic actuators installed in outside wheels have to produce the necessary forces to cancel out the unwanted roll moments, whereas the forces of the two pneumatic actuators in inside wheels are set to zero. Pneumatic force to cancel out roll moment in each corner for counter clockwise steering wheel input is defined as:

$$F'_{pfr} = F'_{prr} \frac{M_s a_y b}{w/2} \quad \text{and} \quad F'_{pfl} = F'_{prl} = 0 \tag{43}$$

Whereas, pneumatic force to cancel out roll moment in each corner for clockwise steering wheel input can be defined as:

$$F'_{prl} = F'_{pfl} = \frac{M_s a_y b}{w/2} \quad \text{and} \quad F'_{pfr} = F'_{prr} = 0 \tag{44}$$

where,

F'_{pfl} = target force of pneumatic system at front left corner produced by inner loop controller

F'_{pfr} = target force of pneumatic system at front right corner produced by inner loop controller

F'_{prl} = target force of pneumatic system at rear left corner produced by inner loop controller

F'_{prr} = target force of pneumatic system at rear right corner produced by inner loop controller

The ideal target forces for each pneumatic actuator are defined as the target forces produced by outer loop controller subtracted with the respective target forces produced by inner loop controller as the following:

$$F_{pfl} = F''_{pfl} - F'_{pfl} \tag{45}$$

$$F_{pfr} = F''_{pfr} - F'_{pfr} \tag{46}$$

$$F_{prl} = F''_{prl} - F'_{prl} \tag{47}$$

$$F_{prr} = F''_{prr} - F'_{prr} \tag{48}$$

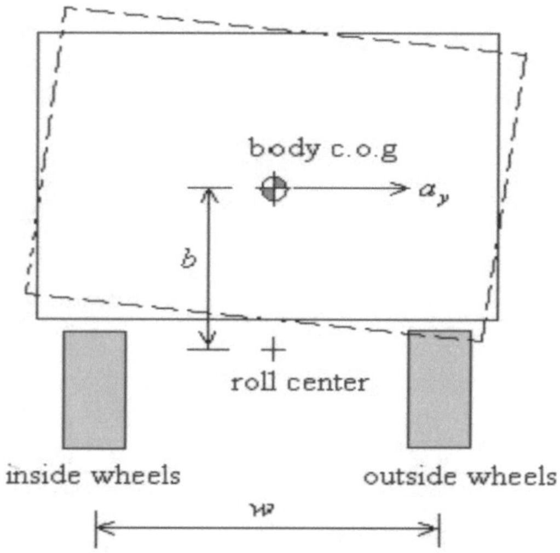

Fig. 4. Roll Moment Generated by Lateral Force

4. Validation of 14-DOF Ride and Handling Model

To verify the full vehicle ride and handling model, experimental works were performed using an instrumented experimental vehicle. This section provides the verification of ride and handling model using visual technique by simply comparing the trend of simulation results with experimental data using the same input conditions. Validation or verification is defined as the comparison of model's performance with a real system. Therefore, the validation is not meant as fitting the simulated data exactly to the measured data, but as gaining confidence that the vehicle handling simulation is giving insight into the behavior of the simulated vehicle. The test data are also used to check whether the input parameters for the vehicle model are reasonable. In general, model validation can be defined as determining the acceptability of a model using some statistical tests for deviance measures or subjectively using visual techniques.

4.1 Instrumented Experimental Vehicle

The data acquisition system (DAS) is installed into the experimental vehicle to obtain the experimental data from the real vehicle reaction to evaluate the vehicle performance in terms of lateral acceleration, body vertical acceleration, yaw rate and roll rate. The DAS uses several types of transducers such as single axis accelerometer to measure the sprung mass and unsprung mass accelerations for each corner, tri-axial accelerometer to measure lateral, vertical and longitudinal accelerations at the body center of gravity, steering wheel sensor and tri-axial gyroscopes for the yaw rate, pitch rate and roll rate. The multi-channel μ-MUSYCS system Integrated Measurement and Control (IMC) is used as the DAS system. Online FAMOS software as the real time data processing and display function is used to ease the data collection. The installation of the DAS and sensors to the experimental vehicle can be seen in Figure 5.

Fig. 5. Instrumented experimental vehicle

4.2 Validation Procedures

The dynamic response characteristics of a vehicle model that include yaw response, lateral acceleration, slip angle in each tire and roll rate can be validated using experimental test through several handling test procedures namely step steer test and double lane change (DLC) test. Step steer test is intended to study transient response of the vehicle under steering wheel input. In this study, step steer tests were performedwith 180 degrees clock wise at 50 km/h. On the other hand, double-lane change test is used to evaluate road holding of the vehicle during crash avoidance. In this study, the speed of 80 km/h was set for the double-lane change test.

4.3 Model Validation Results

In experimental works, all the experimental data were filtered using 5th order Butterworth low pass filter with the cut-off frequency of 5 Hz. It is necessary to note that the measured steering angle from the steering wheel sensorwas used as the input of simulation model. For the simulation model, tire parameters are obtained from Szostak *et al.* (1988) and Singh *et al.* (2000). The results of model verification for 180 degrees step steer test at 50 km/h is shown in Figure 6 to 13.

Figure 6 shows the steering wheel input applied for the step steer test. It can be seen that the trends between simulation results and experimental data are almost similar with acceptable error. The small difference in magnitude between simulation and experimental results is due to the simplification in vehicle dynamics modeling where the effects of anti roll bar were completely ignored. In fact, the anti roll bar plays an important role in reducing the vertical and roll responses of vehicle body. In simulation model, vehicle body is assumed to be rigid. It can be another source of deviation since the body flexibility can influence the roll effects of the vehicle body.

In terms of yaw rate, lateral acceleration and body roll angle, it can be seen that there are quite good comparisons during the initial transient phase as well as during the subsequent steady state phase as shown in Figures 7 to 9. Slip angle responses of the front tires also show satisfying comparison with only small deviation in the transition area between transient and steady state phases as shown in Figures 10 and 11.

It can also be noted that the slip angle responses of all tires in the experimental data are slightly higher than the slip angle data obtained from the simulation particularly for the rear tires as can be seen in Figures 12 and 13. This is due to the fact that it is difficult for the driver to maintain a constant speed during maneuvering. In simulation, it is also assumed that the vehicle is moving on a flat road during step steer maneuver. In fact, it is observed that the road profiles of test field consist of irregular surface. This can be another source of deviation on slip angle response of the tires.

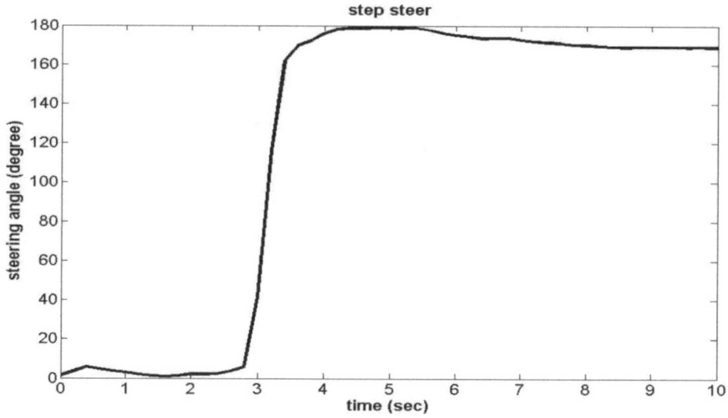

Fig. 6. Steer angle input for 180 deg step steer at 50 km/h

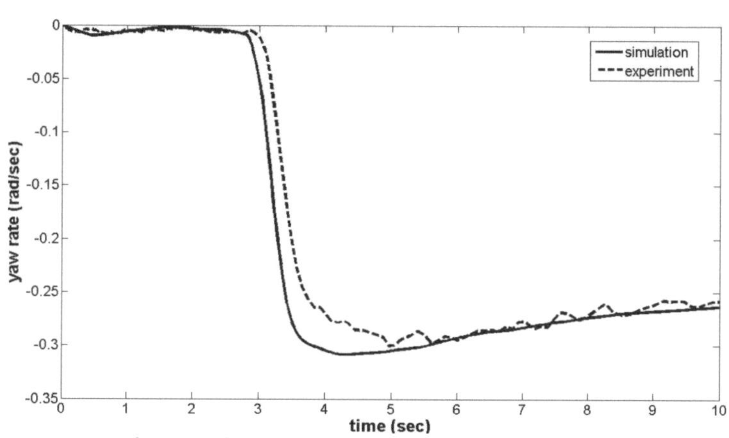

Fig. 7. Yaw rate response for 180 deg step steer at 50 km/h

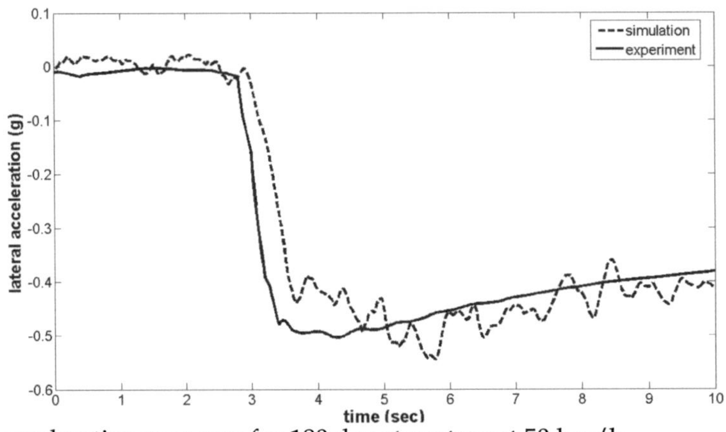

Fig. 8. Lateral acceleration response for 180 deg step steer at 50 km/h

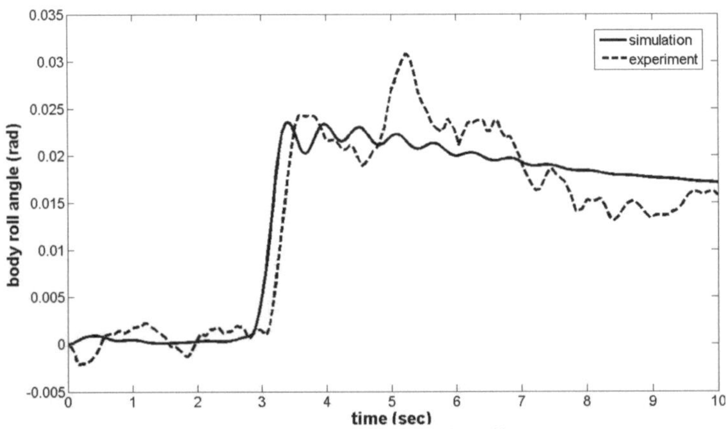

Fig. 9. Roll angle response for 180 deg step steer at 50 km/h

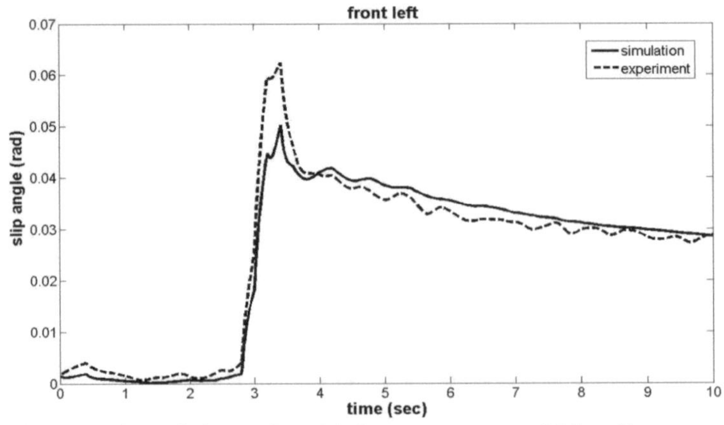

Fig. 10. Slip angle at the front left tire for 180 deg step steer at 50 km/h

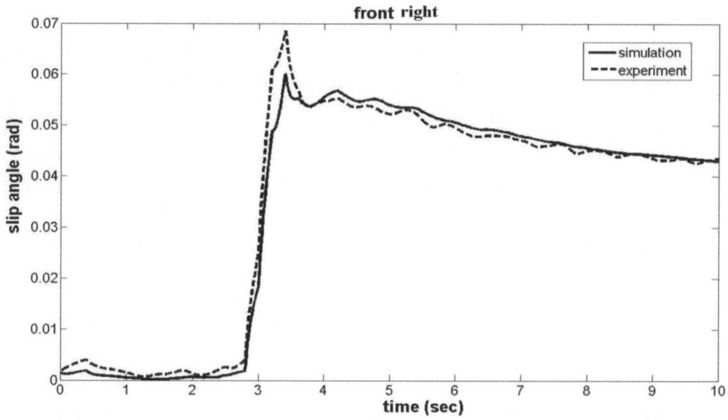

Fig. 11. Slip angle response at front right tire for 180 deg step steer at 50 km/h

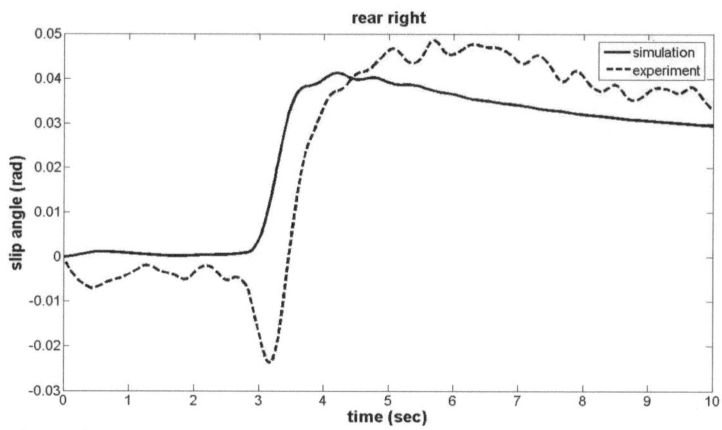

Fig. 12. Slip angle at the rear right tire for 180 deg step steer at 50 km/h

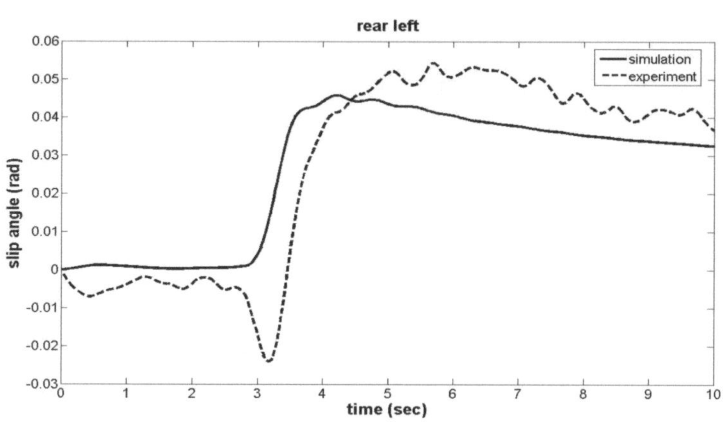

Fig. 13. Slip angle at the rear left tire for 180 deg step steer at 50 km/h

The results of double lane change test indicate that measurement data and the simulation results agree with a relatively good accuracy as shown in Figures 14 to 21. Figure 14 shows the measured steering wheel input from double lane change test maneuver which is also used as the input for the simulation model. In terms of yaw rate, lateral acceleration and body roll angle, it is clear that the simulation results closely follow the measured data with minor difference in magnitude as shown in Figures 15 to 17. The minor difference in magnitude and small fluctuation occurred on the measured data is due to the body flexibility which was ignored in the simulation model. The minor difference in magnitude between measured and simulated data can also be caused by one of the modeling assumptions namely the effects of anti roll bar which is completely ignored in simulation model.

In terms of tire side slip angles, the trends of simulation results have a good correlation with experimental data as can be seen in Figures 18 to 21. Almost similar to the validation results obtained from step steer test, the slip angle responses of all tires in experimental data are higher than the slip angle data obtained from the simulation particularly for the rear tires. Again, this is due to the difficulty of the driver to maintain a constant speed during double lane change maneuver. Assumption in simulation model that the vehicle is moving on a flat road during double lane change maneuver is also very difficult to realize in practice. In fact, road irregularities of the test field may cause the change in tire properties during vehicle handling test. Assumption of neglecting the steering inertia have the possibility in lowering down the magnitude of tire side slip angle in simulation results compared to the measured data.

Overall, it can be concluded that the trends between simulation results and experimental data are having good agreement with acceptable error. The error could be significantly reduced by fine tuning of both vehicle and tire parameters. However, excessive fine tuning works can be avoided since in control oriented model, the most important characteristic is the trend of the model response. As long as the trend of the model response is closely similar with the measured response with acceptable deviation in magnitude, it can be said that the model is valid. The validated model will be used in conjunction with the proposed controller structure of the ARC system in the next section.

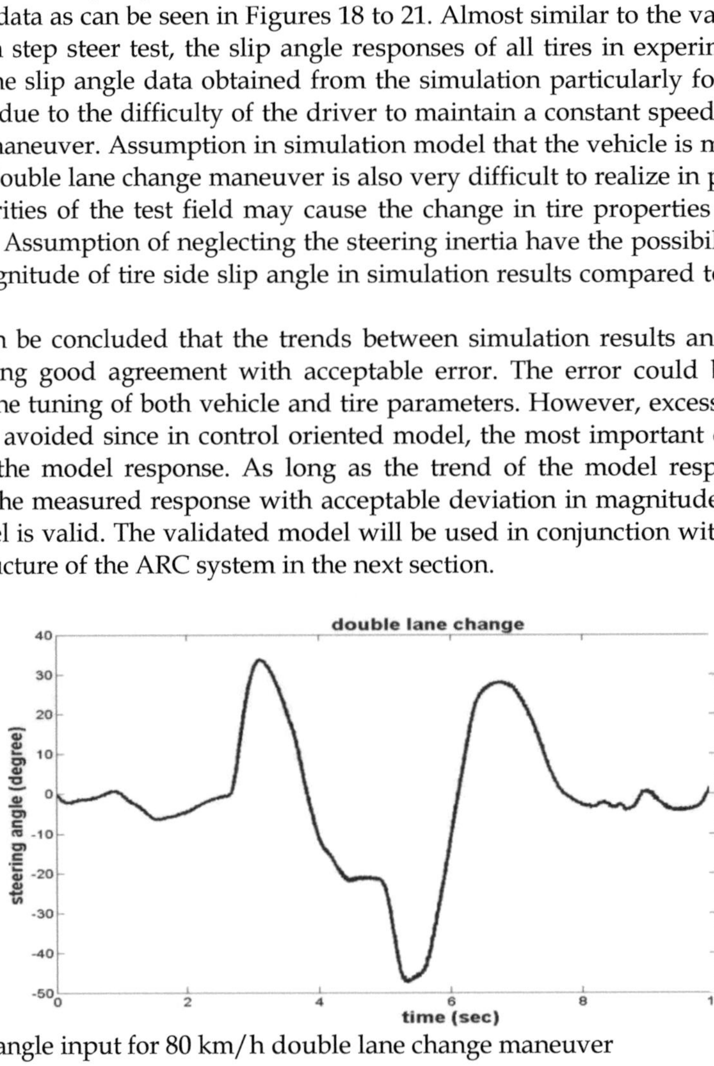

Fig. 14. Steer angle input for 80 km/h double lane change maneuver

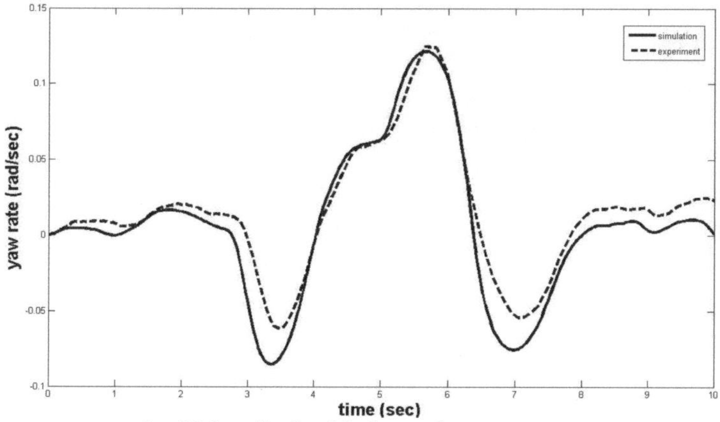

Fig. 15. Yaw rate response for 80 km/h double lane change maneuver

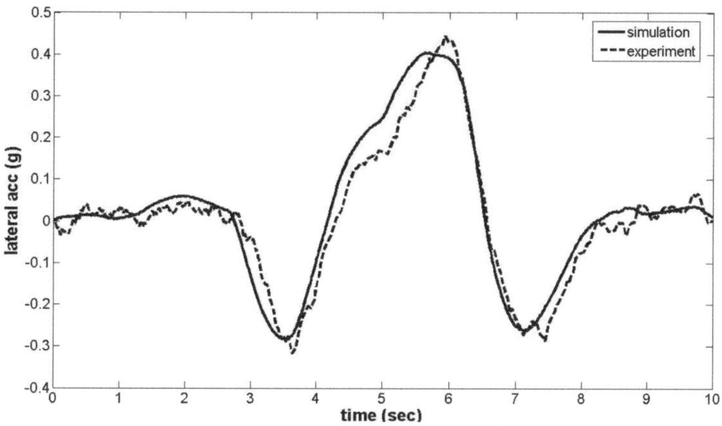

Fig. 16. Lateral acceleration response for 80 km/h double lane change maneuver

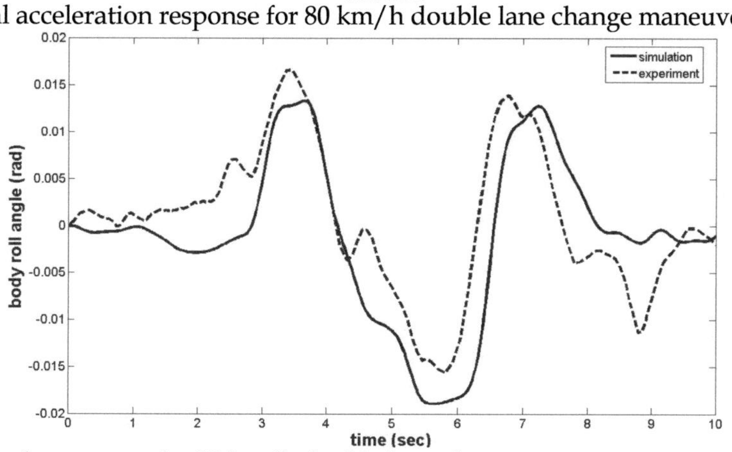

Fig. 17. Roll angle response for 80 km/h double lane change maneuver

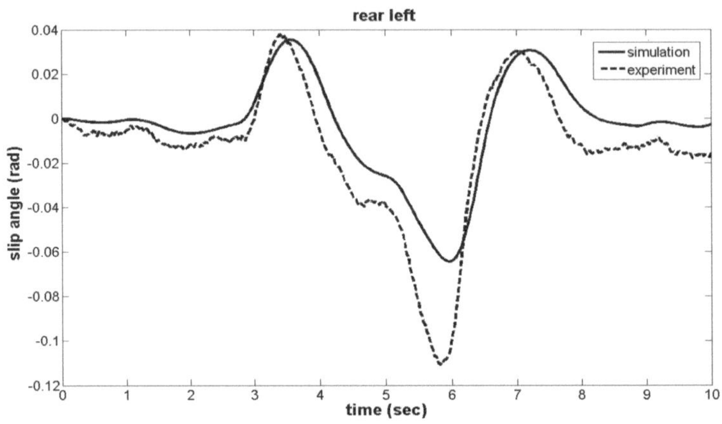

Fig. 18. Slip angle at the front left tire for 80 km/h double lane change maneuver

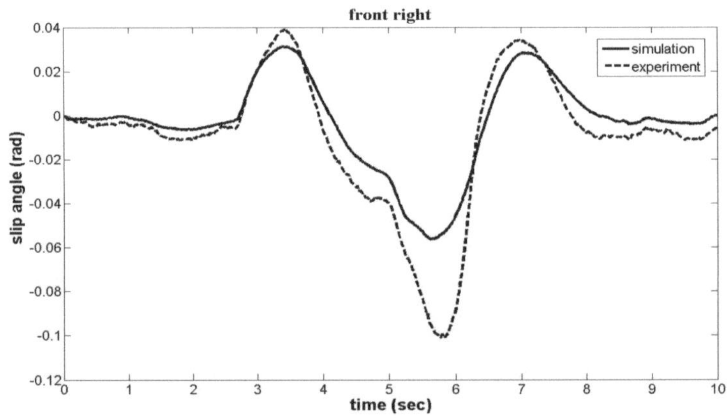

Fig. 19. Slip angle at the front right tire for 80 km/h double lane change maneuver

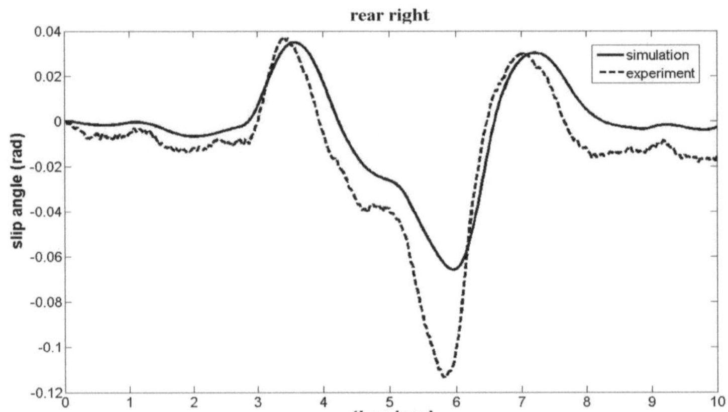

Fig. 20. Slip angle at the rear right tire for 80 km/h double lane change maneuver

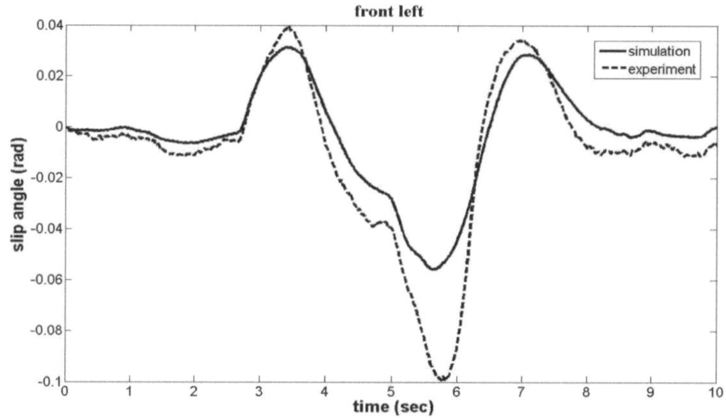

Fig. 21. Slip angle at the rear left tire for 80 km/h double lane change maneuver

5. Performance Assessment of the Proposed Control Structure for ARC System

This section describes the results of performance study of the proposed control structure for the pneumatically actuated ARC system namely PID with roll moment rejection control. Performance of the vehicle with passive system is used as a basic benchmark. To investigate the advantage of additional roll moment rejection loop, the performance of the proposed controller is also compared with PID without roll moment rejection loop. This section begins with introducing all the parameters used in this simulation study, followed by the presentation of the controller performance in step steer and double lane change tests. The PID with roll moment rejection control for ARC system is evaluated for its performance at controlling the lateral dynamics of the vehicle according to the following performance criteria namely body vertical acceleration, body heave, body roll rate and body roll angle.

5.1 Simulation Parameters

The simulation study was performed for a period of 10 seconds using Heun solver with a fixed step size of 0.01 second. The controller parameters are obtained using trial and error technique with some sensitivity studies. The numerical values of the 14-DOF full vehicle model parameters and Calspan tire model parameters as well as the controller parameters are given in the Appendix.

5.2 Performance of ARC System During Step Steer Test

The simulation results of body roll angle and body roll rate at the body centre of gravity on 180 degrees step steer test at 50 km/h are shown in Figures 22 and 23 respectively. It can be seen that the performance of PID control with roll moment rejection loop can outperform its counterpart namely passive system and PID control without roll moment rejection loop. In terms of the roll angle response, it is clear that the additional roll moment rejection loop can effectively reduce the magnitude of the roll angle response. Improvement in roll motion during maneuvering can enhance the stability of the vehicle in lateral direction.
In terms of the roll rate response, PID control with roll moment rejection loop shows significant improvement over passive and PID control without roll moment rejection loop

particularly in the transient response phase area. At steady state response, PID control with roll moment rejection loop shows slight improvement in terms of settling time over PID control without roll moment rejection loop and significant improvement over passive system. Again, the advantage of the additional roll moment rejection loop is shown by reducing the magnitude of the roll rate response. Improvement in both roll rate response and the settling time during maneuvering can increase the stability level of the vehicle in the presence of steering wheel input from the driver.

Body vertical acceleration and body heave responses of the vehicle at the body center of gravity are presented in Figures 24 and 25 respectively. From the body vertical acceleration response, both PID control with and without roll moment rejection loops are able to drastically reduce unwanted vertical acceleration compared to the passive system. It can be seen, the capability of the controller in lowering down the magnitude of body acceleration and in speeding up the settling time. Improvement in vertical acceleration at the body center of gravity will enhance the comfort level of the vehicle as well as avoiding the driver from losing control of the vehicle during maneuvering.

The main goal of ARC system is to keep the vehicle body remain flat in any driving maneuvers. From the body heave response, it is clear that the performance of PID control with roll moment rejection loop is significantly better than that of passive system and PID control without roll moment rejection loop. It means that PID control with roll moment rejection loop shows less vertical displacement during step steer maneuver. This will also enhance the comfort level of the vehicle as well as avoiding the driver from losing control of the vehicle.

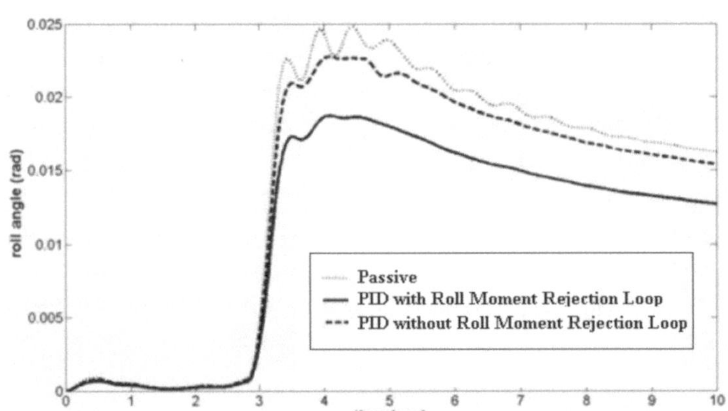

Fig. 22. Roll angle response of ARC System for 180 degrees Step Steer Test at 50 km/h

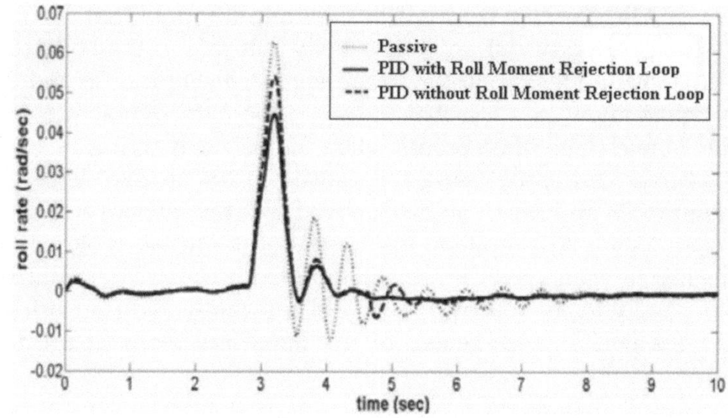

Fig. 23. Roll rate response of ARC System for 180 degrees Step Steer Test at 50 km/h

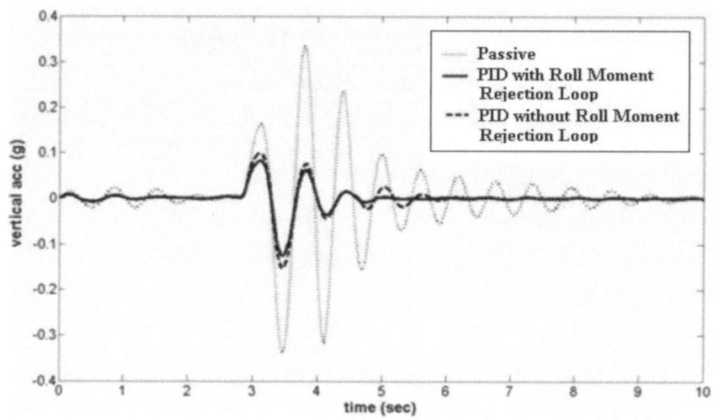

Fig. 24. Vertical acceleration response of ARC System for 180 degrees Step Steer Test at 50 km/h

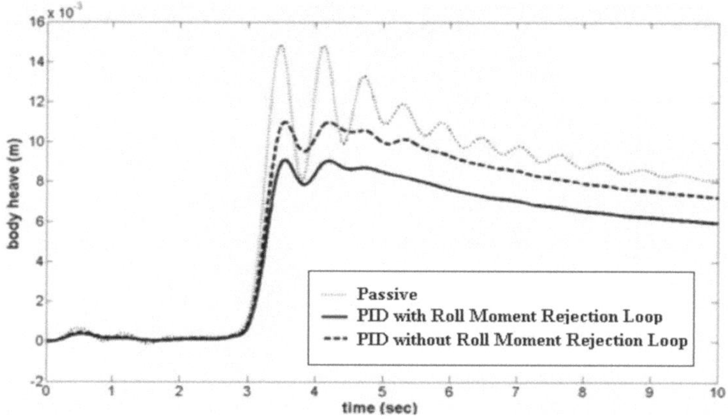

Fig. 25. Vertical displacement response at the body cog of ARC System for 180 degrees Step Steer Test at 50 km/h

5.3 Performance of ARC System During Double Lane Change Test

The simulation results of body roll angle and body roll rate at the body centre of gravity during double lane change test at 80 km/h are shown in Figures 26 and 27 respectively. Double lane-change is know as a test that measures the maneuverability of the vehicle. In real life, a double lane change often occurs when the driver is trying to avoid an accident. This sudden maneuver can easily cause the vehicle to tip on two wheels, resulting in a rollover. From Figures 26 and 27, it can be observed that the maneuverability of the vehicle increases by implementing ARC system. In the case of the driver makes an abrupt swerve like double lane change maneuver, improvement in both roll rate and roll angle responses indicate that the possibility of roll over can be significantly reduced using ARC system. From the figures, the performance benefit of additional roll moment rejection loop is also observed.

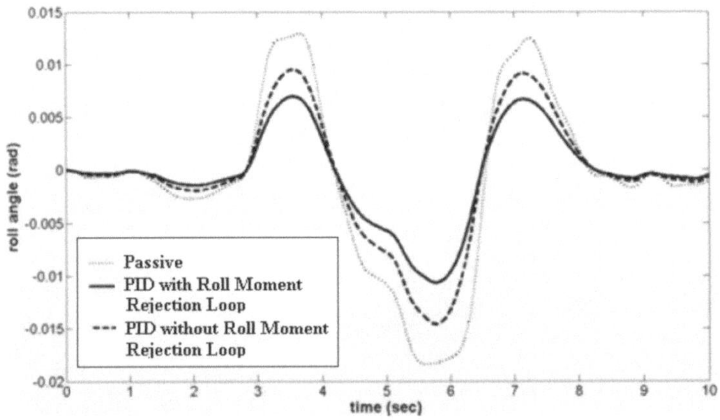

Fig. 26. Roll angle response of ARC System for 80 km/h double lane change

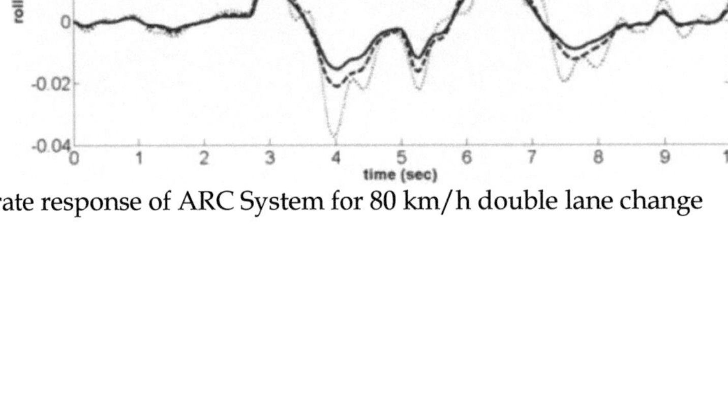

Fig. 27. Roll rate response of ARC System for 80 km/h double lane change

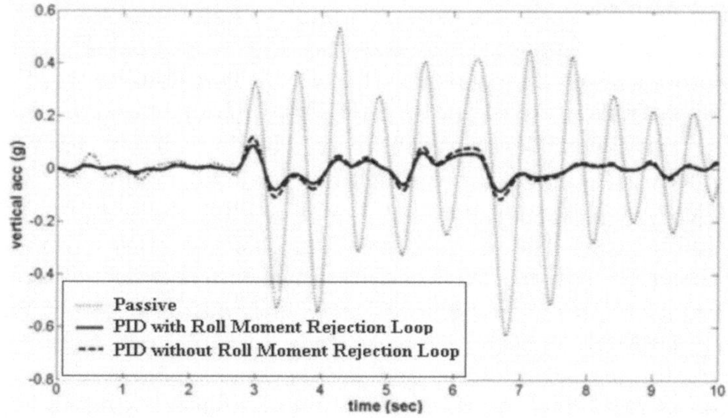

Fig. 28. Vertical acceleration of ARC System for 80 km/h double lane change

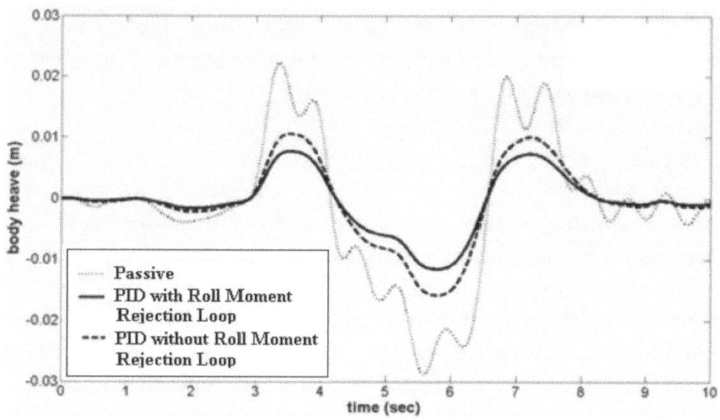

Fig. 29. Vertical displacement response of ARC System for 80 km/h double lane change

Body vertical acceleration and body heave response are presented in Figures 28 and 29. It can be concluded that PID controller with and without roll moment rejection loop for ARC system are able to improvement significantly the ride performance compared to the passive system. Again, the performance benefit of additional roll moment rejection loop is also observed from the figures. Enhancement in ride performance may trim down the rate of driver fatigue and reduce the risk of the driver losing control of the vehicle. It can also be observed from the figures that the performance benefit of additional roll moment rejection loop is minor.

6. Experimental Evaluation of the Proposed Control Structure for ARC System

This section describes the experimental results of ARC system implemented on the instrumented experimental vehicle. Performance of the vehicle equipped with ARC system is compared with passive system in several maneuvers namely step steer and double lane change tests. The response of the passive vehicle is used as a basic benchmark for performance of ARC system. The ARC system is evaluated for its performance at controlling the lateral dynamics of the vehicle according to the following performance criteria namely body vertical acceleration, body vertical displacement, body roll rate and body roll angle.

6.1 Installation of ARC System into the Instrumented Experimental Vehicle

The instrumented experimental vehicle consists of two groups of transducers namely vehicle states sensors and actuator sensors. The vehicle states sensors consist of one unit of K-Beam® Capacitive Triaxial Accelerometer 8393B10 manufactured by Kistler and three units of CRS03 gyro by Silicon Sensing that are installed in the body centre of gravity of the experimental vehicle. The triaxial accelerometer is used to provide measurement data of body vertical, lateral, and longitudinal accelerations while the gyros is used to measure pitch, yaw and roll motions. The vehicle states sensors also consist of one unit of DRS1000 Doppler Radar Speed Sensor manufactured by GMH Engineering to record the real-time vehicle speed during experiment and one unit of Linear Encoder to record the real time steer angle. The actuator sensors consist of four units of LCF451 Load Cells manufactured by Futek to measure the actuator forces. The multi-channel μ-MUSYCS system Integrated Measurement and Control (IMC) is used as the data acquisition system. It is installed into experimental vehicle to collect the experimental data from the transducers to control the vehicle performance in terms of body lateral acceleration, body vertical acceleration, and body roll rate. Online FAMOS software as the real time data processing and display function is used to ease the data collection. More detail specifications of the transducers and the data acquisition system are listed in the appendix.

The pneumatic actuator as the main component of the ARC system consists of 4 unit of pneumatic compact cylinders which are installed in parallel arrangement with passive suspension system. A double acting pneumatic compact cylinder of SDA80x75 is used in this experimental test which has bore size of 80 mm and 75 mm in stroke length. Another components are 5/3 way solenoid valve (center exhaust), 2.5 HP air compressor and the current driver. The 5/3 way solenoid valves of SY7420-5LZD with double coil specification of 24V and 300 mA are installed with the cylinders. The installation of the data acquisition system, sensors and pneumatic system to the experimental vehicle can be seen in Figure 30.

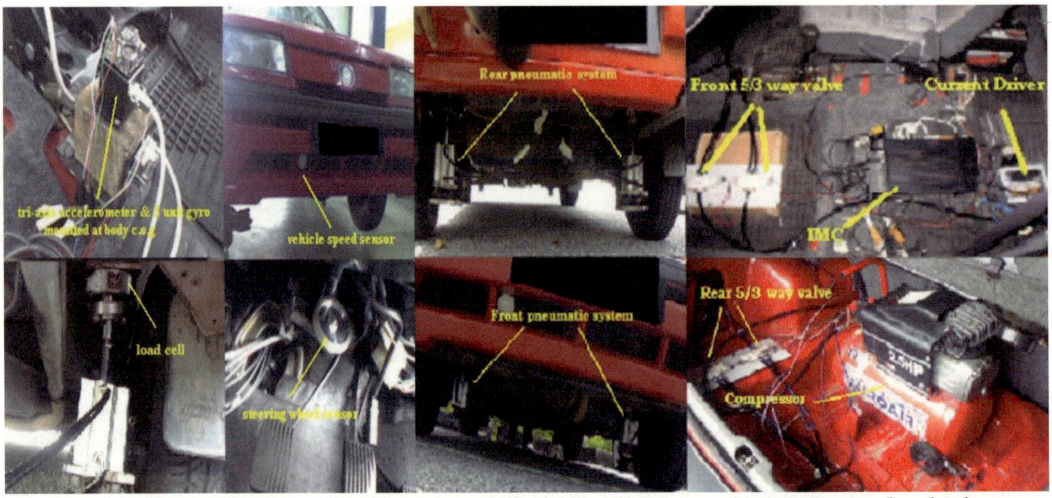

Fig. 30. Four units of pneumatic system installed in instrumented experimental vehicle

6.2 Experimental Parameters

The ARC system is performed in experimental test with two types of maneuver tests namely step steer test and double lane change test. In step steer test, the vehicle begins moving in a straight line with the constant speed of 50 km/h and then the steering suddenly turned 160 degrees clockwise. The double lane change and slalom tests were performed with the constant speed of 50 km/h based on the test track as illustrated in Figure 31.

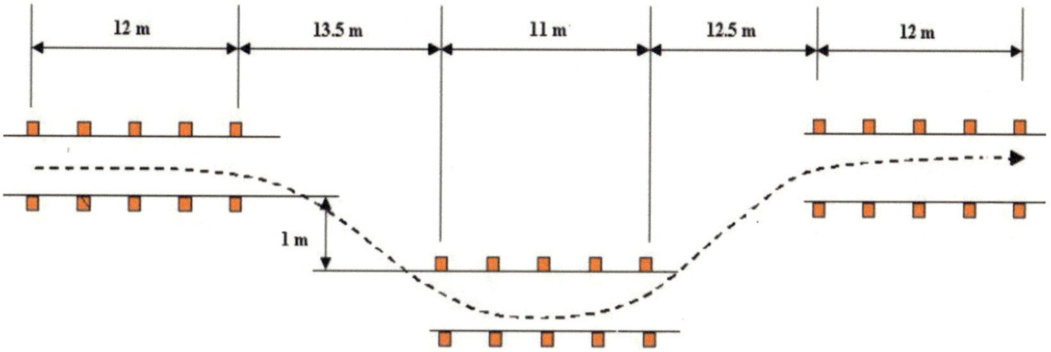

Fig. 31. The track for double lane change test

6.3 Experimental Performance of ARC System during Step Steer Test

Figure 32 shows the visual comparison of experimental results between passive system and vehicle equipped with ARC system during steep steer test. It can be seen that the roll angle of vehicle is reduced for vehicle equipped with ARC system compared to the passive system and able to reduce the possibility of vehicle rollover.

Fig. 32. Visual comparison of passive system and vehicle equipped with ARC system during step steer test

The experimental result of body roll angle at body centre of gravity during step steer test is shown in Figure 33(a). It can be seen that the performance of vehicle equipped with ARC system is better than passive system by reducing the magnitude of body roll angle. The vehicle equipped with ARC system also showing a significant reduction of roll rate at body centre of gravity as compared with passive system as shown in Figure 33(b). The vehicle equipped with ARC system shows an improvement response with respect to passive system by reducing the magnitude of body roll rate.

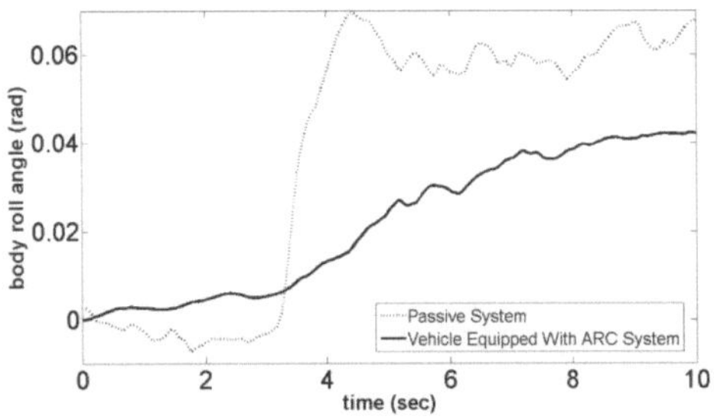

a) Roll angle response at the body center of gravity

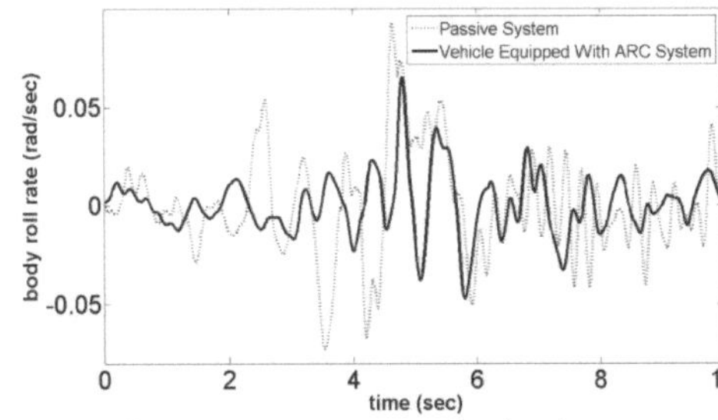

b) Roll rate response at the body center of gravity

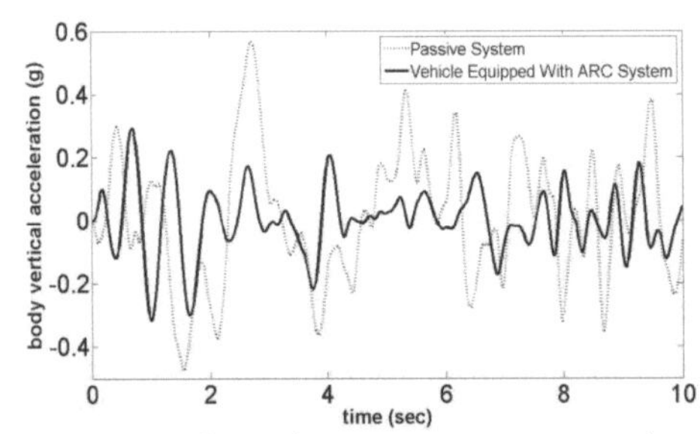

c) Vertical acceleration response at the body center of gravity

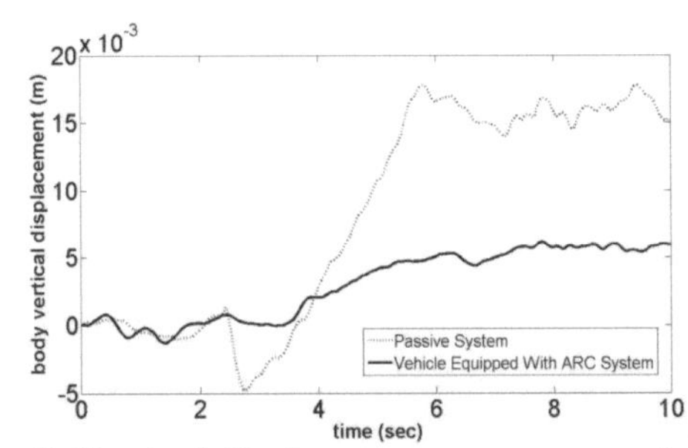

d) Vertical displacement response at the at the body center of gravity

Fig. 33. Experimental results of passive system and vehicle equipped with ARC system for 160 degrees step steer test at 50 km/h

The body vertical displacement performance at body centre of gravity obtained from the experimental result is shown in Figure 33(c). It can be seen that there is an improvement on vertical displacement of vehicle equipped with ARC system over passive system. The experimental result of vehicle equipped with ARC system is having smaller magnitude of vertical displacement than that of passive system. Vehicle equipped with ARC system also offer significant improvement on body vertical acceleration as shown in Figure 33(d). It can be seen that the ARC system is more capable in lowering down the magnitude of body vertical acceleration compared to passive system.

6.4 Experimental Performance of ARC System during Double Lane Change Test

Figure 34 shows the visual comparison of experimental results between passive system and vehicle equipped with ARC system during double lane change test. It can be seen that the stability of the vehicle equipped with ARC system is improved compare to passive system.

Fig. 34. Visual comparison of experimental results between passive system and vehicle equipped with ARC system during double lane change test

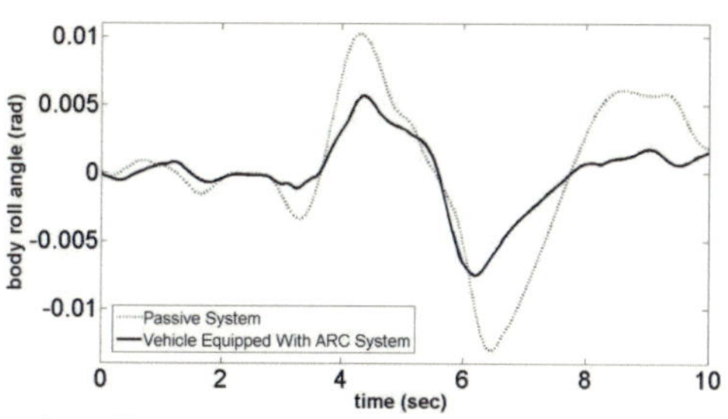

a) Roll angle response at the body center of gravity

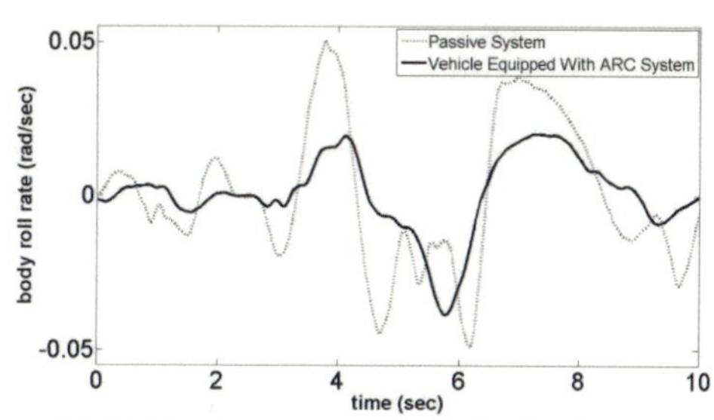

b) Roll rate response at the body center of gravity

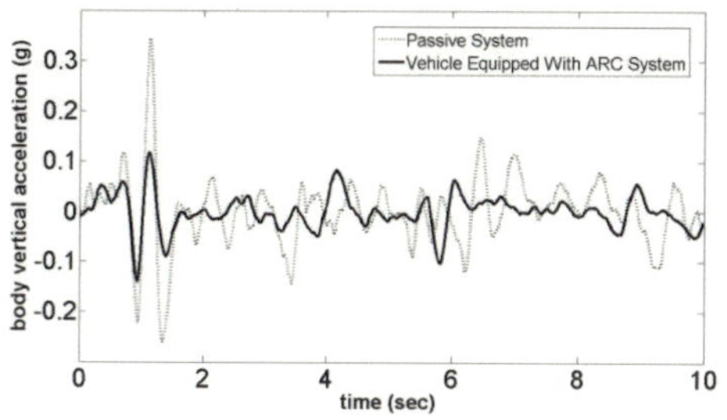

c) Vertical acceleration response at the body center of gravity

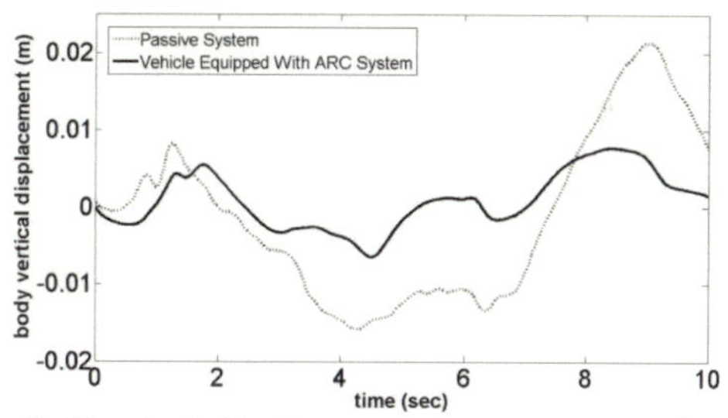

d) Vertical displacement response at the body center of gravity

Fig. 35. Experimental results of passive system and vehicle equipped with ARC system for DLC test at 50 km/h

From Figure 35(a) it can be seen that the body roll angle response of the passive system is higher than the body roll angle response of the vehicle equipped with ARC system. Therefore, it can be said that the vehicle equipped with ARC system is more stable and easier to avoid an obstacle during driving than passive system. The vehicle equipped with ARC system also show more reduction in magnitude in terms of roll rate response at body centre of gravity compared to passive system. The experimental result of roll rate is presented in Figure 35(b). It indicates that the overall vehicle roll rate behavior is improved with the vehicle equipped with ARC system to passive system

The experimental result of body vertical displacement at the body centre of gravity is shown in Figure 35(c). From the result it can be said that in terms of body vertical displacement at the body centre of gravity, the performance of the vehicle equipped with ARC system is better than the passive system. The vertical acceleration at body centre of gravity obtained from the experiment is shown in Figure 35(d). It can be seen that there is an improvement on vertical acceleration of vehicle equipped with ARC system compared to passive system. It can be seen clearly that the ARC system is effective in improving the performance of vehicle body from unwanted body motions namely vertical acceleration. Overall, It can be concluded that the ARC system is able to reduce the unwanted body motion in vertical direction.

7. Conclusions

A 14-DOF full vehicle model for passenger vehicle which consists of ride, handling and Calspan tire subsystems has been developed. An instrumented experimental vehicle has been developed to validate the 14-DOF model with the necessary sensors and data acquisition system installed inside the vehicle. Two types of road tests namely step steer test and double lane change test have been performed and data gathered from the tests were used as the benchmark of the model validation. The wheel steer angle data measured from the test in both step steer and double lane change tests were used as the inputs of the simulation model. Some of the vehicle behaviors to be validated in this works were yaw rate, lateral acceleration, body roll angle and tire slip angle responses. The results of model validation show that the trends between simulation results and experimental data are almost similar with acceptable error. The small difference in magnitude between simulation and experimental results is mainly due to the simplification/idealization in vehicle dynamics modeling and the difficulty of the driver to maintain a constant speed during maneuvering.

From the simulation results, it is clear that the performance of the proposed control structure is proven to outperform the performance of passive system in all the selected performance criteria. The need of additional roll moment rejection loop to the PID controller is also strongly justified. In general, it can be concluded that the proposed PID control with roll moment rejection loop for ARC system significantly enhances the maneuverability of the vehicle by reducing both roll rate and roll angle in the presence of the steering angle input from the driver. Improvement in body acceleration and body heave response indicate that the comfort level of the vehicle can also be improved drastically using the proposed control structure. Improvement in comfort level will avoid the driver from fatigue as well as reduce the possibility of the driver from losing control of the vehicle during maneuvering.

Four units of pneumatic actuators have been installed in parallel with the existing passive suspensions into the instrumented experimental vehicle for the ARC system. A PC-based controller for ARC system using the proposed control structure was then implemented through experimental test on real vehicle situations namely step steer and double lane change tests to investigate the effectiveness of the controller in attenuating the effect of steering input from the driver. The experimental results show that the ARC system is able to reduce the unwanted motions of vehicle body namely body roll angle, body roll rate, body vertical acceleration and body vertical displacement significantly. It can also be concluded that better improvement on vehicle stability was obtained using the ARC system.

Acknowledgement

This work is supported by the Ministry of Higher Education (MoHE) of Malaysia through FRGS project entitled "Development of a Pneumatically Actuated Active Roll Control Suspension System" lead by Dr. Khisbullah Hudha at the Universiti Teknikal Malaysia Melaka. This financial support is gratefully acknowledged.

Appendix

Vehicle Model Parameters:

M (kg)	***lf*** (m)	***lr*** (m)	***w*** (m)	***hcg*** (m)
920	1.34	1.04	1.34	0.5
Iz (kg/m²)	***Jw*** (kg/m²)	***Ir*** (kg/m²)	***Csfl, Csfr, Csrl, Csrr*** (N/msec⁻¹)	***Ksfl, Ksfr, Ksrl, Ksrr*** (N/m)
3190	1.2825	400	750	30000

Tire Parameters:

Parameter	RWD radial
Tire Type	155SR13
Tw	6
Tp	24
FZT	810
C1	1.0
C2	0.34
C3	0.57
C4	0.32
A0	914.02
A1	12.9
A2	2028.24
Ka	0.05
CS/FZ	18.7
μo	0.85

Controller Parameters:

PID	***Kp***	***Ki***	***Kd***
Body Heave Control	30000	0.00033	22500
Body Roll Control	7500	0.00003	3000

8. References

Ahmad, F., Hudha, K., Said, M. R. and Rivai, A. (2008). Development of Pneumatically Actuated Active Stabilizer Bar to Reduce Vehicle Dive. *International Conference on Plant Equipment and Reliability (ICPER).* March 27-28. Kuala Lumpur, Malaysia.

Ahmad, F., Hudha, K. and Jamaluddin, H. (2009a). Modeling Validation and Gain Scheduling PID Control fo Pneumatically Actuated Active Suspension System for Reducing Unwanted Vehicle Motion in Longitudinal Direction. *International Journal of Vehicle Safety (IJVS). 4* (1): 38-45.

Ahmad, F., Hudha, K. and Harun, M. H. (2009b). Pneumatically Actuated Active Suspension System for Reducing Vehicle Dive and Squat. *Jurnal Mekanikal UTM .* (28): 85-114.

Araki, M. (2006). PID Control, in Unbehauen, H. (ed) *Control Systems, Robotics and Automation – Vol. II ,* UNESCO: Encyclopedia of Life Support Systems

Ayat, M. L., Diop, S. and Fenaux, E. (2002a). Development of a Full Active Suspension System. *Proceedings of a15th Triennial World Congress of the International Federation of Automatic Control.* July 21-26. Barcelona, Spain. Paper No. 2658.

Ayat, M. L., Diop, S. and Fenaux, E. (2002b). An Improved Active Suspension Yaw Rate Control. *Proceedings of the American Control Conference.* May 8-10. Anchorage, AK. 2: 863-868.

Ben-Dov, D. and Salcudean, S. E. (1995). A force-controlled pneumatic actuator. *IEEE Transactions on Robotics and Automation.* Vol.11, No.6, pp. 906-911.

Borrelli, F., Bemporad, A., Fodor, M. and Hrovat, D. (2006). An MPC/hybrid System Approach to Traction Control. *IEEE Transactions on Control Systems Technology.* 14(3): 541 – 552.

Bustamante, J., Diong, B. and Wicker, R. (2000). System Identification and Control Design of an Alternative Fuel Engine for Hybrid Power Generation. *Proceeding of the IEEE 35th Intersociety Energy Conversion Engineering Conference and Exhibit, 2000 (IECEC).* July 24-28. Las Vegas, USA. Vol. 1, pp.329-339.

Cabrera, J. A., Ortiz, A., Castillo, J. J. and Simon, A. (2005). A Fuzzy Logic Control for Antilock Braking System Integrated in the IMMa Tire Test Bench. *IEEE Transactions on Vehicular Technology.* 54(6): 1937 – 1949.

Corno, M., Tanelli, M., Savaresi, S. M., Fabbri, L. And Nardo, L. (2008). Electronic Throttle Control for Ride-by-Wire in Sport Motorcycles. *Proceeding of the IEEE International Conference on Control Applications 2008 (CCA'2008).* September 3-5. San Antonio, Taxes, USA. pp. 233-238.

Darling, J. and Ross-Martin, T. J. (1997). A Theoretical Investigation of a Prototype Active Roll Control System. *Journal of Automobile Engineering.* 211(1): 3-12.

Du, H. and Dong, G. (2007). Robust Active Roll Controller Design for Vehicles Considering Variable Speed and Actuator Delay. *SAE Technical Paper Series.* Paper No. 2007-01-0825.

Falcone, P., Borrelli, F., Asgari, J., Tseng, H. E. and Hrovat, D. (2007). Predictive Active Steering Control for Autonomous Vehicle Systems. *IEEE Transactions on Control Systems Technology.* 15(3): 566 – 580.

Hanafi, D. (2010). PID Control Design for Semi-active Car Suspension Based on Model from Intelligent System Identification. *Proceeding of the IEEE Second International Conference on Computer Engineering and Applications (ICCEA 2010)*. March 19-21. Bali Island, Indonesia. pp. 60-63.

Hashemi-Dehkordi, S. M., Mailah, M. and Abu-Bakar, A. R. (2008). A Robust Active Control Method To Reduce Brake Noise. *Proceedings of the IEEE Conference on Robotics and Biometrics*. February 22-25. Bangkok, Thailand. pp. 739-744.

Hudha, K., Jamaluddin, H. Samin, P. M. and Rahman, R. A. (2003). Semi Active Roll Control Suspension System on a New Modified Half Car Model. *SAE Technical Paper Series*. Paper No. 2003-01-2274

Ikenaga, S. A. (2000). *Development of a Real Time Digital Controller: Application to Active Suspension Control of Ground Vehicles*. Michigan University: PhD. Dissertation.

Kadir, Z. A., Hudha, K., Nasir, M. Z. M. and Said, M. R.(2008). Assessment of Tire Models for Vehicle Dynamics Analysis. *Proceedings of the International Conference on Plant Equipment and Reliability*. March 27-28. Kuala Lumpur, Malaysia.

Mammar, S. and Koenig, D. (2002). Vehicle Handling Improvement by Active Steering. *Vehicle System Dynamics*. 38(3): 211-242.

Marino, R., Scalzi, S., Orlando, G. and Netto, M. (2009). A Nested PID Steering Control for Lane Keeping in Vision Based Autonomous Vehicle. *Proceeding of the IEEE American Control Conference Conference Hyatt Regency Riverfront*. June 10-12. Saint Louis, Missouri, USA. pp. 2885-2890.

McCann, R. (2000). Variable Effort Steering for Vehicle Stability Enhancement Using an Electric Power Steering System. *SAE Technical Paper Series*. Paper No. 2000-01-0817.

Messina, A., Giannoccaro, N. I. and Gentile, A. (2005) .Experimenting and modelling the dynamics of pneumatic actuators controlled by the pulse width modulation (PWM) technique. *Mechatronics*. Vol.15, pp. 859-881.

Miege, A. and Cebon, D. (2002). Design and Implementation of an Active Roll Control System for Heavy Vehicles. *Proceedings of the 6th International Symposium on Advanced Vehicle Control (AVEC)*. September 9-13.Hisoshima, Japan.

Mingzhu, Z., Zhili, Z., Jinfa, X. and Zhiqiang, X. (2008). Modeling and Control Simulation for Farm Tractor with Hydro-Mechanical CVT. *Proceeding of the IEEE International Conference on Automationand Logistics 2008 (ICAL 2008)*. September 1-3.Qingdao, China. pp. 908-913. 1.

Mokhiamar, O. and Abe, M. (2002). Effect of Model Response on Model Following Type of Combined Lateral Force and Yaw Moment Control Performance for Active Vehicle Handling Safety. *JSAE Review*. 23: 473-480.

Morita, Y., Torii, K., Tsucida, N., Iwasaki, M., Ukai, H., Matsui, N., Hayashi, T., Ido, W. and Ishikawa, H. (2008). Improvement of Steering Feel of Electric Power Steering System with Variable Gear Transmission System Using Decoupling Control. *Proceeding of the IEEE International Workshop on Advanced Motion Control 2008 (AMC2008)*. March 26-28. Trento, Italy. pp. 417-422,

Richer, E. and Hurmuzlu, Y. (2000). A high performance pneumatic force actuator system Part II –Nonlinear controller design. *Journal of Dynamic Systems, Measurement and Control*. pp.426–434.

Shoubo, L., Chenglin, L., Shanglou, C. and Lifang, W. (2009). Traction Control of Hybrid Electric Vehicle. *Proceeding of the IEEE Vehicle Power and Propulsion Conference, 2009(VPPC'09)*. September 7-10. Dearbon, Michigan, USA. pp. 1535-1540.

Singh, T., Kesavadas, T., Mayne, R., Kim, J. J. and Roy, A. (2002). Design of Hardware/Algorithms for Enhancement of Driver-Vehicle Performance in Inclement Weather Conditions Using a Virtual Environment, *SAE* Paper No. 2002-01-0322.

Situm, Z., Kosic, D. and Essert, M. (2005). Nonlinear Mathematical Model of a Servo Pneumatic System. *Proceedings of Nineth International Research/Expert Conference*, Antalya, Turkey.

Smaoui, M., Brun, X. and Thomasset, D. (2006). A study on tracking position control of an electropneumatic system using backstepping design. *Control Engineering Practice*. Vol.14, No. 8, pp.923-933.

Sorniotti, A. and D'Alfio, N. (2007). Vehicle Dynamics Simulation to Develop an Active Roll Control System. *SAE Technical Paper Series*. Paper No. 2007-01-0828.

Sugisaka, M., Tanaka, H. and Hara, M. (2006). A Control Method of Accelerator of an Electric Vehicle. *Proceedings of the IEEE International Joint Conference (SICE-ICASE,2006)*. October 18-21. Busan, Korea. pp 5300-5303.

Szostak, H. T., Allen, W. R. and Rosenthal, T. J. (1988). Analytical Modeling of Driver Response in Crash Avoidance Maneuvering Volume II: An Interactive Model for Driver/Vehicle Simulation. *US Department of Transportation Report NHTSA DOT HS-807-271*. April.

Takatsu, H., Itoh, T. and Araki, M. (1998). Future Needs for the Control Theory in Industries Report and Topics of the Control Technology Survey in Japanese Industry. *Journal of Process Control*. 8 (5-6): 369-374.

Tan, Y., Robotis, A. and Kanellakopoulos, I. (1999). Speed Control Experiment with an Automated Heavy Vehicle. *Proceeding of the 1999 IEEE International Conference on Control Applications*. August 22-27. Kohala Coast, Hawaii, USA. Vol. 2, pp.1353-1358.

Wang, J., Pu, J. and Moore, P. (1999). Accurate position control of servo pneumatic actuator systems: an application to food packaging. *Control Engineering Practice*. Vol.7, No. 7, pp. 699-706.

Wang, J. and Longoria, R. G. (2006). Coordinated Vehicle Dynamics Control with Control Distribution. *Proceedings of the 2006 American Control Conference*. June 14-16. Minneapolis, Minnesota, USA.

Wang, Y., Kraska, M. and Ortmann, M. (2001). Dyanamic Modeling of a Variable Force and a Cluth for Hydraulic Control in Vehicle Transmission System. *Proceeding of the IEEE 2001 American Control Conference*. August 7. Arlington, Virginia, USA.Vol. 3, pp.1789-1793.

Wang, J., Wilson, D. A., Xu. W. and Crolla, D. A. (2005). Active Suspension Control to Improve Vehicle Ride and Steady State Handling. *Proceedings of the 44th IEEE Conference on Decision and Control*. December 12-15. Seville, Spain. 1982-1987.

Wei, L., Wei, L., Jia, Y. and Gang, J. L. (2010). Simulation of CVT Ratio Control Strategy of Engine Braking. *Proceeding of the IEEE 2nd International Asia Conference on Informatics in Control, Automation and Robotics (CAR)*. March 6-7. Wuhan, China. pp.166-169.

Williams, D. E. and Haddad, W. M. (1995). Nonlinear Control of Roll Moment Distribution to Influence Vehicle Yaw Characteristics. *IEEE Transaction on Control System Technology*. 3(1): 110-116.

Wu, X., Wang, X., Yu, T. and Xie, X. (2008). Control of Electronic Clutch During Vehicles Start. *Proceeding of the IEEE Vehicle Power and Propulsion Conference, 2008 (VPPC'2008)*. September 3-5. Harbin, China. pp.1-5.

Xinpeng, T. and Duan, X. (2007). Simulation and Study of SUV Active Roll Control Based on Fuzzy PID. *SAE Technical Paper Series*. Paper No. 2007-01-3570.

Xu, N., Chen, H., Hu, Y. and Liu, H. (2007). The Integrated Control System in Automatic Transmission. *Proceeding of the IEEE International Conference on Mechatronics and Automation*. August 5-8. Harbin, China. pp. 1655-1659.

Ying, H., Fujun, Z., Fushui, L., Yunshun, G. and Yebao, S. (1999). Gasoline Engine Idle Speed Control System Development Based on PID Algorithm. *Proceeding of the IEEE International Vehicle Electronic Conference 1999 (IVEC'99)*. September 6-9. Changcun, China. Vol. 1, pp.30-31.

Yan, Y. L., Guang, Q. Y. and Feng, L. (2008). Research on Control Strategy and Bench Test of Automobile Steer-by-Wire System. *Proceeding of the IEEE Vehicle Power and Propulsion Conference (VPPC)*. September 3-5. Harbin, China. pp.1-6.

Yuanyuan, Z., Jianbo, S. and Guo, C. (2008). Research on the Modeling and Simulation of the Four-stroke Engine and it's Control. *Proceeding of the IEEE 7th International Conference on System Simulation and Scientific Computing,2008 (ICSC2008)*. October 10-12. Beijing, China. pp.1321-1324.

Zhang, D., Zheng, H., Sun, J., Wang, Q., Wen, Q., Yin, A. and Yang, Z. (1999). Simulation Study for Anti-lock Bracking system of a Light Bus. *Proceeding of the IEEE International Vehicle Electronics Conference,1999(IVEC'99)*. September 6-9. Changcun, China. Vol. 1, pp.70-77.

Zhang, J. D., Qin, G. H., Xu, B,. Hu, H. S. and Chen, Z. X. (2010). Study on Automotive Air Conditioner Control System Based on Incremental-PID. *Journal of Advanced Materials Research*. Vol. 129-131, pp 17-22

Application of Improved PID Controller in Motor Drive System

Song Shoujun and Liu Weiguo
Northwestern Polytechnical University
China

1. Introduction

PID (proportional-integral-derivative) controller has being widely used in motor drive system. More than 90% of industrial controllers are implemented based on PID algorithms (Ang et al., 2005). The structure of PID controller is very simple and its control principle is very clear. It is practical and is very easy to be implemented. What's more, because the functionalities of the three factors in PID controller are very clear, they can be tuned efficiently to obtain desired transient and steady-state responses.
Motor drive systems can be found in many applications, their behaviours can influence the performances of the whole system tremendously. The motor drive system has many distinct features, such as multivariable, strong nonlinearity and strong coupling (Li et al., 2010). Many parameters in the system are time-variant. What's more, in many cases, it's very difficult to get the accurate mathematical model of the motor drive system. All these features make the control of the motor drive system difficult.
PID controller is very popular in the control of the motor drive system. However, since the controller parameters are fixed during control after they have been chosen through a certain optimal method, the conventional PID controller can't always keep satisfying performances. To cope with this problem, the parameters of the controller need to be adjusted dynamically according to the running status of the system. Many on-line tuning algorithms, such as fuzzy logic, neural network and genetic algorithm, have been introduced into PID controller to achieve desired control performances for the entire operating envelope of the motor drive system (Tang et al., 2001; Yu et al., 2009; Lin et al., 2003).
In this chapter, two improved self-tuning PID controllers are given and studied in detail. To verify their validity, two typical motor drive systems, namely switched reluctance motor (SRM) drive system (Chen & Gu, 2010) and brushless DC motor (BLDCM) drive system (Wu et al., 2005), are introduced as examples. Based on the models of these two drive systems, the performances of the improved PID controllers are analyzed in detail.

2. Conventional PID Controller

In analog control system, PID controller is used commonly. The conventional PID (C-PID) controller is a linear control method. It compounds the outputs of proportional, integral and derivative parts linearly to control the system. Fig. 1 shows the block diagram of the C-PID controller.

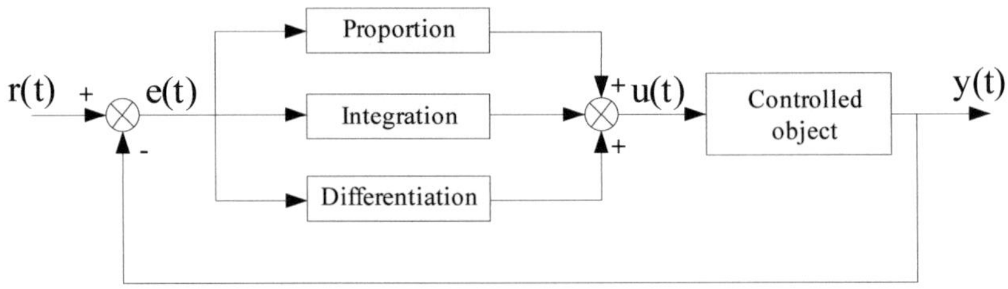

Fig. 1. Block diagram of the C-PID controller

The algorithm of C-PID controller can be given as follows:

$$e(t) = r(t) - y(t) \tag{1}$$

$$u(t) = K_p\left[e(t) + \frac{1}{T_i}\int e(t)dt + T_d\frac{de(t)}{dt}\right] \tag{2}$$

where *y(t)* is the output of the system, *r(t)* is the reference input of the system, *e(t)* is the error signal between *y(t)* and *r(t)*, *u(t)* is the output of the C-PID controller, K_p is proportional gain, T_i is integral time constant and T_d is derivative time constant.
Equation (2) also can be rewritten as (3):

$$u(t) = K_p e(t) + K_i\int e(t)dt + K_d\frac{de(t)}{dt} \tag{3}$$

where K_i is integral gain, K_d is derivative gain, and $K_i=K_p/T_i$, $K_d=K_pT_d$.
In C-PID controller, the relation between PID parameters and the system response specifications is clear. Each part has its certain function as follows (Shi & Hao, 2008):

(1) **Proportion** can increase the response speed and control accuracy of the system. Bigger K_p can lead to faster response speed and higher control accuracy. But if K_p is too big, the overshoot will be large and the system will tend to be instable. Meanwhile, if K_p is too small, the control accuracy will be decreased and the regulating time will be prolonged. The static and dynamic performance will be deteriorated.
(2) **Integration** is used to eliminate the steady-state error of the system. With bigger K_i, the steady-state error can be eliminated faster. But if K_i is too big, there will be integral saturation at the beginning of the control process and the overshoot will be large. On the other hand, if K_i is too small, the steady-state error will be very difficult to be eliminated and the control accuracy will be bad.
(3) **Differentiation** can improve the dynamic performance of the system. It can inhibit and predict the change of the error in any direction. But if K_d is too big, the response process will brake early, the regulating time will be prolonged and the anti-interference capability of the system will be bad.

The three gains of C-PID controller, K_p, K_i and K_d, can be determined conveniently according to the above mentioned function of each part. There are many methods such as NCD (Wei, 2004; Qin et al., 2005) and genetic algorithm can be used to determine the gains effectively.

(1) **NCD** is a toolbox in Matlab. It is developed for the design of nonlinear system controller. On the basis of graphical interfaces, it integrates the functions of optimization and simulation for nonlinear system controller in Simulink mode.

(2) **Genetic algorithm (GA)** is a stochastic optimization algorithm modeled on the principles and concepts of natural selection and evolution. It has outstanding abilities for solving multi-objective optimization problems and finding global optimal solutions. GA can readily handle discontinuous and nondifferentiable functions. In addition, it is easily programmed and conveniently implemented (Naayagi & Kamaraj, 2005; Vasconcelos et al., 2001).

In many conventional applications, the gains of C-PID controller are determined offline by one of the methods mentioned above and then fixed during the whole control process. This control scheme has two obvious shortcomings as follows:

(1) All the methods that can be used to determine the gains of C-PID controller offline are based on the precise mathematical model of the controlled system. However, in many applications, such as motor drive system, it is very difficult to build the precise mathematical model due to the multivariable, time-variant, strong nonlinearity and strong coupling of the real plant.

(2) In many applications, some parameters of the controlled system are not constant. They will be changed according to different operation conditions. For example, in motor drive system, the winding resistance of the motor will be changed nonlinearly along with the temperature. If the gains of C-PID controller are still fixed, the performance of the system will deteriorate.

To overcome these disadvantages, C-PID should be improved. The gains of PID controller should be adjusted dynamically during the control process.

3. Improved PID Controller

There are many techniques such as fuzzy logic control, neural network and expert control (Xu et al., 2004) can be adopted to adjust the gains online according to different conditions. In this chapter, two kinds of Improved PID (I-PID) controller based on fuzzy logic control and neural network are studied in detail.

3.1 Fuzzy Self-tuning PID Controller

Fuzzy logic control (FLC) is a typical intelligent control method which has been widely used in many fields, such as steelmaking, chemical industry, household appliances and social sciences. The biggest feature of FLC is it can express empirical knowledge of the experts by inference rules. It does not need the mathematical model of the controlled object. What's more, it is not sensitive to parameters changing and it has strong robustness. In summary, FLC is very suitable for the controlled object with characteristics of large delay, large inertia, non-linear and time-variant (Liu & Li, 2010; Liu & Song, 2006; Shi & Hao, 2008).

The structure of a SISO (single input single output) FLC is shown in Fig. 2. It can be found that the typical FLC consists of there main parts as follows:

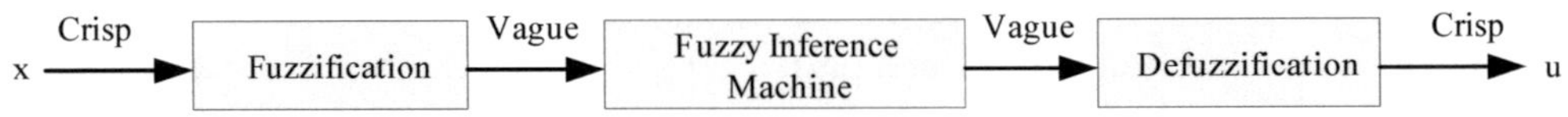

Fig. 2. The structure of SISO FLC

(1) **Fuzzification** comprises the process of transforming crisp inputs into grades of membership for linguistic terms of fuzzy sets. The input values of a FLC consist of measured values from the plant that are either plant output values or plant states, or control errors derived from the set-point values and the controlled variables.
(2) **Fuzzy Inference Machine** is the core of a fuzzy control system. It combines the facts obtained from the fuzzification with the rule base and conducts the fuzzy reasoning process. A proper rule base can be found either by asking experts or by evaluation of measurement data using data mining methods.
(3) **Defuzzification** transforms an output fuzzy set back to a crisp value. Many methods can be used for defuzzification, such as centre of gravity method (COG), centre of singleton method (COS) and maximum methods.

Detailed analyses show that FLC is a nonlinear PD controller. It cannot eliminate steady-state error when the controlled object does not have integral element, so it is a ragged controller. To overcome this disadvantage, FLC is often used together with other controllers. Fig. 3 shows the structure of a controller called Fuzzy_PID compound controller. When the error is big, FLC is used to accelerate the dynamic response, and when the error is small, PID controller is used to enhance the steady-state accuracy of the system (Liu & Song, 2006).

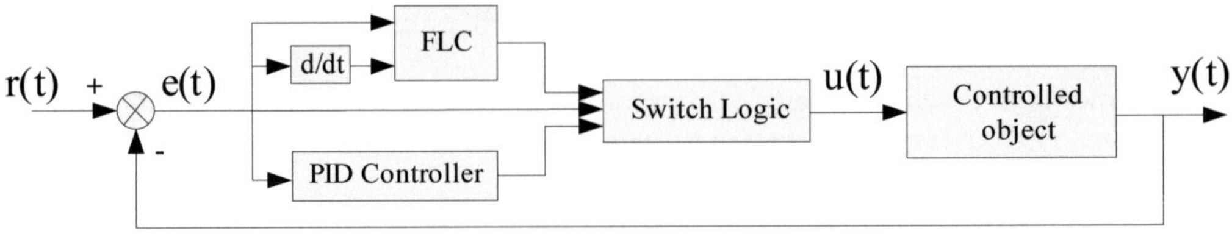

Fig. 3. The structure of Fuzzy_PID compound controller

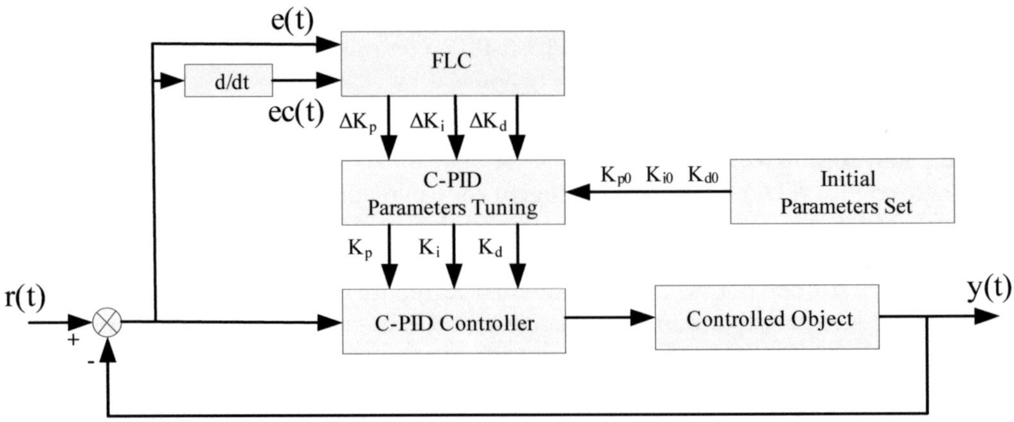

Fig. 4. The structure of FPID controller

In this chapter, an I-PID controller called fuzzy self-tuning PID (FPID) controller is introduced. In this controller, FLC is used to tune the parameters of C-PID controller online according to different conditions. Fig. 4 shows the structure of FPID controller (Liu & Li, 2010).

In FPID controller, the error signal and the rate of change of error are inputted into FLC firstly. After fuzzy inference based on the rule base, the increments of PID control parameters, ΔK_p, ΔK_i and ΔK_d, are obtained, add these increments to initial values of PID control parameters, the actual PID control parameters can be achieved finally. The initial values of PID control parameters, K_{p0}, K_{i0} and K_{d0}, can be obtained by the methods mentioned in the last section.

3.2 Neural Network PID Controller

Neural network (NN) is a mathematical model or computational model inspired by the structure and functional aspects of biological neural systems, such as the brain. It is composed of a large number of highly interconnected processing elements (neurones) working in unison to solve specific problems. Fig. 5 shows the typical structure of a NN. It has one input layer, one output layer and several hidden layers. In each layer, there are a certain number of nodes (neurons). The neurons in adjacent layers are connected together, while there are no connections between neurons in the same layer. Just like the biological neural systems, the NN also can learn by itself. During the learning phase, the connection strength (weights) between neurons can be adjusted by certain algorithms automatically based on external or internal information that flows through the network. (Tao, 2002; Liu, 2003; Wang et al., 2007)

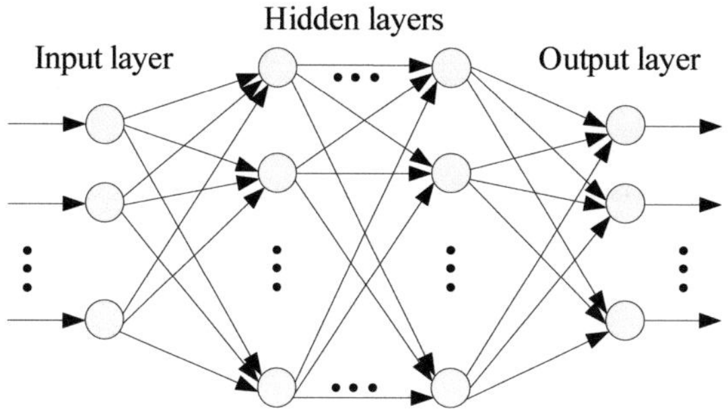

Fig. 5. The structure of a typical NN

The greatest advantage of NN is its ability to be used as an arbitrary function approximation mechanism which 'learns' from observed data. There are many other remarkable advantages of NN as follows:

(1) **Adaptive learning:** An ability to learn how to do tasks based on the data given for training or initial experience.
(2) **Real time operation:** NN can process massive data and information in parallel. Special hardware devices are being designed and manufactured which take advantage of this capability.
(3) **Fault tolerance:** Some capabilities of NN can be retained even with major network damage.

BP (backpropagation) neural network (BPNN) is the most popular neural network for practical applications. It adopts the backpropagation learning algorithm which can be divided into two phases: data feedforward and error backpropagation.

(1) **Data feedforward:** In this phase, the data, such as the error of the controlled system, inputted into the input layer is fed into the hidden layer and then into the output layer. Finally, the output of the BPNN can be obtained from the output layer. It is the function of the connection weights between neurons.

(2) **Error backpropagation:** In this phase, the actual output value of the network obtained in the last phase is compared with a desired value. The error between them is propagated backward. The connection weights between neurons are adjusted by some means, such as gradient descent algorithm, based on the error.

These two phases are repeated continuously until the performance of the network is good enough.

In this chapter, BPNN is used to tune the parameters of C-PID controller online. Fig. 6 shows the structure of this I-PID controller named NNPID controller.

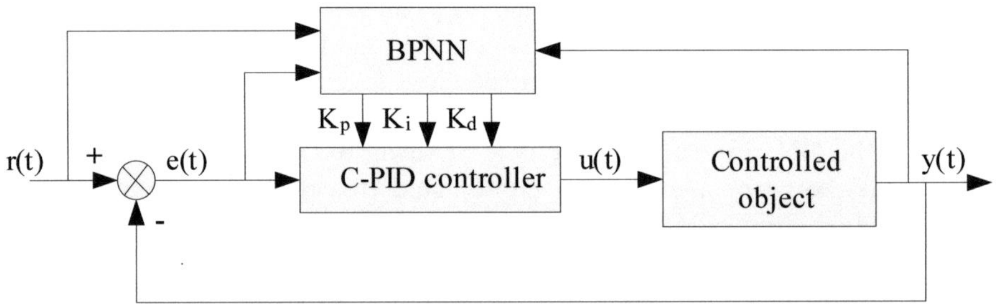

Fig. 6. The structure of NNPID controller

It can be seen that NNPID controller consists of C-PID controller and BPNN. C-PID controller is used to control the plant directly. Its output, u(t), can be obtained by (3). In order to optimize the performance of the system, BPNN is used to adjust the three parameters of C-PID controller online based on some state variables of the system.

4. Motor Drive System

Motor is the main controlled object in motor drive system. In practical applications, there are many kinds of motors. In this chapter, the brushless DC motor (BLDCM) and switched reluctance motor (SRM) are studied as examples. Their mathematical models are built to simulate the performance of different control methods.

4.1 Brushless DC Motor

In BLDCM, electronic commutating device is used instead of the mechanical commutating device. Because BLDCM has many remarkable advantages, such as high efficiency, silent operation, high power density, low maintenance, high reliability and so on, it has been widely used in many industrial and domestic applications.

The voltage equation for one phase in BLDCM can be written as:

$$u = e + i_a R_a + L_a \frac{di_a}{dt} \tag{4}$$

where u, i_a, R_a and L_a are the voltage, current, resistance and inductance of one phase, respectively. e is the back EMF (electromotive force) which can be calculated by

$$e = C_e \Phi n = k_v \omega \tag{5}$$

where ω is the angular speed of the rotor, k_v is a constant which can be calculated by

$$k_v = \frac{30}{\pi} C_e \Phi = 9.55 C_e \Phi \tag{6}$$

where C_e is the EMF constant and Φ is the flux per pole.
The torque equation can be given as

$$T_{em} = T_L + B\omega + J\frac{d\omega}{dt} \tag{7}$$

where T_{em} is electromagnetic torque, T_L is load torque, B is damping coefficient and J is rotary inertia.
T_{em} also can be obtained by

$$T_{em} = C_T \Phi i_a = k_t i_a \tag{8}$$

where k_t is a constant which can be calculated by

$$k_t = C_T \Phi \tag{9}$$

where C_T is the torque constant.
Based on all above equations, the state space equation of BLDCM can be obtained as

$$\frac{d}{dt}\begin{bmatrix} i_a \\ \omega \end{bmatrix} = \begin{bmatrix} -\frac{R_a}{L_a} & -\frac{k_v}{L_a} \\ \frac{k_t}{J} & -\frac{B}{J} \end{bmatrix} \begin{bmatrix} i_a \\ \omega \end{bmatrix} + \begin{bmatrix} \frac{1}{L_a} & 0 \\ 0 & -\frac{1}{J} \end{bmatrix} \begin{bmatrix} u \\ T_L \end{bmatrix} \tag{10}$$

The Laplace transform of (10) can be written as two equations as follows:

$$\begin{cases} I_a(s) = \dfrac{-k_v \omega(s) + U(s)}{L_a s + R_a} \\ \omega(s) = \dfrac{k_t I_a(s) - T_L}{Js + B} \end{cases} \tag{11}$$

According to (11), the simulation model of BLDCM can be built in Matlab/simulink as shown in Fig. 7.

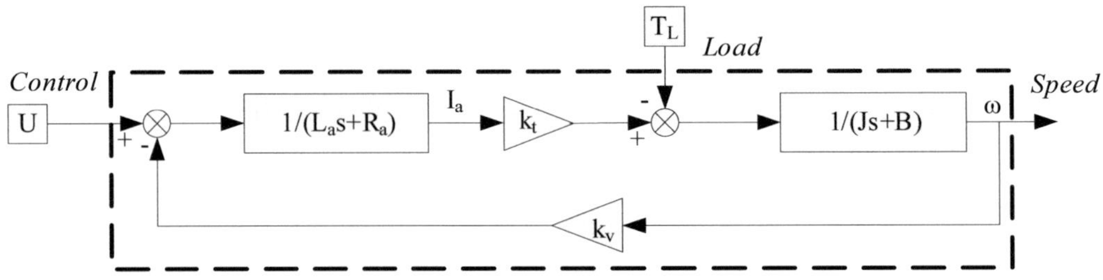

Fig. 7. The simulation model of BLDCM

4.2 Switched Reluctance Motor

The SRM is a brushless synchronous machine with salient rotor and stator teeth. There are concentrated phase windings in the stator, and no magnets and windings in the rotor. It has many remarkable advantages such as simple magnetless and rugged construction, simple control, ability of extremely high speed operation, relatively wide constant power capability, minimal effects of temperature variations offset, low manufacturing cost and ability of hazard-free operation. These advantages make the SRM very suitable for applications in more/all electric aircraft (M/A EA), electric vehicle (EV) and wind power generation.

Because the nonlinear model of SRM is very complex, people generally use its quasi-linear model to design and analyze control methods.

According to the quasi-linear model of SRM, the average torque equation can be obtained as (12) when the phase current is flat topped (Wang, 1999).

$$T_{av} = \frac{mN_r U_s^2}{2\pi\omega_r^2}\left(\theta_{off} - \theta_1\right)\left[\frac{\theta_1 - \theta_{on}}{L_{min}} - \frac{1}{2}\frac{\theta_{off} - \theta_1}{L_{max} - L_{min}}\right] \tag{12}$$

where T_{av} is the average torque, m is the number of motor phase, N_r is the number of rotor tooth, U_s is the power supply voltage, ω_r is angular speed of the rotor, θ_{on} is the angle of starting the excitation, θ_{off} is the angle of switching off the excitation, θ_1 is the starting angle of the phase inductance increasing, L_{max} and L_{min} are the maximum and minimum value of phase inductance, respectively.

Based on (12), the total differential equation of T_{av} can be written as (He et al., 2004)

$$dT_{av} = \frac{\partial T_{av}}{\partial U_s}dU_s + \frac{\partial T_{av}}{\partial \theta_{on}}d\theta_{on} + \frac{\partial T_{av}}{\partial \theta_{off}}d\theta_{off} + \frac{\partial T_{av}}{\partial \omega_r}d\omega_r \tag{13}$$

According to the linearization theory, the differential of each variable in (13) can be replaced by corresponding increment. If voltage PWM control is adopted, θ_{on} and θ_{off} are fixed. The simplified small-signal torque equation can be obtained as

$$\Delta T_{av} = k_u \Delta U_s - k_\omega \Delta\omega_r \tag{14}$$

The increment of the average torque also can be indicated as

$$\Delta T_{av} = J\frac{d\Delta\omega_r}{dt} + B\Delta\omega_r + \Delta T_L \tag{15}$$

where J is rotary inertia, B is damping coefficient, T_L is load torque.
The voltage chopping can be treated as a sampling process of the controller's output ΔU_{ASR}, and the amplification factor is K_c. The small-signal model of power inverter can be given as

$$\Delta U_s(s) = k_c \cdot \frac{1-e^{-Ts}}{s}\Delta U_{ASR}(s) = k_c \cdot \frac{T}{1+Ts}\Delta U_{ASR}(s) \tag{16}$$

The feedback of angular speed can be treated as a small inertial element.

$$G_\omega(s) = k_n/(1+T_\omega s) \tag{17}$$

where K_n is feedback coefficient and T_ω is time constant of the measurement system.
Based on above analysis, the simplified small-signal model of SRM can be got as shown in Fig. 8.

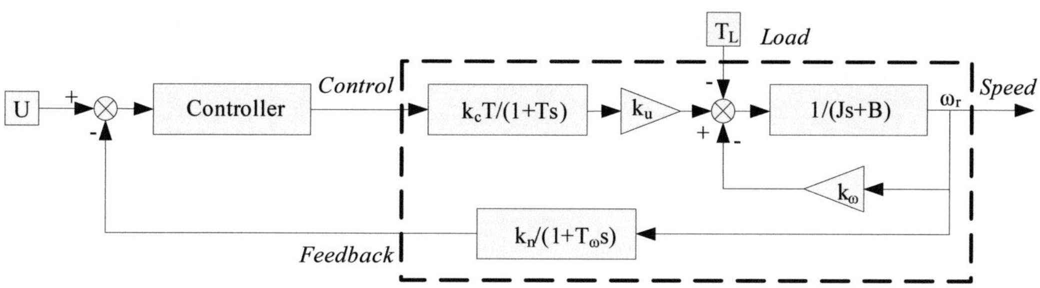

Fig. 8. The simulation model of SRM

5. Design of I-PID Controller

5.1 FPID Controller for SRM

Based on Fig.1, Fig. 4 and Fig. 8, the simulation model of C-PID and FPID for SRM can be obtained as shown in Fig. 9 and Fig. 10. The internal structure of the module marked "SRM" is the part that enclosed by dashed box in Fig. 8.
It can be found that the three parameters of the PID controller in FPID control can be obtained by

$$\begin{cases} K_p = K_{p0} + \Delta K_p \\ K_i = K_{i0} + \Delta K_i \\ K_d = K_{d0} + \Delta K_d \end{cases} \tag{18}$$

Where K_{p0}, K_{i0} and K_{d0} are the initial PID parameters obtained by NCD or GA. ΔK_p, ΔK_i and ΔK_d are provided by FLC. They are used to adjust the three parameters online. In other words, the parameters of C-PID can be dynamically tuned by FLC according to different operation conditions. Fig. 11 shows the structure of the FLC used in FPID controller. It has two input variables and three output variables.

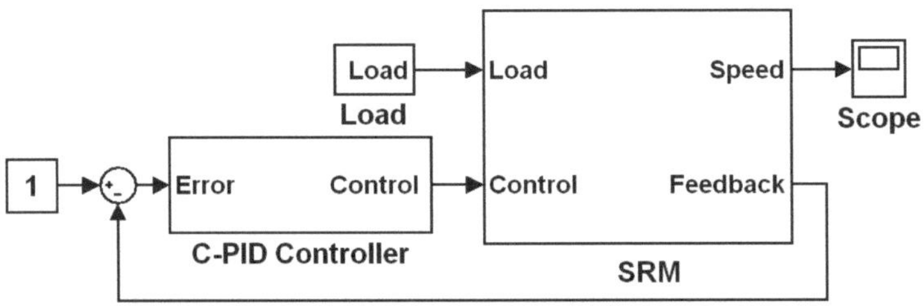

Fig. 9. The simulation model of C-PID controlled SRM system

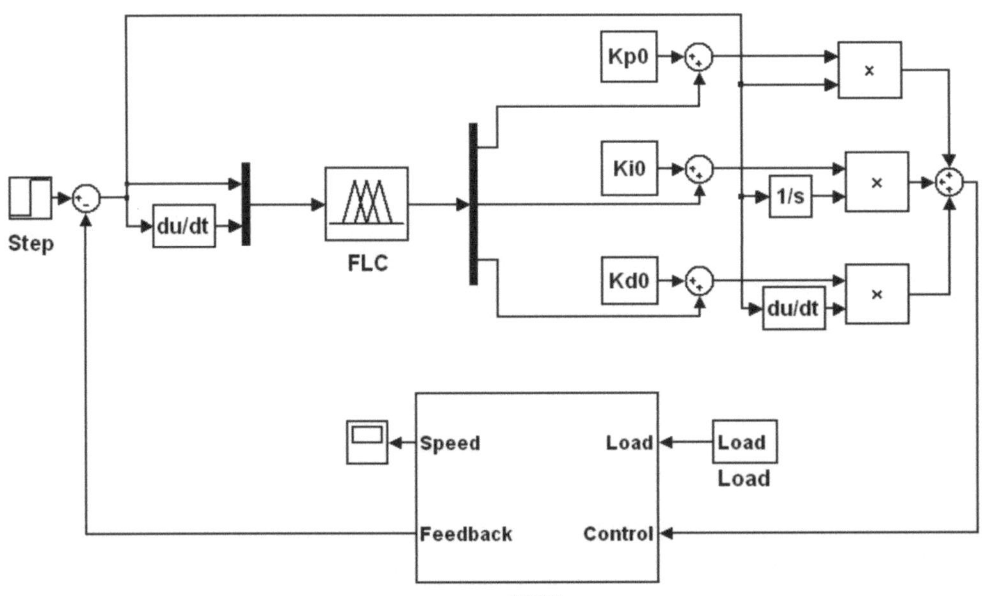

Fig. 10. The simulation model of FPID controlled SRM system

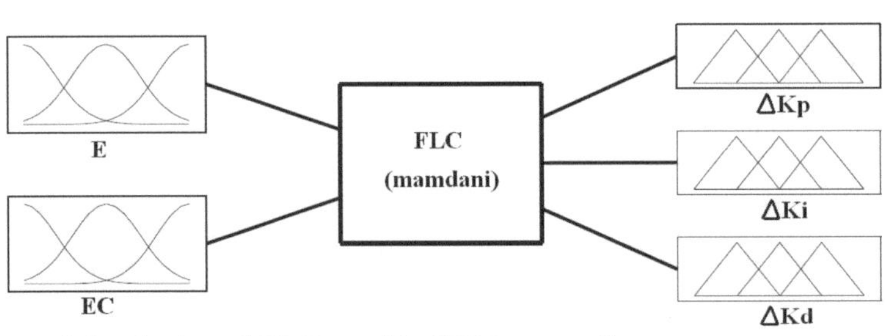

Fig. 11. Structure of the designed FLC used in FPID controller

The most important thing for the design of FPID controller is the determination of the fuzzy rule base. According to the functions of each PID parameter mentioned in section 2, the principles for their adjustment can be summarized as follows (Shi & Hao, 2008):

(1) When the absolute value of the system error, |e(t)|, is relatively big: K_p is increased to get faster tracking speed, K_i is reduced to avoid overshoot.

(2) When |e(t)| is relatively small: K_p and K_i are increased to enhance the tracking precision, K_d should be proper to avoid steady-state oscillation.

(3) When |e(t)| is medium: K_p is reduced to avoid overshoot, K_i is increased slightly to enhance the steady-state precision, and K_d should be proper to guarantee the stability of the system.

Based on above principles and consider the change rate of the system error, ec(t), the fuzzy rule base of the three parameters can be obtained. As an example, Table.1 shows the fuzzy rule base for ΔK_p.

ΔK_p ec / e	NB	NM	NS	ZO	PS	PM	PB
NB	NB	NB	NM	PB	PB	PB	ZO
NM	NB	NB	NM	PM	PB	ZO	PS
NS	NB	NM	NS	PS	ZO	NM	NS
ZO	NB	NM	NS	ZO	NS	NM	NB
PS	NS	NM	ZO	PS	NS	NM	NB
PM	PS	ZO	PB	PM	NM	NB	NB
PB	ZO	PB	PB	PB	NM	NB	NB

Table 1. Fuzzy rule base of ΔK_p

It can be seen that there are totally 49 fuzzy rules and they are represented by fuzzy linguistic terms, such as *if* e=NB and ec=NB *then* ΔK_p=NB, ΔK_i=NB, ΔK_d=PB.

In this chapter, all the variables are described by seven linguistic terms. They are negative big (NB), negative middle (NM), negative small (NS), zero (ZO), positive small (PS), positive middle (PM) and positive big (PB). The universe of input variables, e and ec, is {-3 -2 -1 0 1 2 3}. The universe of output variables, ΔK_p, ΔK_i and ΔK_d, is {-0.6 -0.4 -0.2 0 0.2 0.4 0.6}.

Fig. 12 and 13 show the membership function of each variable.

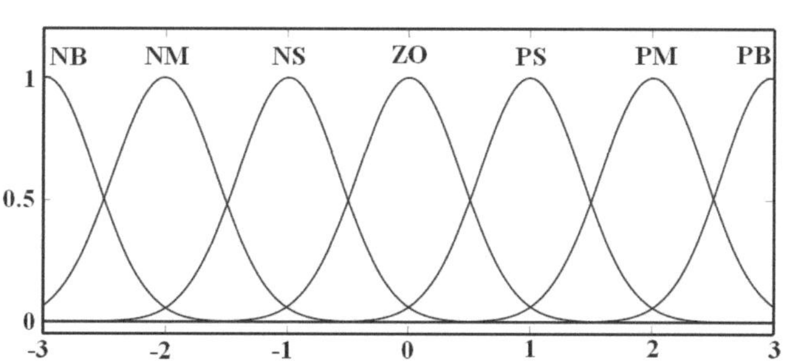

Fig. 12. Membership function of e and ec

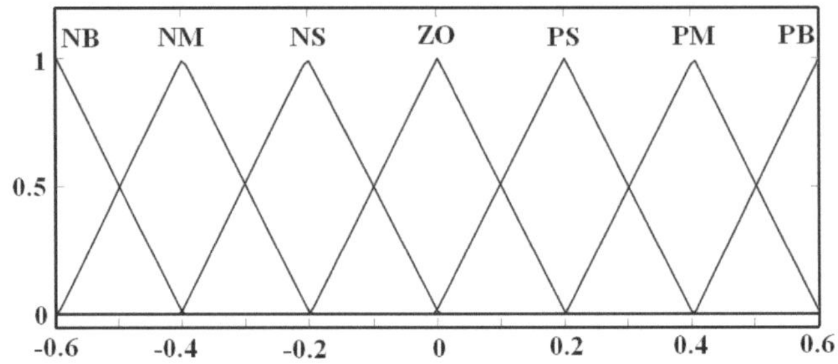

Fig. 13. Membership function of ΔK_p, ΔK_i and ΔK_d

In this chapter, the MAX-MIN method is used for fuzzy inference and centroid is used for defuzzification.

5.2 NNPID Controller for BLDCM

Based on Fig.1 and Fig. 7, the simulation model of C-PID for BLDCM can be obtained as shown in Fig. 14. The internal structure of the module marked "BLDCM" is the part that enclosed by dashed box in Fig. 7.

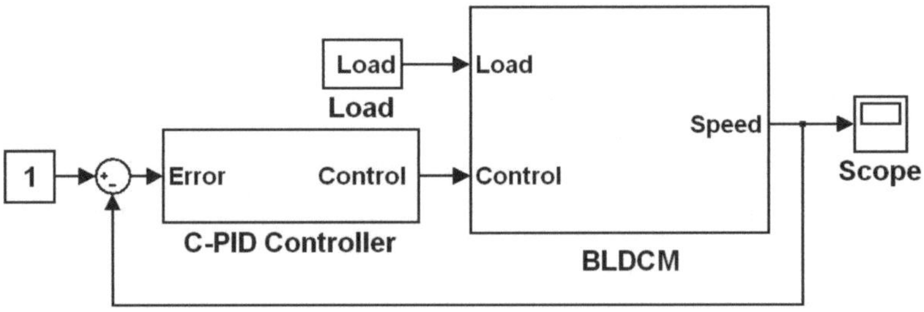

Fig. 14. The simulation model of C-PID controlled BLDCM system

Based on Fig.5 and Fig. 6, the structure of the BPNN used in the NNPID controller is shown in Fig.15.

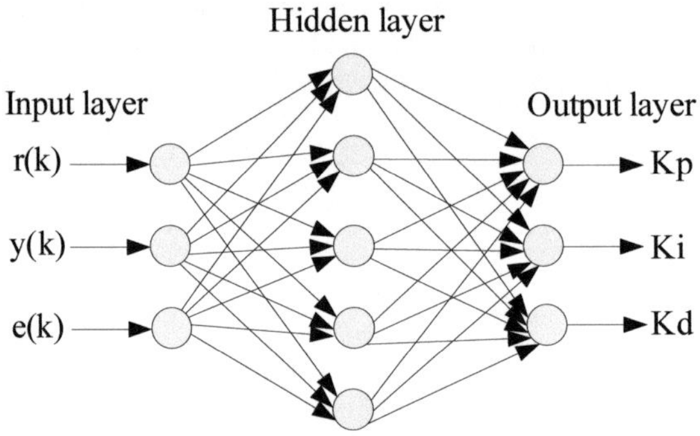

Fig. 15. The structure of the BPNN used in NNPID controller

It can be seen that the adopted BPNN has three layers: one input layer, one hidden layer and one output layer. There are three input variables and three output variables. r(k) is the reference input of the system, y(k) is the real output of the system and e(k) is the error between them. K_p, K_i and K_d are the three parameters of the C-PID controller. There are five nodes (neurones) in the hidden layer.

During operation, the connection strength (weights) between neurons can be adjusted automatically through learning based on the input information. The three output variables of NN, K_p, K_i and K_d, will be changed along with the adjustment of the connection weights. Finally, the performance of the system can be improved.

The output of nodes in input layer equals to their input. The input and output of nodes in hidden layer and output layer can be represented as (Liu, 2003)

$$\text{Hidden} \quad \begin{cases} in_i^{(2)}(k) = \sum_{j=1}^{3} w_{ij}^{(2)} out_j^{(1)}(k) \\ out_i^{(2)}(k) = f\left[in_i^{(2)}(k)\right] \end{cases} \quad i = 1,2,3,4,5 \tag{19}$$

$$\text{Output} \quad \begin{cases} in_l^{(3)}(k) = \sum_{i=1}^{5} w_{li}^{(3)} out_i^{(1)}(k) \\ out_l^{(3)}(k) = g\left[in_l^{(3)}(k)\right] \end{cases} \quad l = 1,2,3 \tag{20}$$

where $w_{ij}^{(2)}$ is connection weight between input and hidden layer, $w_{li}^{(3)}$ is connection weight between hidden and output layer, $f[\cdot]$ and $g[\cdot]$ are activation functions. In this chapter, the activation function of hidden layer is sigmoid function. Because the output variables of NN, K_p, K_i and K_d, can't be negative, the activation function of output layer is nonnegative sigmoid function, that is

$$\begin{cases} f[x] = \tanh(x) = \dfrac{e^x - e^{-x}}{e^x + e^{-x}} \\ g[x] = \dfrac{(1 + \tanh(x))}{2} = \dfrac{e^x}{e^x + e^{-x}} \end{cases} \tag{21}$$

In this chapter, the output variables of NN are the three parameters of C-PID controller, that is

$$\begin{cases} out_1^{(3)}(k) = K_p \\ out_2^{(3)}(k) = K_i \\ out_3^{(3)}(k) = K_d \end{cases} \tag{22}$$

With (19) ~ (22), NN completes the feedforward of the information. The output of the C-PID controller can be got easily based on the three updated parameters, and then the output of the system, y(k), can be obtained. The next step is the backpropagation of the error.

To minimize the error between y(k) and r(k), a performance index function is introduced as

$$J(k)=\frac{1}{2}(r(k)-y(k))^2=\frac{1}{2}e(k)^2 \quad (23)$$

Typically, the connection weights are adjusted by steepest descent method. To increase the convergence speed, an inertia term is added.

$$\Delta w_{li}^{(3)}(k)=-\eta\frac{\partial J(k)}{\partial w_{li}^{(3)}}+\alpha\Delta w_{li}^{(3)}(k-1) \quad (24)$$

where η is learning rate, is inertia coefficient. In this chapter, η=0.001 and α=0.05.
Based on (19) ~ (24), the connection weights can be tuned dynamically. In this chapter, the NNPID controller for BLDCM is implemented by M-File in Matlab.

6. Performance Verification

6.1 FPID Controller for SRM

In this chapter, the parameter values of the SRM (see Fig. 8) are as follows:
k_c=45, T=0.5ms, k_u=22.45, J=1kg·m^2, B=1, k_n=1, T_ω=1.5ms, k_ω=0.05.
The initial values of the three parameters in C-PID are
K_p=5, K_i=7, K_d=2.
Fig. 16 shows the step response of the SRM with C-PID and FPID controller, respectively. The reference angular speed is 100rad/s. The motor is started without load, and at 10s a 50Nm load is added.

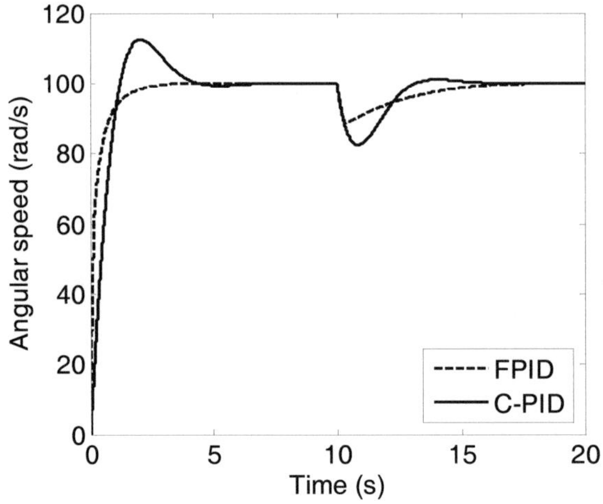

Fig. 16. The step response of the SRM

It can be seen that compared with C-PID controller, the FPID controller can improve the performance of the system significantly. It has advantages of no overshoot, shorter adjusting time. Moreover, when add load torque suddenly at 10s, the drop of the angular speed is smaller and the transition time is shorter.

Fig. 17 shows the adjustment of the three parameters, K_p, K_i and K_d, in FPID controller.

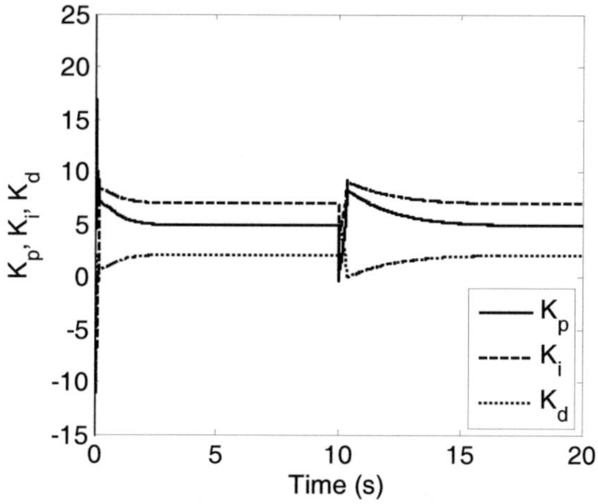

Fig. 17. The adjustment of K_p, K_i and K_d in FPID controller

It can be found that during the adjustment process of the angular speed, the three parameters are tuned dynamically as well. When the system reaches its steady-state, the angular speed is constant and the three parameters are also changed into their initial values.

6.2 NNPID Controller for BLDCM

The discrete form of the BLDCM used in this chapter is

$$y(k) = 0.417 * y(k\text{-}1) + 0.102 * y(k\text{-}2) + 3.058 * u(k\text{-}1) \tag{25}$$

where u is the output of the C-PID controller.

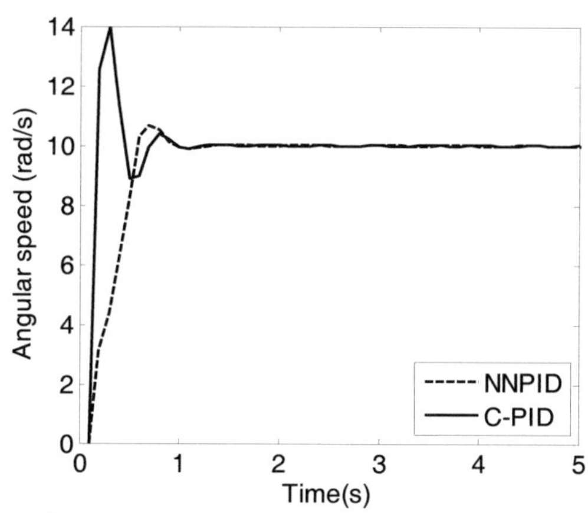

Fig. 18. The step response of the BLDCM

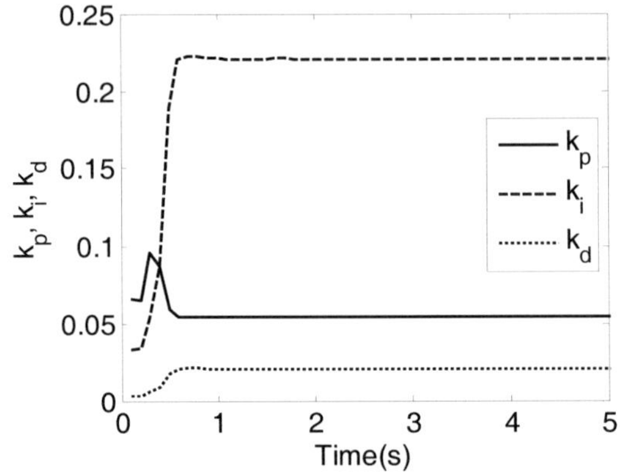

Fig. 19. The adjustment of K_p, K_i and K_d in NNPID controller

Fig. 18 shows the step response of the BLDCM with C-PID and NNPID controller, respectively. The reference angular speed is 10rad/s. The motor is started without load.
It can be seen that compared with C-PID controller, the NNPID controller can improve the performance of the system significantly. The overshoot of the system is nearly eliminated. However, because NNPID controller needs time to train NN, the adjusting times of the system with two controllers are almost the same.
Fig. 19 shows the adjustment of the three parameters, K_p, K_i and K_d, in NNPID controller.
It can be found that during the adjustment process of the angular speed, the three parameters are tuned dynamically. When the system reaches its steady-state, the angular speed is constant and the three parameters are constant as well.

7. Conclusion

In this chapter, the structure and operation principle of C-PID controller are introduced firstly. According to the shortcomings of C-PID controller, two improved PID controllers, namely FPID and NNPID controller, are studied. The structure and operation principle of them are analyzed. Then, the BLDCM and SRM drive system are introduced and their mathematical models are built. Based on the models, FPID and NNPID controller are designed in detail. Finally, the performances of the designed controllers are tested by simulation. The simulation results show that compared with C-PID controller, both FPID and NNPID controller can improve the performance of the system significantly.

8. References

Ang K. H.; Chong G. & Li Y. (2005). PID Control System Analysis, Design, and Technology. *IEEE Transactions on Control Systems Technology*, Vol.13, No.4, July 2005, 559-576, ISSN 1063-6536.

Li X.; Mao X. & Lin W. (2010). Permanent Magnet Synchronous Motor Decoupling Control Study Based on the Inverse System, *Proceedings of Progress In Electromagnetics Research Symposium*, pp. 300-303, ISBN 978-1-934142-14-1, Cambridge, July 2010, Electromagnetics Academy, Cambridge.

Tang K. S.; Man K. F.; Chen G. & Kwong S. (2001). An Optimal Fuzzy PID Controller. *IEEE Transactions on Industrial Electronics*, Vol.48, No.4, August 2001, 757-765, ISSN 0278-0046.

Yu H.; Zhang X. & Hu Q. (2009). Application of NN-PID Control in Linear Elevator, *Proceedings of Fifth International Conference on Natural Computation*, pp. 35-37, ISBN 978-0-7695-3736-8, Tianjian, August 2009, IEEE Computer Society, Los Alamitos.

Lin C.; Jan H. & Shieh N. (2003). GA-Based Multiobjective PID Control for a Linear Brushless DC Motor. *IEEE/ASME Transactions on Mechatronics*, Vol.8, No.1, March 2003, 56-65, ISSN 1083-4435.

Chen H. & Gu J. J. (2010). Implementation of the Three-Phase Switched Reluctance Machine System for Motors and Generators. *IEEE/ASME Transactions on Mechatronics*, Vol.15, No.3, June 2010, 421-432, ISSN 1083-4435.

Wu H.; Cheng S. & Cui S. (2005). A Controller of Brushless DC Motor for Electric Vehicle. *IEEE Transactions on Magnetics*, Vol.41, No.1, January 2005, 509-513, ISSN 0018-9464.

Shi X. & Hao Z. (2008). *Fuzzy Control and Its Simulation in Matlab*, Tsinghua University Press, ISBN 978-7-81123-029-1, Beijing.

Wei W. (2004). *MATLAB control engineering toolbox technique manual*, National Defence Industrial Press, ISBN 7-118-03205-0, Beijing.

Qin G.; Yao W. & Zhang W. (2005). Design of Nonlinear Optimization PID Controller for BLDCM Based on Neuro-Fuzzy Identified Model, *Proceedings of the Eighth International Conference on Electrical Machines and Systems*, pp. 1524-1527, ISBN 7-5062-7407-8, Nanjing, September 2005, Beijing World Publishing Corporation, Beijing.

Naayagi R. T. & Kamaraj V. (2005). Shape Optimization of Switched Reluctance Machine for Aerospace Applications, *Proceedings of 31st Annual Conference of IEEE Industrial Electronics Society*, pp. 1748-1751, ISBN 0-7803-9252-3, Raleigh, November 2005, IEEE Industrial Electronics Society, Los Alamitos.

Vasconcelos J. A.; Ramirez J. A.; Takahashi R. H. C. & Saldanha R. R. (2001). Improvements in Genetic Algorithms. *IEEE Transactions on Magnetics*, Vol.37, No.5, September 2001, 3414-3417, ISSN 0018-9464.

Xu J.; Zhao J.; Luo L. & Wan S. (2004). Expert PID Control for AC/DC Converter, *Proceedings of The Fifth World Congress on Intelligent Control and Automation*, pp. 5586-5590, ISBN 0-7803-8273-0, Hangzhou, June 2004, IEEE Industrial Electronics Chapter, Singapore.

Liu X. & Li X. (2010). Research on Tension Control System Based on Fuzzy Self-tuning PID Control, *Proceedings of 2010 Chinese Control and Decision Conference*, pp. 3385-3390, ISBN 978-1-4244-5182-1, Xuzhou, May 2010, IEEE Industrial Electronics Chapter, Singapore.

Liu W. & Song S. (2006). Application of Fuzzy Control in Switched Reluctance Motor Speed Regulating System, *Proceedings of International Conference on Computational Intelligence for Modelling, Control and Automation*, pp. 72-72, ISBN 0-7695-2731-0, Sydney, November 2006, Patrick Kellenberger, Sydney.

Tao Y. (2002). *Novel PID Control and Its Application*, China Machine Press, ISBN 7-111-06299-X, Beijing.

Liu J. (2003). *Novel PID Control and Its Simulation in Matlab*, Publishing House of Electronic Industry, ISBN 7-5053-8427-9, Beijing.

Wang J.; Zhang C.; Jing Y. & An D. (2007). Study of Neural Network PID Control in Variable-frequency Air-conditioning System, *Proceedings of IEEE International Conference on Control and Automation*, pp. 317-322, ISBN 1-4244-0818-0, Guangzhou, May 2007, IEEE Control Systems Chapter, Singapore.

Wang H. (1999). *Switched Reluctance Motor Speed Adjusting System*, China Machine Press, ISBN 7111-04623-4, Beijing.

He L.; Wan P. & Xiao H. (2004). Application of PID-Fuzzy Controller in Switched Reluctance Motor Drives. *Middle and Small Motor*, Vol.31, No.4, August 2004, 32-35, ISSN 1001-8085.

5

PID control with gravity compensation for hydraulic 6-DOF parallel manipulator

Chifu Yang*, Junwei Han
O.Ogbobe Peter, and Qitao Huang
State Key Laboratory of Robotics and System,, Harbin Institute of Technology, Harbin, Heilongjiang, 150001, China

Abstract

A novel model-based controller for 6 degree-of-freedom (DOF) hydraulic driven parallel manipulator considering the nonlinear characteristic of hydraulic systems-proportional plus derivative with dynamic gravity compensation controller is presented, in order to improve control performance and eliminate steady state errors. In this paper, 6-DOF parallel manipulator is described as multi-rigid-body systems, the dynamic models including mechanical system and hydraulic driven system are built using Kane method and hydromechanics methodology, the numerical forward kinematics and inverse kinematics is solved with Newton-Raphson method and close-form solutions. The model-based controller is developed with feedback of actuator length, desired trajectories and system states acquired by forward kinematics solution as the input and servovalve current as its output. The hydraulic system is decoupled by local velocity compensation in inner control loop prerequisite for the controller. The performance revolving stability, accuracy and robustness of the proposed control scheme for 6-DOF parallel manipulator is analyzed in theory and simulation. The theoretical analysis and simulation results indicate the controller can improve the control performance and eliminate the steady state errors of 6-DOF hydraulic driven parallel manipulator.
Keywords: Parallel manipulator; Proportional-derivative control; Hydraulic servo-systems; Dynamic compensation

---

*Correspondence to: C.F Yang, State Key Laboratory of Robotics and System, Harbin Institute of Technology, Harbin, Heilongjiang, 150001, China.
E-mail: ycf1008@163.com; Tel: +86-451-86412548(325); Fax: +86-451-86412258

Contract/grant sponsor: China Academy of Space Technology;
contract/grant number: HgdJG00401D04
Contract/grant sponsor: State Key Laboratory of Robotics and System (HIT);
contract/grant number: SKLRS200803B

1. Introduction

Hydraulic driven 6-DOF parallel manipulator with long stroke actuators and heavy load is applied in most of the current high fidelity simulators, which is used to simulate various motions in different environments by exporting varying position and orientation. There are several advantages in the application of hydraulic driven parallel manipulator which includes large output force and torque, higher rigidity and accuracy due to the parallel path and averaged link to end effectors error, compared with serial manipulator [1-2]. A classical proportional plus integral plus derivative controller is applied in hydraulic driven 6-DOF parallel manipulator continually due to easy to implementation [3], nevertheless the existence of large steady state errors and dynamic errors in virtue of the influence of system gravity taken no account of hydraulic and mechanical systems in classical proportional plus integral plus derivative control system may degrade the control performance. Well-known facts, it is very difficult for the classical PID control to satisfy the requirements, less steady state error and superior dynamic performance simultaneously. With respect to hydraulic 6-DOF parallel manipulator with heavy payload, the system gravity, the uppermost turbulence to control system for slow motion, results in large steady and dynamic errors in gravitational direction. Therefore, the design and realization of proportional plus derivative controller with dynamic gravity compensation in hydraulic 6-DOF parallel manipulator is of critical importance for improving system control performance of hydraulic driven 6-DOF parallel manipulator especially for the parallel manipulator with heavy payload.

Parallel manipulator has been extensively studied due to its high force-to-weight ratio and widespread application [4]. 6-DOF parallel manipulator is named Stewart platform after Stewart illustrated the use of such parallel structure [5], it is also referred to as Gough-platform who presented the practical use of such a system [6]. Hunt [7] researched the kinematics of parallel manipulators based on screw theory and enumerated promising kinematics structures. Do and Yang [8] used the Newton-Euler approach to solve the inverse dynamics for Stewart platform assuming the joints as frictionless and legs asymmetrical. The control strategies for parallel manipulator may be largely divided into two schemes, joint-space control developed in joint space coordinates [9-11], and workspace control designed based on the workspace coordinates [12-14]. The joint space control scheme can be readily implemented as a collection of multiple, independent single-input single-output control system using data on each actuator length only. A classical proportional plus integral plus derivative control in joint space has been employed in industry, but it does not always guarantee a high performance for parallel robots [15]. This novel joint space control approaches have been proposed to improve control performance by rejecting the uncertainty and nonlinear effects in motion equations. Kim proposed a robust nonlinear control scheme in joint space for a hydraulic parallel system based on Lyapunov redesign method [10], yet the pressure closed loop control is hard to implemented for real hydraulic system due to the effect of pipeline pressure transient and frication force. Nguyen et al [11] developed a joint-space adaptive control scheme applied to an electromechanically driven Stewart platform using Lyapunov direct method. Su presented a robust auto-disturbance rejection controller in joint space for 6-DOF parallel manipulator [16]. Kim et al [12] discussed robust nonlinear task space control with a friction estimator for dynamoelectric Gough-Stewart platform. Burdet et al [17] investigated a nonlinear controller with dynamic compensation which depended on system state and velocity of 6-DOF parallel manipulator. Noriega et al [18] presented a neural network control scheme and showed its superiority over a kinematics control. Kim et al [19] researched and applied a high speed tracking control for 6-DOF electric driven Stewart platform using an enhanced sliding mode control approach. Cervantes et al [20] studied tracking problem of robot manipulator based on multi-rigid body models with revolute joints via PID control. Although the above advanced model-based control strategies are effective for 6-DOF parallel manipulator, the characteristics of hydraulic driven system is not taken into account. Davliakos et al [21] developed operational error joint feedback control scheme embedding the forward kinematics in the feedback control loop for 6-DOF electrohydraulic Parallel manipulator platforms. However, only simulation is investigated for the model-based control scheme. Besides, the influences of dynamical gravity to system control performance are not analyzed and attracted attention for the proposed controller.

In this paper, a proportional plus derivative controller with dynamic gravity compensation (PDGC) is developed to improve the control performance including steady and dynamic precision via compensating steady state errors and reducing dynamic errors for a 6-DOF hydraulic driven parallel manipulator with symmetric joint locations. This paper begins with a practical strategy to obtain 6-DOF hydraulic driven Gough-Stewart platform essential to the developed controller. The dynamics models of the 6-DOF platform system are built using Kane method, considering the Gough-Stewart platform as 13 rigid bodies, and the hydraulic driven system are established in terms of hydromechanics theory. The desired actuator length is calculated by a closed-solution inverse kinematics, and the system states of 6-DOF Gough-Stewart platform are obtained by a numerical forward kinematics, the forward kinematics and inverse kinematics models are described with Newton-Raphson method and closed-form solution, respectively. The proportional plus derivative with dynamics compensation control scheme is gained, combing the kinematics control and inverse dynamics method, the proposed controller employs rigid body and actuation dynamic and yields the input current vector of the servovalve, the dynamic gravity term including the gravity of platform, load and hydraulic cylinders is used to compensate the influence of gravity of 6-DOF Gough-Stewart platform, and the decoupling of hydraulic system is implemented by local velocity compensation in inner control loop. The performance including stability, precision and robustness of the proposed controller is analyzed in theory and simulation. The proportional plus derivative with dynamic compensation control scheme is studied to improve the performance of control system for 6-DOF hydraulic driven Gough-Stewart platform.

2. System model

The kinematics of 6-DOF Gough-Stewart platform has been studied extensively [22, 23]. Therefore, the kinematics models of 6-DOF Gough-Stewart are briefly described in the paper. Fig.1 depicts the configuration of the 6-DOF Gough-Stewart; Fig.2 explains the two Cartesian coordinate systems; the {**B**} coordinate system is the body coordinate system fixed to the movement platform, while the {**L**} coordinate system is the base coordinate system for the inertial frame. The linear motions denotes as surge (q_1), sway (q_2), and heave (q_3) are along the X_L-Y_L-Z_L axis for base coordinate system, and the angular motions labeled as roll (q_4), pitch (q_5), and yaw (q_6) are Euler angles of platform at X_L, Y_L, Z_L axis. The body coordinate system {**B**} and the base coordinate system {**L**} are superposition in the initial state q_i=0, i=1,…,6.

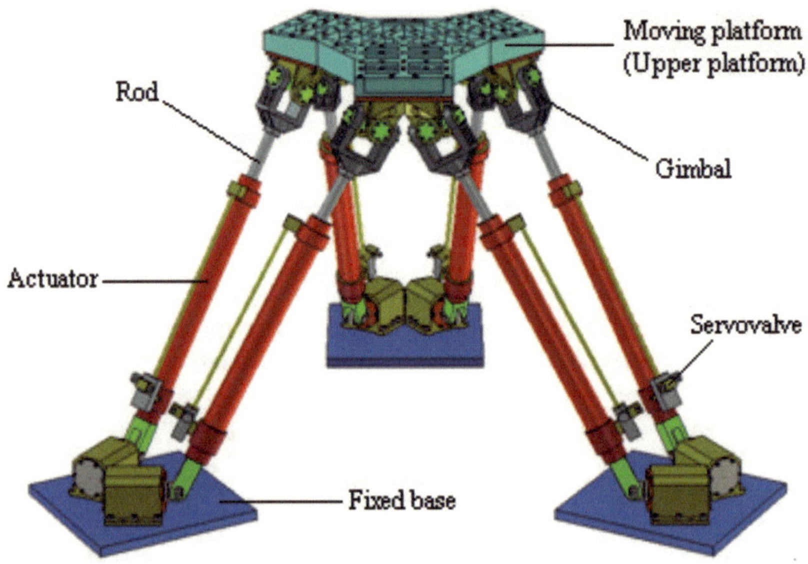

Fig. 1. Configuration of 6-DOF Gough-Stewart platform

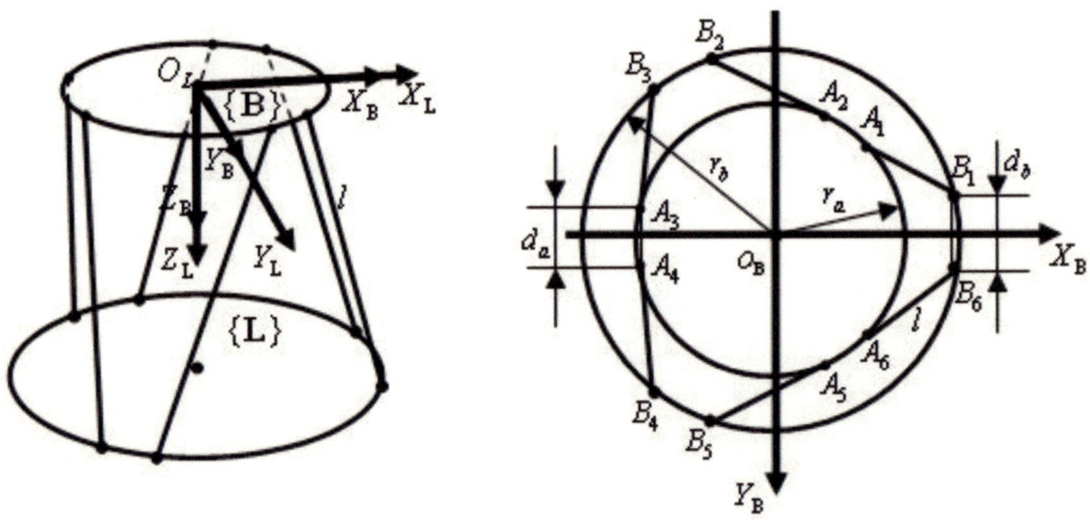

Fig. 2. Definition of the Cartesian coordination systems and vectors in dynamics and kinematics equations of

6-DOF Gough-Stewart platform

For the movement including the linear and angular motions of Gough-Stewart platform, the inverse kinematics model is derived using closed-form solution [22].

$$\widetilde{\mathbf{l}} = (\mathbf{R} \cdot \widetilde{\mathbf{A}} - \widetilde{\mathbf{B}}) + \widetilde{\mathbf{c}} \tag{1}$$

where $\widetilde{\mathbf{l}}$ is a 3×6 actuator length matrix of platform, $\mathbf{R}$ is a 3×3 rotation matrix of transformation from body coordinates to base coordinates, $\widetilde{\mathbf{A}}$ is a 3×6 matrix of upper gimbal points, $\widetilde{\mathbf{B}}$ is a 3×6 matrix of lower gimbal points, and $\widetilde{\mathbf{c}}$ is position 3×1 vector of platform, $\widetilde{\mathbf{c}} = (q_1, q_2, q_3)^{\mathrm{T}}$. The rotation matrix under Z-Y-X order is given by

$$\mathbf{R} = \begin{bmatrix} \cos q_5 \cdot \cos q_6 & \cos q_6 \cdot \sin q_5 \cdot \sin q_4 - \sin q_6 \cdot \cos q_4 & \sin q_6 \cdot \sin q_4 + \cos q_6 \cdot \sin q_5 \cdot \cos q_4 \\ \sin q_6 \cdot \cos q_5 & \cos q_6 \cdot \cos q_4 + \sin q_6 \cdot \sin q_5 \cdot \sin q_4 & \sin q_6 \cdot \sin q_5 \cdot \cos q_4 - \cos q_6 \cdot \sin q_4 \\ -\sin q_5 & \cos q_5 \cdot \sin q_4 & \cos q_5 \cdot \cos q_4 \end{bmatrix} \quad (2)$$

The forward kinematics is used to solve the output state of platform for a measured length vector of actuators; it is formulated with Newton-Raphson method [23].

$$\widetilde{\mathbf{\Theta}}_{j+1} = \widetilde{\mathbf{\Theta}}_j + \mathbf{J}_{l,\widetilde{\Theta}}^{-1} \cdot (\widetilde{\mathbf{l}}_0 - \widetilde{\mathbf{l}}_j) \quad (3)$$

where $\widetilde{\mathbf{\Theta}}$ is a 6×1 state vector of the platform generalized coordinates, $\widetilde{\mathbf{\Theta}} = (q_1, q_2, q_3, q_4, q_5, q_6)^{\mathrm{T}}$, j is the iterative numbers, $\widetilde{\mathbf{l}}_0$ is the initial measured length 6×1 vector of actuator of the platform, $\widetilde{\mathbf{l}}_j$ is the 6×1 solving actuator vector during the iterative calculation, $\mathbf{J}_{l,\widetilde{\Theta}}$ is a Jacobian 6×6 matrix, which is one of the most important variables in the Gough-Stewart platform, relating the body coordinates to be controlled and used as basic model coordinates, and the actuator lengths, which can be measured.

The dynamic model for motion platform as a rigid body can be derived using Newton-Euler and Kane method [24, 25].

$$\widetilde{\boldsymbol{\tau}} + \widetilde{\mathbf{G}}(\widetilde{\mathbf{\Theta}}) = \widetilde{\mathbf{M}}(\widetilde{\mathbf{\Theta}}) \cdot \ddot{\mathbf{\Theta}} + \widetilde{\mathbf{V}}(\widetilde{\mathbf{\Theta}}, \dot{\mathbf{\Theta}}) \cdot \dot{\mathbf{\Theta}} \quad (4)$$

where $\widetilde{\mathbf{M}}(\widetilde{\mathbf{\Theta}})$ is a 6×6 mass matrix, $\widetilde{\mathbf{V}}(\widetilde{\mathbf{\Theta}}, \dot{\mathbf{\Theta}})$ is a 6×6 matrix of centrifugal and Coriolis terms, $\widetilde{\mathbf{G}}(\widetilde{\mathbf{\Theta}})$ is a 6×1 vector of gravity terms, see Appendix A, $\widetilde{\boldsymbol{\tau}}$ is a 6×1 vector of generalized applied forces, $\dot{\mathbf{\Theta}}$ is a 6×1 velocity vector, which is given by

$$\dot{\mathbf{\Theta}} = (\dot{\widetilde{\mathbf{c}}} \quad \widetilde{\boldsymbol{\omega}})^{\mathrm{T}} \quad (5)$$

where $\widetilde{\boldsymbol{\omega}}$ is a 3×1 angular velocity vector in base coordinate system, $\widetilde{\boldsymbol{\omega}} = (\omega_x \quad \omega_y \quad \omega_z)^{\mathrm{T}}$. Note that $\dot{\mathbf{\Theta}} \neq \dot{\widetilde{\mathbf{\Theta}}}$.

The applied forces $\boldsymbol{\tau}$ can be transformed from mechanism actuator forces, which is given by

$$\widetilde{\boldsymbol{\tau}} = \mathbf{J}_{l,\widetilde{\Theta}}^{\mathrm{T}} \cdot \mathbf{F}_{\mathrm{a}} \quad (6)$$

where $\mathbf{F}_a$ is a 6×1 vector representing actuator forces, $\mathbf{F}_{\rm a}=(f_{\rm a1}\ \ f_{\rm a2}\ \ \cdots\ \ f_{\rm a6})^{\rm T}$, $f_{\rm ai}$ (i=1,…,6) is actuator output force.
The rotation of actuator around itself is ignored, thus the dynamic model for each hydraulic actuator (piston rod and cylinder) using Newton-Euler and Kane method is described as

$$(\mathbf{J}_{\rm uc,ai}\mathbf{J}_{{\rm ai},\tilde{\Theta}})^{\rm T} m_{\rm u}\cdot\mathbf{g}+(\mathbf{J}_{\rm tc,ai}\mathbf{J}_{{\rm ai},\tilde{\Theta}})^{\rm T} m_{\rm t}\cdot\mathbf{g}=\overline{\mathbf{F}}_{\rm i} \tag{7(a)}$$

$$\begin{aligned}\overline{\mathbf{F}}_{\rm i}&=(\mathbf{J}_{\rm uc,ai}\mathbf{J}_{{\rm ai},\tilde{\Theta}})^{\rm T} m_{\rm u}\cdot\dot{\overline{\mathbf{v}}}_{\rm uc}+(\mathbf{J}_{\rm uc,ai}\mathbf{J}_{{\rm ai},\tilde{\Theta}})^{\rm T}(\mathbf{I}_{\rm a}\dot{\overline{\boldsymbol{\omega}}}_i+\tilde{\boldsymbol{\omega}}\times\mathbf{I}_{\rm a}\overline{\boldsymbol{\omega}}_i)+\mathbf{J}_{\rm tc,ai}\mathbf{J}_{{\rm ai},\tilde{\Theta}})^{\rm T} m_{\rm t}\cdot\dot{\overline{\mathbf{v}}}_{\rm tc}\\&+(\mathbf{J}_{\rm tc,ai}\mathbf{J}_{{\rm ai},\tilde{\Theta}})^{\rm T}(\mathbf{I}_{\rm b}\dot{\overline{\boldsymbol{\omega}}}_i+\tilde{\boldsymbol{\omega}}\times\mathbf{I}_{\rm b}\overline{\boldsymbol{\omega}}_i)\end{aligned} \tag{7(b)}$$

where $\mathbf{J}_{\rm uc,ai}$, $\mathbf{J}_{\rm tc,ai}$ are 3×3 Jacobian matrix, , $\mathbf{J}_{{\rm ai},\tilde{\Theta}}$ is 3×6 Jacobian matrix, $m_{\rm u}$ is the mass of piston rod of a actuator, $m_{\rm t}$ is the mass of cylinder of a actuator, $\overline{\boldsymbol{\omega}}_i$ is the angular velocity of actuator relative to relevant lower gimbal point, $\overline{\mathbf{v}}_{\rm uc}$, $\overline{\mathbf{v}}_{\rm tc}$ are the linear velocity of the mass center of piston rod and cylinder, respectively, $\mathbf{I}_{\rm a}$, $\mathbf{I}_{\rm b}$ are the inertia of piston rod and cylinder, respectively, $\mathbf{g}$ is acceleration vector of gravity, $\mathbf{g}=(0\ 0\ g)^{\rm T}$.
Combining Eqs.(4), (5),(6) and (7), the dynamics model of 6-DOF Gough-Stewart platform as thirteen rigid body is obtained with Kane method, given by

$$\tilde{\boldsymbol{\tau}}+\mathbf{G}^*(\tilde{\boldsymbol{\Theta}})=\mathbf{M}^*(\tilde{\boldsymbol{\Theta}})\cdot\ddot{\boldsymbol{\Theta}}+\mathbf{V}^*(\tilde{\boldsymbol{\Theta}},\dot{\boldsymbol{\Theta}})\dot{\boldsymbol{\Theta}} \tag{8}$$

where, $\mathbf{M}^*(\tilde{\boldsymbol{\Theta}})$ is a mass matrix, $\mathbf{V}^*(\tilde{\boldsymbol{\Theta}},\dot{\boldsymbol{\Theta}})$ is a matrix of centrifugal and Coriolis terms, $\mathbf{G}^*(\tilde{\boldsymbol{\Theta}})$ is a vector of gravity terms, see Appendix B.
The hydraulic systems are studied in depth for symmetrical servovalve and actuator [26], it is assumed that Coulomb frictions are zero (Coulomb friction $F_{\rm ci} << B_c\dot{l}_i$, not zero, practically) the hydraulic system mathematical models of symmetric and matched servovalve and symmetrical actuator are given as

$$q_{{\rm L}i}=C_{\rm d}\cdot w\cdot x_{{\rm v}i}\sqrt{\frac{1}{\rho}(p_{\rm s}-{\rm sign}(x_{{\rm v}i})p_{{\rm L}i})} \tag{9}$$

$$q_{{\rm L}i}=A\cdot\dot{l}_i+C_{\rm te}\cdot p_{{\rm L}i}+\frac{V_{\rm t}}{4E}\dot{p}_{{\rm L}i} \tag{10}$$

$$A\cdot p_{{\rm L}i}=f_{\rm ai}+f_{fi} \tag{11}$$

where $q_{{\rm L}i}$ is load flow of the $i^{\rm th}$ hydraulic actuator, w is area grads, $x_{{\rm v}i}$ is position of the $i^{\rm th}$ servovalve, ρ is fluid density, p_s is supply pressure of servosystem, p_{Li} is load pressure of the $i^{\rm th}$ actuator, A is effective acting area of piston, $C_{\rm te}$ is the leakage coefficient, $V_{\rm t}$ is actuator cubage, E is bulk modulus of fluid, l_i is the length of the $i^{\rm th}$ actuator, $C_{\rm d}$ is flow coefficient, f_{fi} is joint space friction force in the ith actuator. A number of methods can be used to model the friction $\mathbf{F}_f$ [21, 27]. A widely method for modeling friction as

$$\mathbf{F}_f(\dot{l}) = \mathbf{F}_v(\dot{l}) + \mathbf{F}_c(\dot{l}) + \mathbf{F}_s \tag{12}$$

where $\mathbf{F}_f$ is total friction vector, $\mathbf{F}_f = [f_{f1} \quad \cdots \quad f_{f6}]^{\mathrm{T}}$, $\mathbf{F}_v$, $\mathbf{F}_c$ and $\mathbf{F}_s$ are the viscous, Coulomb and static friction vector, respectively, with elements

$$f_{vi}(\dot{l}_i) = \begin{cases} B_c \dot{l}_i, & \dot{l}_i \neq 0 \\ 0, & \dot{l}_i = 0 \end{cases} \qquad i=1,2,\ldots,6 \tag{13}$$

$$f_{ci}(\dot{l}_i) = \begin{cases} f_{c0,i}\mathrm{sign}(\dot{l}_i), & \dot{l}_i \neq 0 \\ 0, & \dot{l}_i = 0 \end{cases} \qquad i=1,2,\ldots,6 \tag{14}$$

$$f_{si}(\dot{l}_i) = \begin{cases} f_{\mathrm{ext},i}, & |f_{\mathrm{ext},i}| < f_{s0,i} \quad \dot{l}_i = 0, \ddot{l}_i = 0 \\ f_{s0,i}\,sign(\dot{l}_i), & |f_{\mathrm{ext},i}| > f_{s0,i} \quad \dot{l}_i = 0, \ddot{l}_i \neq 0 \\ 0, & \dot{l}_i \neq 0 \end{cases} \qquad i=1,2,\ldots,6 \tag{15}$$

where B_c is viscous damping coefficient, $f_{c0,i}$ is the element of Coulomb friction, $f_{\mathrm{ext},i}$ is the external force element, $f_{s0,i}$ is the breakaway force element.

3. Control design

In this section, the inverse dynamic methodology [20] is adopted to derive a proportional plus derivative controller with dynamic gravity compensation for 6-DOF hydraulic driven Gough-Stewart platform in the case in which the system parameters are known, the PDGC control scheme are described in Fig.3.

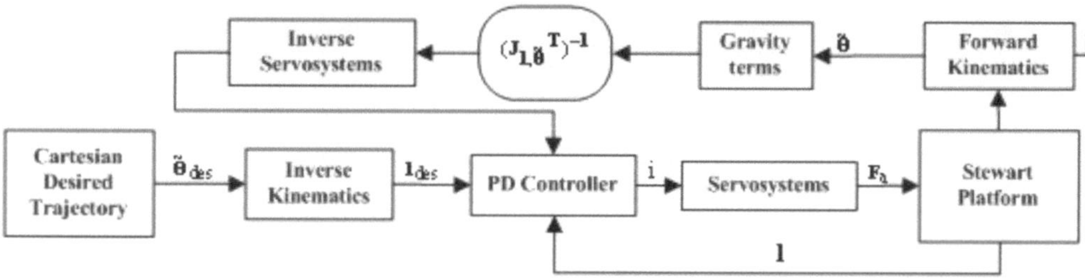

Fig. 3. Control block diagram for PDGC

The PDGC controller considered the dynamic characteristic of parallel manipulator embedded the forward kinematics, dynamic gravity terms and inverse of transfer function from the input position of servovalve to the output force of actuator and Jacobian matrix $(\mathbf{J}_{l\Theta}{}^{\mathrm{T}})^{-1}$ in inverse of transpose form in inner control loop. It is should be noted that

the friction force, zero bias and dead zone of servovalve also affect the steady and dynamic precision as well as system gravity. However, the valve with high performance index may be chosen to avoid the effect of dead zone of control valve. In fact, the dead zone of servovalve in hydraulic system is very small, which can achieve 0.01mm even a general servovalve. The zero bias of servovalve may be measured and compensated for control system. For large hydraulic parallel manipulator with heavy payload, the system gravity is much more that the maximal friction even that no payload exist in hydraulic 6-DOF parallel manipulator. Therefore, the dynamical gravity, the most chief influencing factor of steady precision, and viscous friction is taken into account for designing of the developed control scheme without considering Column and static friction in this paper. Besides, the classical PID is widely applied in hydraulic 6-DOF parallel manipulator in practice, then the considered system gravity is associated with PID control to improve the steady and dynamic precision without destroy the steadily of the original control system.
The nature frequency of servovalve is higher than the mechanical and hydraulic commix system, so Eqs.(9) can be linearized using Taylor formulation, rewritten by

$$q_{Li} = K_q x_{vi} - K_c p_{Li} \tag{16}$$

With Eqs.(10)-(13), (10) and (11) are rewritten in the form of La-transformation.

$$Q_{Li} = A \cdot sL_i + C_{te} \cdot P_{Li} + \frac{V_t}{4E} \cdot sP_{Li} \tag{17}$$

$$A \cdot P_{Li} = B_c \cdot sL_i + F_{ai} \tag{18}$$

The input current of servovalve is direct proportion to position of servovalve, so

$$i_i = K_0 x_{vi} \tag{19}$$

where, K_0 is a constant.

Substituting the Eqs.(16),(17) and (19) in Eqs.(17), the output of inverse servosystem, given by

$$\tilde{I}_i = \{\frac{1}{A}(F_{ai} + B_c \cdot sL_i)(K_c + C_{te} + \frac{V_t}{4E}s) + A \cdot sL_i\} \cdot \frac{K_0}{K_q} \tag{20}$$

where,

$$F_{ai} = \{(\mathbf{J}_{l,\tilde{\Theta}}{}^{\mathrm{T}})^{-1} \cdot \mathbf{G}^*(\tilde{\mathbf{\Theta}})\}_i$$

The developed controller is extended to model-based control scheme allowing tracking of the reference inputs for platform. Desired position vector of hydraulic cylinders and actual position vector of hydraulic cylinders are used as input commands of the controller, and the controller provides the current sent to the servovale, the closed-loop control law can be shown as

$$u_i = f_i = (K_0 k_p e_i + K_0 k_d \dot{e}_i - \tilde{i}_i) \cdot G \tag{21}$$

where u_i is the output of actuator, k_p and k_d are control gain of system, G is the transfer function of the output current of servovalve to the actuator output forces, e is actuator length error of the platform, $e_i=l_{ides}-l_i$, l_{ides} is the desired hydraulic cylinders length, l_i is the feedback hydraulic cylinder length.
Using Eqs.(20), the Eqs.(21) can be rewritten by

$$\mathbf{u}=(k_p\mathbf{e}+k_d\dot{\mathbf{e}})K_0\cdot G-(\mathbf{J}_{l,\tilde{\Theta}}{}^{\mathrm{T}})^{-1}\cdot\mathbf{G}^*(\tilde{\mathbf{\Theta}}) \tag{22}$$

where,

$$\mathbf{u}=(u_1,u_2,\dots,u_6)^{\mathrm{T}},$$
$$\mathbf{e}=(e_1,e_2,\dots,e_6)^{\mathrm{T}}.$$

Combining Eqs.(8), (22), an system equation of the 6-DOF parallel manipulator with PDGC controller can be obtained, which can be shown as

$$\mathbf{J}_{l,\tilde{\Theta}}{}^{\mathrm{T}}\cdot\mathbf{u}+\mathbf{G}^*(\tilde{\mathbf{\Theta}})=\mathbf{M}^*(\tilde{\mathbf{\Theta}})\cdot\ddot{\mathbf{\Theta}}+\mathbf{V}^*(\tilde{\mathbf{\Theta}},\dot{\mathbf{\Theta}})\dot{\mathbf{\Theta}} \tag{23}$$

According to Eqs.(23), the system error dynamics for pointing control can be written as

$$\mathbf{M}^*(\tilde{\mathbf{\Theta}})\cdot\ddot{\mathbf{e}}+[\mathbf{V}^*(\tilde{\mathbf{\Theta}},\dot{\mathbf{\Theta}})+k_d]\cdot\dot{\mathbf{e}}+k_p\mathbf{e}=0 \tag{24}$$

The Lyapunov function is chosen for PDGC control scheme, and the rest of stability proof is identical to the one in [28].

$$V=\frac{1}{2}\dot{\mathbf{e}}^{\mathrm{T}}\mathbf{M}^*(\tilde{\mathbf{\Theta}})\dot{\mathbf{e}}+\frac{1}{2}\mathbf{e}^{\mathrm{T}}k_p\mathbf{e} \tag{25}$$

The error term $(\dot{\mathbf{e}},\ddot{\mathbf{e}})$ and the generalized coordinates term $(\dot{\mathbf{\Theta}},\ddot{\mathbf{\Theta}})$ in Eqs.(24) are zero in steady state, so the steady state error vector $\mathbf{e}$ converge to zero, the actual actuator length $\mathbf{l}$ can converge asymptotical to the desired actuator length $\mathbf{l}_{des}$ without errors.

4. Experiment results

The control performance including steady state precision, stability and robustness of the proposed PDGC is evaluated on a hydraulic 6-DOF parallel manipulator in Fig.4 via experiment, which features (1) six hydraulic cylinders, (2) six MOOG-G792 servo-valves, (3) hydraulic pressure power source, (4) signal converter and amplifier, (5) D/A ACL-6126 board, (6) A/D PCL-816/818 board, (7) position and pressure transducer, (8) a real-time industrial computer for real-time control, and (9) a supervisory control computer. The control program of the parallel manipulator is programmed with Matlab/Simulink and compiled to gcc code executed on target real-time computer with QNX operation system using RT-Lab. The sampling time for the control system is set to 1 ms, and the parameters of the hydraulic 6-DOF parallel manipulator are summarized in Table 1.

Parameters	Value
Maximal/Maximal stroke of cylinder, l_{min}/l_{max} (m)	-0.37/0.37
Initial length of cylinder, l_0 (m)	1.830
Upper joint spacing, d_u (m)	0.260
Lower joint spacing, d_d (m)	0.450
Upper joint radius, R_u (m)	0.560
Lower joint radius, R_d (m)	1.200
Mass of upper platform and payload, m_p (Kg)	2940
Moment of inertia of upper platform and payload, I_{xx}, I_{yy}, I_{zz} (Kg·m^2)	217.37, 217.37, 266.75

Table 1. Parameters of hydraulic 6-DOF parallel manipulator

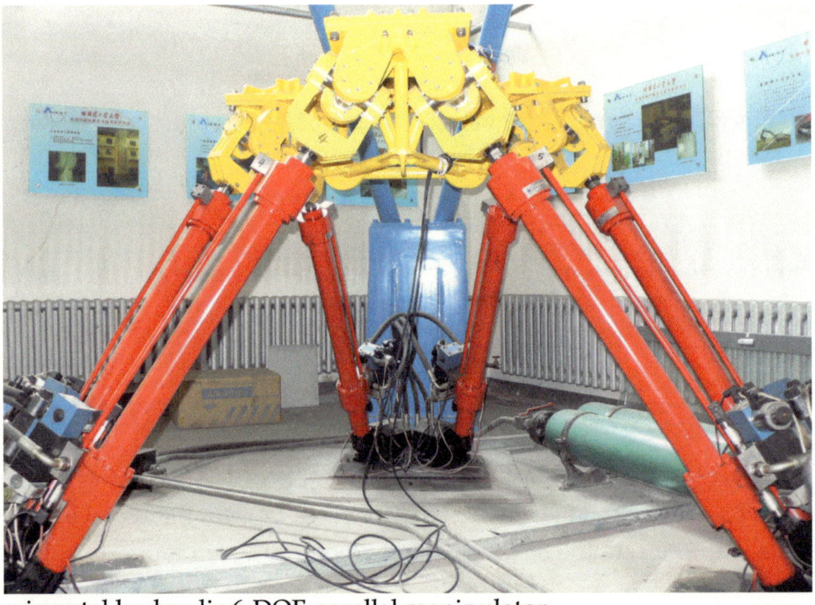

Fig. 4. Experimental hydraulic 6-DOF parallel manipulator

The spatial states of parallel manipulator are critical to determine the control input for compensating system gravity, turbulence for the control system of hydraulic 6-DOF parallel manipulator. Fortunately, the real-time forward kinematics for estimating system states has been investigated and implemented with high accuracy (less than 10^{-7}m) and sample 1-2ms [29]. It is should be noted that the steady state error in principle of control system mainly results from system gravity of the 6-DOF parallel manipulator especially for hydraulic parallel manipulator with heavy payload, even though the friction always exists in the system under position control, since the gravity of the payload and upper platform is much more than friction.

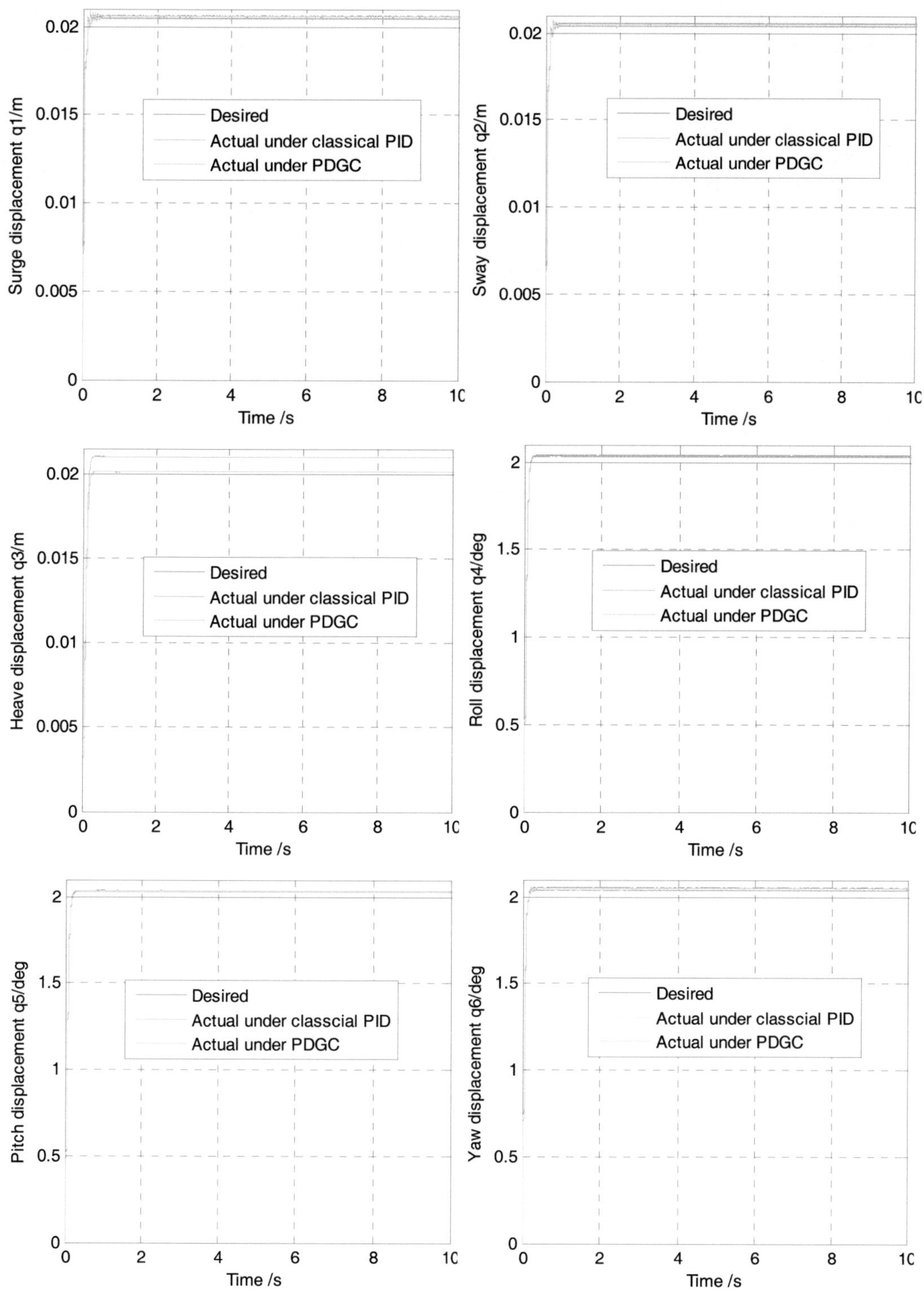

Fig. 5. Responses to desired step trajectories of classical PID and PDGC controller

With online forward kinematics available, the proposed PDGC strategy is implemented in a real 6-DOF hydraulic parallel manipulator. The classical PID control scheme is also applied to the parallel manipulator as benchmarking for that the classical PID control is a typical control strategy in theory and practice, particularly in industrial hydraulic 6-DOF parallel manipulator with heavy payload. It is should be noted that the proposed PDGC control is an improved PID control with dynamical gravity compensation to improve the control performance involving both steady and dynamic precision of hydraulic 6-DOF parallel manipulator, the control strategy with gravity compensation also may be incorporated with other advanced control scheme to derive better control performance. The classical PID gain Kp is experimental tuned to 40, which is identical with the proposed PDGC gains. All six DOFs step signals (Surge: 0.02m, Sway: 0.02m, Heave: 0.02m, roll: 2deg, Pitch: 2deg, Yaw: 2deg) are applied to the actual control system, respectively. Fig.5 shows the responses to the desired step trajectory of experimental hydraulic parallel manipulator.

As shown in Fig.5, the PDGC control scheme can respond to the desired step trajectories promptly and steadily in all DOFs. Moreover, the proposed PDGC shows superior control performance in steady precision to those of the classical PID control along all six DOFs directions. The maximal steady state error is 0.41mm in linear motions and 0.04deg in angular motions under the PDGC, 1.01mm in linear motions and 0.052deg in angular motions under the classical PID. The maximal steady state error chiefly influenced by system gravity appeared in heave direction motion for all 6 DOFs motions under the classical PID control, which was compensated via the proposed PDGC control, depicted in Fig.6. Compared with the PDGC controller, the maximal steady state error in angular motions presented in yaw direction under classical PID control is also shown in Fig.6. The steady state error is 0.1mm in heave and 0.03deg in yaw with PDGC, 1.01mm in heave and 0.052deg in yaw with classical PID. Additionally, the responses to the step trajectories also illustrate that the control system, both PDGC and classical PID control, is steady.

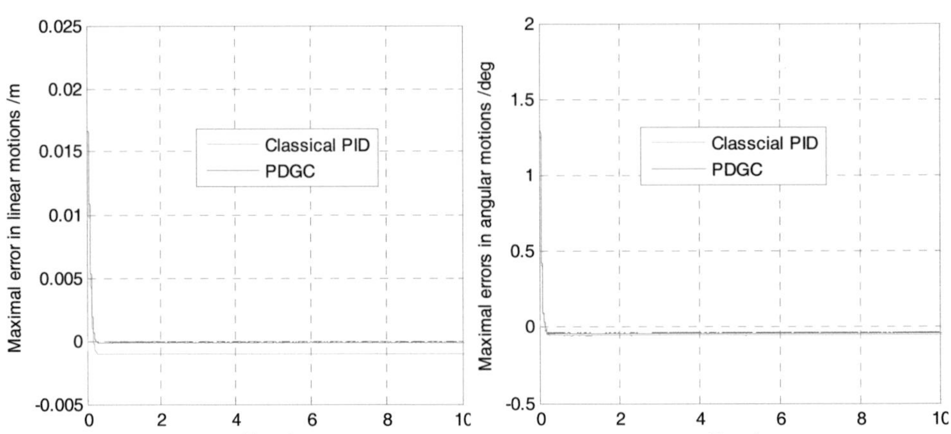

Fig. 6. The maximal errors of PDGC and classical PID controller in position and orientation

With a view of evaluating the dynamic control performance of the PDGC, the desired sinusoidal signals are inputted to the hydraulic parallel manipulator. Under sinusoidal inputs along six directions: surge (0.01m/1Hz), sway (0.01m/2Hz), heave (0.01m/1Hz), roll (1deg/1Hz), pitch (1deg/2Hz), and yaw (1deg/1Hz), the trajectory tracking for the PDGC control and the classical PID control scheme are shown in Fig. 7.

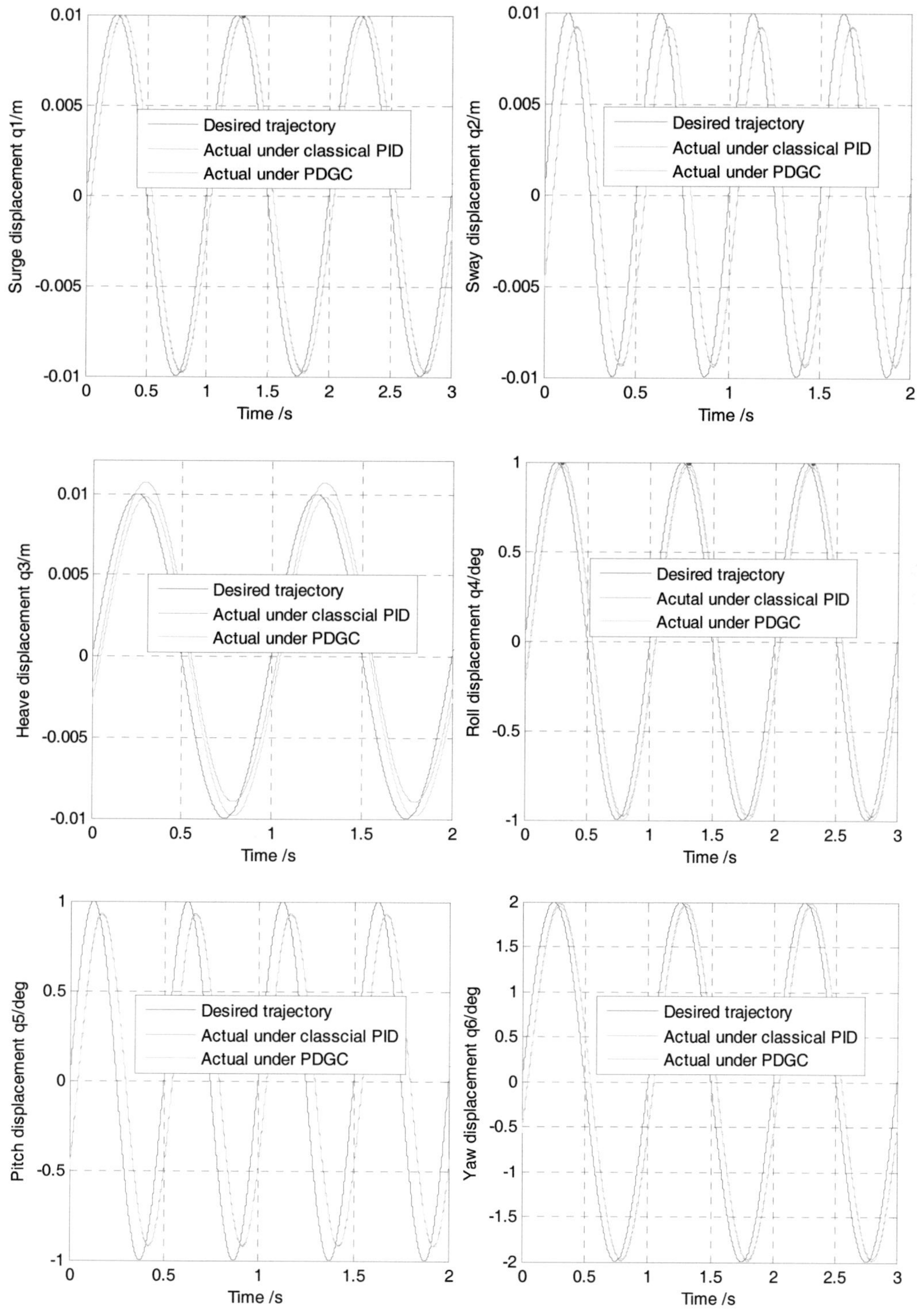

Fig. 7. Responses to desired sinusoidal trajectories of classical PID and PDGC controller

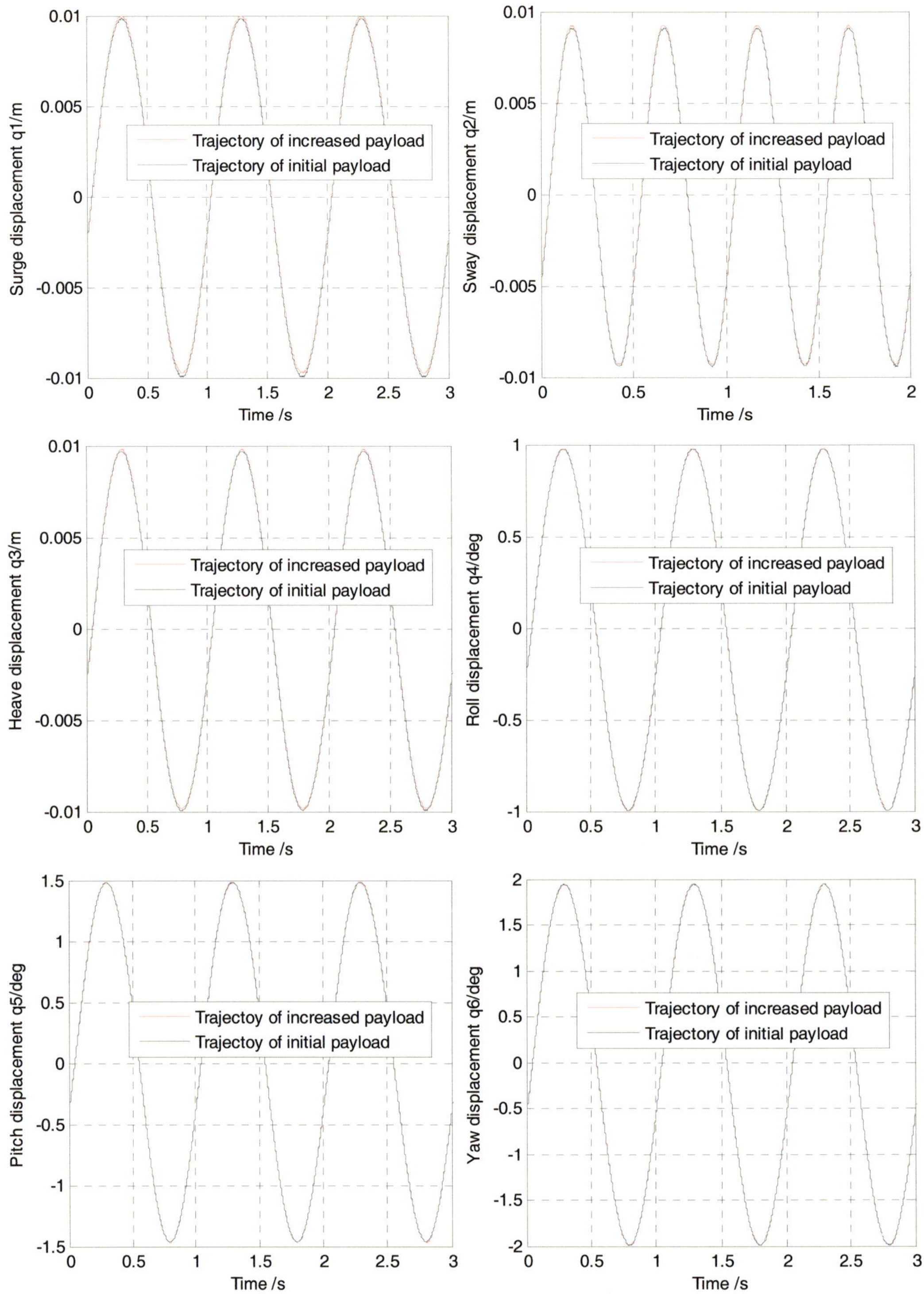

Fig. 8. Experimental results for different mass of payload

As can be deduced form Fig. 5-7, the hydraulic 6-DOF Gough-Stewart platform with PDGC, lead the systems to the desired location with smaller steady state error neglected in large hydraulic 6-DOF parallel manipulator, while the classical proportional plus integral plus derivative control scheme exist large steady state errors in the system, and the PDGC control system can implement trajectory tracking of sine wave with excellent performance in all DOFs motions, which is better than classical proportional plus integral plus derivative controller especially in heave direction motion.
The influence of platform load variable during the motion of 6-DOF parallel manipulator and the robustness of the controller can be illustrated by applied the controller to the system in the case of the platform load increase by 12%, the experimental results are shown in the Fig.8.

Comparison of results demonstrate that the maximal amplitude fading with increased mass of payload is 0.644dB in linear motions (q1, q2, q3), 0.154dB in angular motions (q4, q5, q6), and it is 0.661dB in linear motions and 0.153dB in angular motions for initial mass of payload, the maximal phase delay of PDGC controller with 112% of initial mass is 0.14rad relative to initial mass in linear motions, while it is 0.023rad phase delay than it was with initial mass in angular motions. Consequently, the proposed control still has excellent performance (robustness) with incorrect mass of payload which is 112% of initial mass. Moreover, the experimental results display that the proposed PDGC control scheme can improve the steady precision and reduce system dynamic errors of hydraulic 6-DOF parallel manipulator even 12% uncertainty exists in gravity, especially for 6-DOF parallel manipulator with heavy payload.

5. Conclusions

In this paper, a proportional plus derivative control with dynamic gravity compensation is studied for 6-DOF parallel manipulator. The system models are derived, including the dynamics model of 6-DOF Gough-Stewart platform and actuators using Kane method and the forward kinematics with Newton-Raphson method and the inverse kinematics in closed-form solution, and the hydraulic systems based on hydromechanics theory. The control law of proportional plus derivative control with dynamic gravity compensation is developed in the paper, the inner loop feedback controller employed dynamic gravity term, forward kinematics and Jacobian matrix and yield servovalve currents, and the dynamics of hydraulic systems are decoupled by local velocity compensation in inverse servosystem, the outer loop implement the position control of actuator length. The direct estimation method for the system states required in the proposed control based on the forward kinematics are employed in order to realize the control scheme in the base coordinate systems instead of the state observer with the actuator length output. The performances with respect to stability, precision and robustness are analyzed. The theoretical analysis and simulation results demonstrate that the proposed controller represent excellent performance for the 6-DOF hydraulic driven Gough-Stewart platform, it is stable, the steady state errors of the system due to gravity of the systems are converge asymptotically to zero, and the controller reveal superexcellent robustness. Furthermore, the effective PDGC control for the hydraulic 6-DOF parallel manipulator with heavy payload is obtained in this paper; it can not only be used in hydraulic driven 6-DOF parallel manipulator for improving classical PID control performance, but also can be associated with other advanced control scheme to get better control performance and applied in other systems.

Acknowledgements

This research was supported by 921 Manned Space Project from China Academy of Space Technology and Self-Planned Task (No.SKLRS200803B) of State Key Laboratory of Robotics and System (HIT). The authors would like to thank CAST, HIT, Prof S. J. Li of Department of Mechanical and Electrical Engineering, Harbin Institute of Technology, and to thank the Editor, Associate Editors, and anonymous reviewers for their constructive comments.

Appendix A.

The 6×6 mass matrix $\tilde{\mathbf{M}}(\tilde{\mathbf{\Theta}})$, 6×6 centrifugal and Coriolis matrix $\tilde{\mathbf{V}}(\tilde{\mathbf{\Theta}},\dot{\mathbf{\Theta}})$, and 6×1 vector of gravity terms $\tilde{\mathbf{G}}(\tilde{\mathbf{\Theta}})$ in Eqs.(4) are given by

$$\tilde{\mathbf{M}}(\tilde{\mathbf{\Theta}}) = \begin{bmatrix} m_{\mathrm{p}}\mathbf{I}_{\mathrm{e}} & 0 \\ 0 & \mathbf{I}_L \end{bmatrix} \tag{A.1a}$$

$$\tilde{\mathbf{V}}(\tilde{\mathbf{\Theta}},\dot{\mathbf{\Theta}}) = \begin{bmatrix} 0_{3\times3} & 0 \\ 0 & \mathbf{\Omega}\cdot\mathbf{I}_L \end{bmatrix} \tag{A.1b}$$

$$\tilde{\mathbf{G}}(\tilde{\mathbf{\Theta}}) = \begin{bmatrix} 0 & 0 & g & 0 & 0 & 0 \end{bmatrix}^{\mathrm{T}} \tag{A.1c}$$

where $\mathbf{I}_{\mathrm{e}}$ is unit 3×3 matrix, $\mathbf{I}_{\mathrm{L}}$ is a 3×3 inertia matrix of upper platform in base coordinates system, m_{p} is the mass of upper platform.

$$\mathbf{I}_{\mathrm{e}} = \begin{bmatrix} 1 & 0 & 0 \\ 0 & 1 & 0 \\ 0 & 0 & 1 \end{bmatrix} \tag{A.2a}$$

$$\mathbf{I}_L = \mathbf{R}\cdot\mathbf{I}_{\mathrm{p}}\cdot\mathbf{R}^{\mathrm{T}} \tag{A.2b}$$

$$\mathbf{\Omega} = \begin{bmatrix} 0 & -\omega_z & \omega_y \\ \omega_z & 0 & -\omega_x \\ -\omega_y & \omega_x & 0 \end{bmatrix} \tag{A.2c}$$

where $\mathbf{I}_{\mathrm{p}}$ is 3×3 inertia matrix relative to its symmetrical axis system, $\mathbf{I}_{\mathrm{p}} = \mathrm{diag}\{I_{xx}, I_{yy}, I_{zz}\}$.

Appendix B.

The mass matrix $\mathbf{M}^*(\tilde{\boldsymbol{\Theta}})$, matrix of centrifugal and Coriolis term $\mathbf{V}^*(\tilde{\boldsymbol{\Theta}},\dot{\boldsymbol{\Theta}})$, and gravity terms $\mathbf{G}^*(\tilde{\boldsymbol{\Theta}})$ in Eqs.(8) are given by

$$\mathbf{G}^*(\tilde{\boldsymbol{\Theta}}) = \tilde{\mathbf{G}}(\tilde{\boldsymbol{\Theta}}) + \sum_{i=1}^{6}[(\mathbf{J}_{\mathrm{uc,ai}}\mathbf{J}_{\mathrm{ai},\tilde{\Theta}})^{\mathrm{T}} m_{\mathrm{u}} \cdot \mathbf{g} + (\mathbf{J}_{\mathrm{tc,ai}}\mathbf{J}_{\mathrm{ai},\tilde{\Theta}})^{\mathrm{T}} m_{\mathrm{t}} \cdot \mathbf{g}] \tag{B.1a}$$

$$\mathbf{M}^*(\tilde{\boldsymbol{\Theta}}) = \tilde{\mathbf{M}}(\tilde{\boldsymbol{\Theta}}) + \sum_{i=1}^{6} \mathbf{J}^{\mathrm{T}}{}_{\mathrm{ai},\tilde{\Theta}}(\mathbf{M}_{\mathrm{li}} + \mathbf{M}_{\mathrm{agi}})\mathbf{J}_{\mathrm{ai},\tilde{\Theta}} \tag{B.1b}$$

$$\mathbf{V}^*(\tilde{\boldsymbol{\Theta}},\dot{\boldsymbol{\Theta}}) = \tilde{\mathbf{V}}(\tilde{\boldsymbol{\Theta}},\dot{\boldsymbol{\Theta}}) + \sum_{i=1}^{6}\{\mathbf{J}^{\mathrm{T}}{}_{\mathrm{ai},\tilde{\Theta}} \cdot \mathbf{V}_{\mathrm{li}} + \mathbf{V}_{\mathrm{ani}}\} \tag{B.1c}$$

where,

$$\mathbf{M}_{\mathrm{li}} = \mathbf{J}^{\mathrm{T}}{}_{\mathrm{uc,ai}} m_{\mathrm{u}} \mathbf{J}_{\mathrm{uc,ai}} + \mathbf{J}^{\mathrm{T}}{}_{\mathrm{tc,ai}} m_{\mathrm{t}} \mathbf{J}_{\mathrm{tc,ai}} \tag{B.2a}$$

$$\mathbf{M}_{\mathrm{agi}} = \frac{(\mathbf{I}_a + \mathbf{I}_b)}{l_i^2}(\mathbf{I} - \mathbf{l}_{\mathrm{n}} \cdot \mathbf{l}_{\mathrm{n}}{}^{\mathrm{T}}) \tag{B.2b}$$

$$\mathbf{V}_{\mathrm{li}} = \mathbf{J}^{\mathrm{T}}{}_{\mathrm{uc,ai}} m_{\mathrm{u}} (\mathbf{J}^{\mathrm{T}}{}_{\mathrm{uc,ai}} \dot{\mathbf{J}}_{\mathrm{ai},\tilde{\Theta}} + \dot{\mathbf{J}}_{\mathrm{uc,ai}} \mathbf{J}_{\mathrm{ai},\tilde{\Theta}}) \tag{B.2c}$$

$$\mathbf{V}_{\mathrm{ani}} = -\frac{2(\mathbf{I}_a + \mathbf{I}_b)}{l_i^2}(\mathbf{l}_{\mathrm{ni}}{}^{\mathrm{T}} \cdot \mathbf{J}_{\mathrm{ai},\tilde{\Theta}}\dot{\boldsymbol{\Theta}})(\mathbf{I} - \mathbf{l}_{\mathrm{n}} \cdot \mathbf{l}_{\mathrm{n}}{}^{\mathrm{T}})\mathbf{J}_{\mathrm{ai},\tilde{\Theta}} \tag{B.2d}$$

6. References

[1] J.P. Merlet, Parallel robots, Kluwer Academic Publisher, Netherlands, 2000.

[2] J. Gallardo, J.M. Rico, A. Frisoli, D. Checcacci, M. Bergamasco, Dynamic of parallel manipulators by means of screw theory, Mechanism and Machine Theory 38 (2003) 1113-1131

[3] A. Lopes, F. Almeida, A force-impedance controlled industrial robot using an active robotic auxiliary device, Robotics and Computer-Integrated Manufacturing 24 (2008) 299-309.

[4] Y.N. Wu, C. Gosselin, Design of reactionless 3-DOF and 6-DOF parallel manipulators using parallelepiped mechanisms, IEEE Transactions on Robotics 21 (5) (2005) 821-833

[5] D. Stewart, A platform with six degree of freedom: A new form of mechanical linage which enables a platform to move simultaneously in six degree of freedom developed by Elliot-Automation, Aircraft Engineering and Aerospace Technology 38 (4) (1965) 30-35.

[6] J.P. Merlet, Parallel manipulators: state of the art and perspectives, Advanced Robotics 8 (6) (1994) 589-596.

[7] K.H. Hunt, Structural kinematics of in-parallel-actuated robot-arms, ASME Journal of Mechanisms, Transmissions and Automation in Design 105(4) (1983) 705-712.

[8] W.Q.D. Do, D.C.H. Yang, Inverse dynamic analysis and simulation of a platform type of robot, Journal of Robotic Systems 5 (1988) 209-227.
[9] M. Honegger, R. Brega, G. Schweizer, Application of a nonlinear adaptive controller to a 6 dof parallel manipulator, In Proceeding of the 2000 IEEE International Conference on Robotics and Automation, San Francisco, CA, USA, (2000) pp. 1930-1935.
[10] D.H. Kim, J.Y. Kang, K-II. Lee, Robust tracking control design for a 6 DOF parallel manipulator, Journal of Robotics Systems 17(10) (2000) 527-547.
[11] C.C. Nguyen, S.S. Antrazi, Z.L. Zhou, C.E. Campbell, Adaptive control of a Stewart platform-based manipulator, Journal of Robotics Systems 10(5) (1992) 657-687.
[12] H.S. Kim, Y.M. Cho, K-II. Lee, Robust nonlinear task space control for a 6 DOF parallel manipulator, Automatica 41 (2005) 1591-1600.
[13] Y. Ting, Y.S. Chen, H.C. Jar, Modeling and control for a Gough-Stewart platform CNC machine, Journal of Robotics System 21(11) (2004) 609-623.
[14] J.L. Chen, W.D. Chang, Feedback linearization control of a two-link robot using a multi-crossover genetic algorithm, Expert Systems with Applications 36 (2009) 4154-4159.
[15] K.J. Astrom, T. Hagglund. PID Controllers: Theory, Design, and Tuning. Instrument Society of America: NC, 1995.
[16] Y.X Su, B.Y. Duan, C.H. Zheng, Y.F. Zhang, G.D. Chen, J.W. Mi, Disturbance-rejection high-precision motion control of a Stewart platform, IEEE Transactions on Control Systems Technology 12(3) (2004) 364-374.
[17] E. Burdet, M. Honegger, A. Codourey, Controllers with desired dynamic compensation and their implementation on a 6 DOF parallel manipulator, Proceeding of the 2000 IEEE/RSJ International Conference on Intelligent Robots and Systems, 2000, pp. 39-45.
[18] J.R. Noriega, H. Wang, A direct adaptive neural-network control for unknown nonlinear systems and its application, IEEE Transactions on Neural Networks 9 (1) (1998) 27-34.
[19] N.I. Kim, C.W. Lee, High speed tracking control of Stewart platform manipulator via enhanced sliding mode control, In Proceeding of the 1998 IEEE Conference on Robotics and Automation, Leuven, Belgium, (1998) pp. 2716-2721.
[20] I. Cervantes, J.A. Ramirez, On the PID tracking control of robot manipulators, Systems & Control Letters 42 (2001) 37-46.
[21] I. Davliskos, E. Papadopoulous, Model-based control of 6-DOF electrohydraulic parallel manipulator platform, Mechanism and Machine Theory 43 (2008) 1385-1400.
[22] F. Behi, Kinematic analysis for a six-degree-of-freedom 3-PRPS parallel mechanism, IEEE Journal of Robotics and Automation 4(5) (1988) 561-565.
[23] D.M. Ku, Direct displacement analysis of a Stewart platform mechanism. Mechanism and Machine Theory 34 (1999) 453-465.
[24] B. Dasgupta, T.S. Mruthyunjaya, A Newton-Euler formulation for the inverse dynamics of the Stewart platform manipulator, Mechanism and Machine Theory 33(8) (1998) 1135-1152.
[25] S.H. Koekebakker, Model based control of a flight simulator motion system. Ph.D Thesis. Netherlands: Delft University of Technology, 2001.
[26] H.E. Merrit, Hydraulic Control Systems, Wiley, 1967.
[27] D. Rowell, D.N. Wormley, System Dynamics: An Introduction, Prentice Hall, 1997.
[28] R. Gorez, Globally stable PID-like control of mechanical systems, Systems & Control Letters 38 (1999) 61-72.
[29] C.F. Yang, J.F. He, J.W. Han, X.C. Liu, Real-time state estimation for spatial six-degree-of-freedom linearly actuated parallel robots. Mechatronics 19(6) (2009) 1026-1033.

6

Sampled-Data PID Control and Anti-aliasing Filters*

Marian J. Blachuta and Rafal T. Grygiel
Department of Automatic Control, Silesian University of Technology
Poland

1. Introduction

Consider a typical configuration of the sampled-data control system. It consists of the plant to be controlled, a sampler, a discrete-time controller and a zero-order hold. Disturbance can be seen as an integral part of the plant so that the plant is characterized by the control path responsible for control signal influence on the output and the disturbance. The system output is usually sensed by sensors whose output signal can be corrupted by noise. Sometimes analog filters are put between the analog sensor output signal and sampler. In the control literature

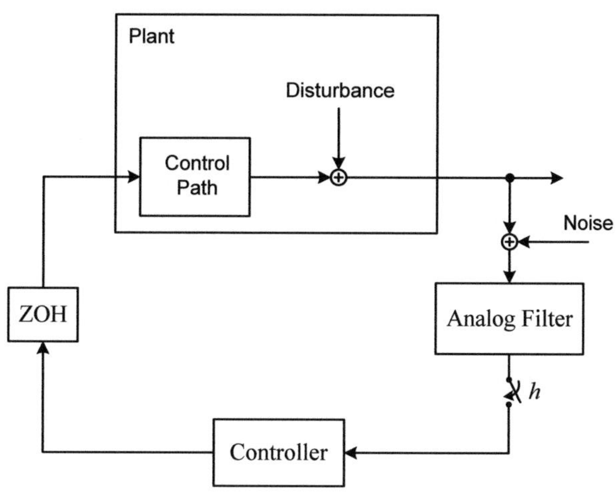

Fig. 1. General control system diagram

(Åström and Wittenmark, 1997; Feuer and Goodwin, 1996) strong belief is expressed, that filters are necessary prior to sampling to guarantee correct digital signal processing and control. This belief is usually supported by heuristic speculations based on Shannon-Kotelnikov Reconstruction Theorem, e.g. (Jerri, 1977), which states that in order to reconstruct the signal $s(t)$ from its samples $s(ih), -\infty < i < \infty$, the sampling frequency should be at least twice the highest frequency component in the signal. Since the spectra of physical signals often stretch on infinite frequency range, this gives rise to the idea of so called anti-aliasing filters that cut off the portion of frequency spectrum lying outside the region determined by that theorem.

*This work has been granted by the Polish Ministry of Science and Higher Education from funds for years 2008-2011

It should, however, be stressed that no proofs are available concerning the necessity of anti-aliasing filters in sampled-data systems, and no statements can be found with regard to the consequences of the lack of such filters.
Anti-aliasing filters usually take the form of Butterworth filters whose cutoff frequency equals to the so called Nyquist frequency $\omega_N = \pi/h$, which is depending solely on sampling period h. As an alternative, so called integrating or averaging samplers are considered (Blachuta & Grygiel, 2008a;b; Feuer and Goodwin, 1996; Goodwin et al., 2001; Steinway and Melsa, 1971; Shats and Shaked, 1989).
In (Blachuta & Grygiel, 2008a;b) we studied the impact of antialiasing filters for pure signal processing, while in (Blachuta & Grygiel, 2009b) the context of discrete-time LQG control was discussed. The statement was made, that there is no reason for using them in the noiseless case, and practically they find no use in the case of noisy measurements. The best results in the latter case are obtained when the continuous-time output is passed through a continuous-time Kalman filter, which depends rather on disturbance and noise characteristics than the sampling period, before being sampled. Similar results were observed in PID control systems (Blachuta & Grygiel, 2009a;b;c)and (Blachuta & Grygiel, 2010)
In this chapter we summarize these results and compare them with LQG minimum-variance benchmark control using simple, but representative examples.

2. Analog part of the system

2.1 Plant, disturbance and noise model

The model of system displayed in Fig. 1 is presented in Fig. 2, where $K_c(s)$ is the transfer function of control path of the plant, while $K_d(s)$ and $K_n(s)$ represent filters forming stochastic disturbance and noise, respectively. $K_f(s)$ stands for a continuous-time filter.

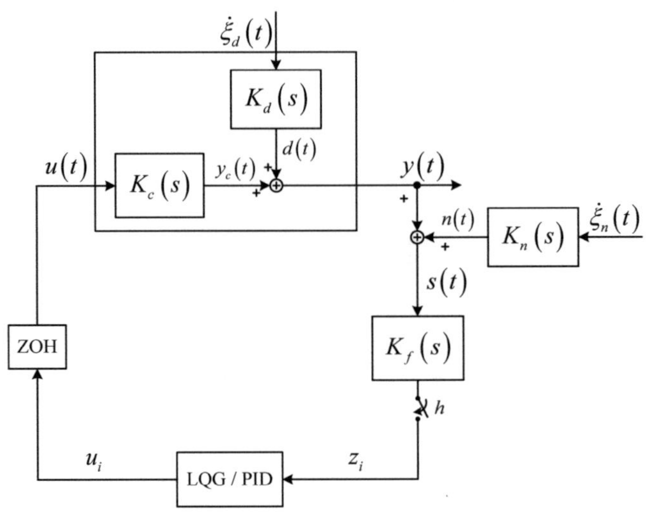

Fig. 2. Control system

The entire continuous-time system can be modeled in state-space as follows:

$$\dot{\boldsymbol{x}}(t) = \boldsymbol{A}\boldsymbol{x}(t) + \boldsymbol{b}u(t) + \boldsymbol{C}\dot{\boldsymbol{\xi}}(t), \tag{1}$$

$$y(t) = \boldsymbol{d}_y'\boldsymbol{x}(t), \tag{2}$$

$$s(t) = \boldsymbol{d}_s'\boldsymbol{x}(t), \tag{3}$$

$$z(t) = \boldsymbol{d}'\boldsymbol{x}(t), \tag{4}$$

where:

$$\boldsymbol{A} = \begin{bmatrix} \boldsymbol{A}_c & \boldsymbol{0} & \boldsymbol{0} & \boldsymbol{0} \\ \boldsymbol{0} & \boldsymbol{A}_d & \boldsymbol{0} & \boldsymbol{0} \\ \boldsymbol{0} & \boldsymbol{0} & \boldsymbol{A}_n & \boldsymbol{0} \\ \boldsymbol{b}_f\boldsymbol{d}_c' & \boldsymbol{b}_f\boldsymbol{d}_d' & \boldsymbol{b}_f\boldsymbol{d}_n' & \boldsymbol{A}_f \end{bmatrix}, \quad \boldsymbol{C} = \begin{bmatrix} \boldsymbol{0} & \boldsymbol{0} \\ \boldsymbol{c}_d & \boldsymbol{0} \\ \boldsymbol{0} & \boldsymbol{c}_n \\ \boldsymbol{0} & \boldsymbol{0} \end{bmatrix},$$

$$\boldsymbol{b} = \begin{bmatrix} \boldsymbol{b}_c \\ \boldsymbol{0} \\ \boldsymbol{0} \\ \boldsymbol{0} \end{bmatrix}, \quad \boldsymbol{d}_y = \begin{bmatrix} \boldsymbol{d}_c \\ \boldsymbol{d}_d \\ \boldsymbol{0} \\ \boldsymbol{0} \end{bmatrix}, \quad \boldsymbol{d}_s = \begin{bmatrix} \boldsymbol{d}_c \\ \boldsymbol{d}_d \\ \boldsymbol{d}_n \\ \boldsymbol{0} \end{bmatrix}, \quad \boldsymbol{d} = \begin{bmatrix} \boldsymbol{0} \\ \boldsymbol{0} \\ \boldsymbol{0} \\ \boldsymbol{d}_f \end{bmatrix},$$

$$\boldsymbol{x}(t) = \begin{bmatrix} \dot{\boldsymbol{x}}_c(t) \\ \dot{\boldsymbol{x}}_d(t) \\ \dot{\boldsymbol{x}}_n(t) \\ \dot{\boldsymbol{x}}_f(t) \end{bmatrix}, \quad \dot{\boldsymbol{\xi}}(t) = \begin{bmatrix} \dot{\xi}_d(t) \\ \dot{\xi}_n(t) \end{bmatrix}.$$

Processes $\dot{\xi}_d(t)$ and $\dot{\xi}_n(t)$ are independent continuous-time white noises with zero means and covariance functions defined as unit Dirac pulse functions, i.e.:

$$\mathrm{E}\left[\dot{\xi}_d(t)\right] = 0, \qquad \mathrm{E}\left[\dot{\xi}_d(t)\dot{\xi}_d(\tau)\right] = \delta(t-\tau); \tag{5}$$

$$\mathrm{E}\left[\dot{\xi}_n(t)\right] = 0, \qquad \mathrm{E}\left[\dot{\xi}_n(t)\dot{\xi}_n(\tau)\right] = \delta(t-\tau). \tag{6}$$

2.2 Analog Filters

In the paper two types of filters are considered: Butterworth filter as the anti-aliasing filter, as well as a continuous-time Kalman filter as a filter based on signals spectra.

2.2.1 Butterworth Filter

Transfer function of the Butterworth filter has the form:

$$K_f(s) = \frac{1}{B_n\left(\frac{s}{\omega_o}\right)}, \tag{7}$$

where $B_n(*)$ is the nth-degree Butterworth's polynomial and ω_o is called the cutoff frequency. In this paper ω_o will be assumed as Nyquist frequency $\omega_o = \omega_N = \frac{\pi}{h}$. The first Butterworth's polynomials are definded as follows:

$$B_1(x) = x + 1; \quad B_2(x) = x^2 + \sqrt{2}\cdot x + 1. \tag{8}$$

2.2.2 Kalman Filter

Kalman filter is the one that provides the best noise filtering under assumptions of our model. Since the noise added to the measured output is not white, the classical Kalman filter for a system consisting of disturbance and noise becomes singular. One way to overcome the problem is to replace the continuous-time filter with a discrete-time one working at a high enough sampling frequency $1/h_f$. The output of such filter could be re-sampled at lower frequency if necessary.
Very often the power spectrum $S_n(\omega)$ of noise $n(t)$, defined by transfer function $K_n(s)$, is much wider than that of the signal of interest $y(t)$. In such case it can be modeled as white noise $n(t)$

$$\mathrm{E}\left[n(t)\right] = 0, \qquad \mathrm{E}\left[n(t)n(\tau)\right] = \eta^2\delta(t-\tau); \tag{9}$$

with constant spectral density η^2 independent of frequency ω. The model of disturbances is then simplified to

$$\dot{\boldsymbol{x}}_d(t) = \boldsymbol{A}_d\boldsymbol{x}_d(t) + \boldsymbol{c_d}\dot{\xi}_d(t), \tag{10}$$

$$y_{dn}(t) = \boldsymbol{d}_d'\boldsymbol{x}_d(t) + \eta\dot{\xi}_n(t), \tag{11}$$

with

$$\eta = |K_n(0)| = |\boldsymbol{d}_n'\boldsymbol{A}_n^{-1}\boldsymbol{c}_n| \tag{12}$$

The continuous-time Kalman filter is then defined by:

$$\dot{\boldsymbol{x}}_f(t) = \boldsymbol{A}_d\boldsymbol{x}_f(t) + \boldsymbol{k}_c^f\left[y_{dn}(t) - \boldsymbol{d}_d'\boldsymbol{x}_f(t)\right] \tag{13}$$

where:

$$\boldsymbol{k}_c^f = \frac{\boldsymbol{P}\boldsymbol{d}_d}{\eta^2}; \quad \boldsymbol{A}_d\boldsymbol{P} + \boldsymbol{P}\boldsymbol{A}_d' - \frac{\boldsymbol{P}\boldsymbol{d}_d\boldsymbol{d}_d'\boldsymbol{P}}{\eta^2} + \boldsymbol{c}_d\boldsymbol{c}_d' = 0. \tag{14}$$

We use this filter in the system to pass the signal $y_2(t)$ through it, i.e. we substitute $y_{dn}(t) = y_2(t)$ and receive $z(t) = \boldsymbol{d}_d'\boldsymbol{x}_f(t)$
Since only a rough characterization of noise is required and filter equations are of lower order equal to the order of disturbance model, analog filtering is greatly simplified.

3. Control algorithms

The aim of the control system is to keep the output of the system close to the reference value $y^r(t) = 0$, i.e. to make the error $e(t) = y^r(t) - y(t)$ small. Since standard deviation is a good measure of the expected magnitude, the quality of the control systems will be assessed based on standard deviation of output and control signals. To this end, appropriate variations should be calculated.

3.1 PID controller

Discrete-time PID controller defined by transfer function:

$$K_{reg}(z) = \frac{U(z)}{E(z)} = k_P\left(1 + \frac{h}{T_I}\frac{z}{z-1} + \frac{T_D}{h}\frac{z-1}{z}\right) \tag{15}$$

can be presented in the state-space form, assuming $e_i = -z_i$, as follows:

$$\boldsymbol{x}_{i+1}^r = \boldsymbol{F}_r\boldsymbol{x}_i^r - \boldsymbol{g}_r z_i, \tag{16}$$

$$u_i = \boldsymbol{d}_r'\boldsymbol{x}_i^r - e_r z_i, \tag{17}$$

P	$k_P = \frac{T}{k \cdot L}$	–	–
PI	$k_P = 0.9\frac{T}{k \cdot L}$	$T_I = 3.33 \cdot L$	–
PID	$k_P = 1.2\frac{T}{k \cdot L}$	$T_I = 2 \cdot L$	$T_D = 0.5 \cdot L$

Table 1. QDR PID controller settings

where:

$$\boldsymbol{F}_r = \begin{bmatrix} 1 & 0 \\ 0 & 0 \end{bmatrix}, \quad \boldsymbol{g}_r = \begin{bmatrix} 1 \\ 1 \end{bmatrix}, \quad \boldsymbol{d}_r = \begin{bmatrix} k_P \frac{h}{T_I} \\ -k_P \frac{T_D}{h} \end{bmatrix}, \quad e_r = k_p \left[1 + \frac{h}{T_I} + \frac{T_D}{h} \right] \tag{18}$$

3.1.1 QDR controller settings

There are several methods to find continuous-time PID controller settings. Perhaps the simplest one is the so called QDR (Quarter Decay Ratio) method, which is based on lag-delay approximation of the plant. We adapt this method to sampled-data controller using a continuous-time approximation of the discrete-time system consisting of ZOH, plant, filter and sampler. Moreover, a lag-delay approximation $G_{OL}(s)$ of the control path including respective filter, $K_{OL}(s) = K_c(s)K^f(s)$, is used.

$$G_{OL}(s) = \frac{k}{Ts+1} e^{-s\tau}. \tag{19}$$

The parameters of $G_{OL}(s)$ can be determined by several methods based on the step response of $K_{OL}(s)$. One of them, called "two points method", relies on two time instants, t_1 and t_2, at which the step response reaches the values 63.2% and 28.3% of the steady state, respectively. We then have:

$$T = 1.5\,(t_1 - t_2)\,, \qquad \tau = t_1 - T. \tag{20}$$

Then the QDR settings (Goodwin et al., 2001) are taken from Table 1 where L accounts for ZOH and sampler as follows:

$$L = \tau + \frac{h}{2}, \tag{21}$$

which corresponds to the $h/2$ delay approximation of ZOH.

3.1.2 Optimal PID controller

QDR controller settings do not depend on disturbance and noise characteristics. Therefore optimal controllers settings $\hat{\boldsymbol{p}} = \begin{bmatrix} \hat{k}_P & \hat{T}_I^j & \hat{T}_D^j \end{bmatrix}'$ will be chosen as the ones minimizing the output variance of the controlled system:

$$\hat{\boldsymbol{p}} = \arg\min_{\boldsymbol{p}} \operatorname{var}\{y_i\} \tag{22}$$

where the variance $\operatorname{var}\{y_i\}$ is determined by the formuale in (24) - (28) that take disturbance and noise characteristics into account. Denoting $\hat{\boldsymbol{p}}^j = \begin{bmatrix} \hat{k}_P^j & \hat{T}_I^j & \hat{T}_D^j \end{bmatrix}'$ at j-th stage of the minimization procedure, the computation stops when:

$$\|\hat{\boldsymbol{p}}^j - \hat{\boldsymbol{p}}^{j-1}\| < \varepsilon \text{ where } \varepsilon = 0.01 \tag{23}$$

In the above, Powell method of extremum seeking, amended with a procedure determining the range of stable values of parameters at each direction, can be used. The parameters resulting from QDR tuning can then be chosen as an initial guess.

3.1.3 PID Control System Assessment

The output and control variances are as follows:

$$\sigma_y^2 = \text{var}\{y_i\} = \boldsymbol{d}_y' \boldsymbol{V}^p \boldsymbol{d}_y, \tag{24}$$

$$\sigma_u^2 = \text{var}\{u_i\} = \boldsymbol{d}_r' \boldsymbol{V}^r \boldsymbol{d}_r + e_r \boldsymbol{d}' \boldsymbol{V}^p \boldsymbol{d} e_r - \boldsymbol{d}_r' \boldsymbol{V}^{rp} \boldsymbol{d} e_r - e_r \boldsymbol{d}' \boldsymbol{V}^{pr} \boldsymbol{d}_r, \tag{25}$$

where the covariance matrix $\boldsymbol{V}$

$$\boldsymbol{V} = \mathrm{E}\left\{ \begin{bmatrix} \boldsymbol{x}_i \\ \boldsymbol{x}_i^r \end{bmatrix} \begin{bmatrix} \boldsymbol{x}_i' & \boldsymbol{x}_i^{r\prime} \end{bmatrix} \right\} = \begin{bmatrix} \boldsymbol{V}_i^p & \boldsymbol{V}_i^{pr} \\ \boldsymbol{V}_i^{rp} & \boldsymbol{V}_i^r \end{bmatrix} \tag{26}$$

is a solution of

$$\boldsymbol{V} = \boldsymbol{\Phi} \boldsymbol{V} \boldsymbol{\Phi}' + \boldsymbol{\Lambda} \boldsymbol{W} \boldsymbol{\Lambda}' \tag{27}$$

with

$$\boldsymbol{\Phi} = \begin{bmatrix} (\boldsymbol{F} - \boldsymbol{g} e_r \boldsymbol{d}') & \boldsymbol{g} \boldsymbol{d}_r' \\ -\boldsymbol{g}_r \boldsymbol{d}' & \boldsymbol{F}_r \end{bmatrix}, \quad \boldsymbol{\Lambda} = \begin{bmatrix} \boldsymbol{I} \\ 0 \end{bmatrix} \tag{28}$$

3.2 MV LQG control law

The best control accuracy is achieved when using the optimal Minimum-Variance sampled-data LQG controller which will be used as a benchmark to assess PID control quality.

3.2.1 Controller

LQG control problem with a continuous performance index J is formulated, where

$$J = \lim_{N \to \infty} \mathrm{E} \frac{1}{Nh} \int_0^{Nh} \{y^2(t) + \lambda u^2(t)\} dt. \tag{29}$$

Setting $\lambda = 0$ defines a MV sampled-data LQG problem. Since noise influences only state estimate $\hat{\boldsymbol{x}}_{i|i}$ and not the control law, being itself a linear function of $\hat{\boldsymbol{x}}_{i|i}$ the above sampled data control problem can be reformulated as follows.

The problem defined by modulation equation

$$u(t) = u_i, \text{ for } t \in (ih, ih+h], \, i = 0, 1, \ldots, \tag{30}$$

state equation

$$\dot{\boldsymbol{x}}_p(t) = \boldsymbol{A}_p \boldsymbol{x}_p(t) + \boldsymbol{b}_p u(t) + \boldsymbol{c}_p \dot{\xi}(t), \tag{31}$$

$$y(t) = \boldsymbol{d}_p' \boldsymbol{x}_p(t), \tag{32}$$

where:

$$\boldsymbol{A}_p = \begin{bmatrix} \boldsymbol{A}_c & \boldsymbol{0} \\ \boldsymbol{0} & \boldsymbol{A}_d \end{bmatrix}, \quad \boldsymbol{b}_p = \begin{bmatrix} \boldsymbol{b}_c \\ \boldsymbol{0} \end{bmatrix}, \quad \boldsymbol{c}_p = \begin{bmatrix} \boldsymbol{0} \\ \boldsymbol{c}_d \end{bmatrix},$$

$$\boldsymbol{d}_p = \begin{bmatrix} \boldsymbol{d}_c \\ \boldsymbol{d}_d \end{bmatrix}, \quad \boldsymbol{x}_p(t) = \begin{bmatrix} \boldsymbol{x}_c(t) \\ \boldsymbol{x}_d(t) \end{bmatrix}, \quad \dot{\xi}(t) = \dot{\xi}_d(t),$$

and feedback signal z_i, is equivalent with the following discrete-time problem

$$\boldsymbol{x}_{i+1}^p = \boldsymbol{F}_p \boldsymbol{x}_i^p + \boldsymbol{g}_p u_i + \boldsymbol{w}_i^p, \tag{33}$$

$$z_i = \boldsymbol{d}_p' \boldsymbol{x}_i^p, \tag{34}$$

$$J = \lim_{N\to\infty} \mathrm{E} \frac{1}{N} \sum_{i=0}^{N-1} \{\boldsymbol{x}_i^{p\prime} \boldsymbol{Q}_1 \boldsymbol{x}_i^p + 2\boldsymbol{x}_i^{p\prime} \boldsymbol{q}_{12} u_i + q_2 u_i^2 + q_w\}, \tag{35}$$

where

$$\boldsymbol{Q}_1 = \frac{1}{h}\int_0^h \boldsymbol{F}_p'(\tau)\boldsymbol{M}\boldsymbol{F}_p(\tau)d\tau, \qquad \boldsymbol{M} = \boldsymbol{d}_p\boldsymbol{d}_p',$$

$$\boldsymbol{q}_{12} = \frac{1}{h}\int_0^h \boldsymbol{F}_p'(\tau)\boldsymbol{M}\boldsymbol{g}_p(\tau)d\tau,$$

$$q_2 = \frac{1}{h}\int_0^h \boldsymbol{g}_p'(\tau)\boldsymbol{M}\boldsymbol{g}_p(\tau)d\tau + \lambda,$$

$$q_w = \boldsymbol{d}_p' \left\{ \int_0^h \int_0^\tau \boldsymbol{F}_p(\tau - s)\boldsymbol{c}_p\boldsymbol{c}_p'\boldsymbol{F}_p'(\tau - s)ds d\tau \right\} \boldsymbol{d}_p,$$

$$\boldsymbol{F}_p(\tau) = e^{\boldsymbol{A}_p\tau}, \qquad \boldsymbol{F}_p = \boldsymbol{F}_p(h), \tag{36}$$

$$\boldsymbol{g}_p(\tau) = \int_0^\tau e^{\boldsymbol{A}_p v}\boldsymbol{b}_p dv, \qquad \boldsymbol{g}_p = \boldsymbol{g}_p(h) \tag{37}$$

and $\boldsymbol{w}_i^p$ is a zero mean vector Gaussian noise with $\mathrm{E}\{\boldsymbol{w}_i^p \boldsymbol{w}_i^{p\prime}\} = \boldsymbol{W}_p$, and

$$\boldsymbol{W}_p = \int_0^h \mathrm{e}^{\boldsymbol{A}_p s}\boldsymbol{c}_p\boldsymbol{c}_p'\mathrm{e}^{\boldsymbol{A}_p' s}ds. \tag{38}$$

Vectors $\boldsymbol{x}_0^p$ and $\boldsymbol{w}_i^p$ are independent for all $i \geq 0$. The optimal control law minimizing the performance index (35) for the discrete stochastic system (33)-(34) is a linear function

$$u_i = -\boldsymbol{k}_x'\hat{\boldsymbol{x}}_{i|i}^p, \tag{39}$$

where $\hat{\boldsymbol{x}}_{i|i}^p$ denotes the Kalman filter estimate of $\boldsymbol{x}_i^p$ based on available information up to and including i from (47)-(48).The feedback gain $\boldsymbol{k}_x$,

$$\boldsymbol{k}_x' = \frac{\boldsymbol{q}_{12} + \boldsymbol{F}_p'\boldsymbol{K}\boldsymbol{g}_p}{q_2 + \boldsymbol{g}_p'\boldsymbol{K}\boldsymbol{g}_p} \tag{40}$$

depends on the positive definite solution $\boldsymbol{K}$ of the following algebraic Riccati equation:

$$\boldsymbol{K} = \boldsymbol{Q}_1 + \boldsymbol{F}_p'\boldsymbol{K}\boldsymbol{F}_p - \frac{(\boldsymbol{q}_{12} + \boldsymbol{F}_p'\boldsymbol{K}\boldsymbol{g}_p)(\boldsymbol{q}_{12} + \boldsymbol{F}_p'\boldsymbol{K}\boldsymbol{g}_p)'}{q_2 + \boldsymbol{g}_p'\boldsymbol{K}\boldsymbol{g}_p}.$$

3.2.2 Discrete-time Kalman filter

Simple instantaneous sampling with sampling period h consists in taking the values of the sampled signal at discrete time instants $t_i = ih, i = 0, 1, \ldots$. Available measurements z_i are expressed as

$$z_i = y_2(t_i). \tag{41}$$

The problem defined by measurement equation $z_i = z(ih)$ and state equation (1) is equivalent to the following discrete-time system:

$$\boldsymbol{x}_{i+1} = \boldsymbol{F}\boldsymbol{x}_i + \boldsymbol{g}u_i + \boldsymbol{w}_i, \tag{42}$$

$$z_i = \boldsymbol{d}'\boldsymbol{x}_i, \tag{43}$$

where:

$$\boldsymbol{F}(\tau) = e^{\boldsymbol{A}\tau}, \qquad \boldsymbol{F} = \boldsymbol{F}(h), \tag{44}$$

$$\boldsymbol{g}(\tau) = \int_0^{\tau} e^{\boldsymbol{A}v}\boldsymbol{b}dv, \qquad \boldsymbol{g} = \boldsymbol{g}(h) \tag{45}$$

and $\boldsymbol{w}_i$ is a zero mean vector Gaussian noise with $\mathrm{E}\{\boldsymbol{w}_i\boldsymbol{w}_i'\} = \boldsymbol{W}$, and

$$\boldsymbol{W} = \int_0^h \mathrm{e}^{\boldsymbol{A}s}\boldsymbol{C}\boldsymbol{C}'\mathrm{e}^{\boldsymbol{A}'s}ds. \tag{46}$$

Vectors $\boldsymbol{x}_0$ and $\boldsymbol{w}_i$ are independent for all $i \geq 0$.
The limiting Kalman filter, (Anderson & Moore, 1979), that provides ($\hat{\boldsymbol{x}}_{i|i} = \mathrm{E}[\boldsymbol{x}_i|\vec{z}_i]$) for the discrete-time system in (42)-(43) as $i \to \infty$ has the form:

$$\hat{\boldsymbol{x}}_{i+1|i+1} = \hat{\boldsymbol{x}}_{i+1|i} + \boldsymbol{k}^f(z_{i+1} - \boldsymbol{d}'\hat{\boldsymbol{x}}_{i+1|i}), \tag{47}$$

$$\hat{\boldsymbol{x}}_{i+1|i} = \boldsymbol{F}\hat{\boldsymbol{x}}_{i|i} + \boldsymbol{g}u_i, \qquad \boldsymbol{x}_{0|-1} = 0, \tag{48}$$

where

$$\boldsymbol{k}^f = \frac{\boldsymbol{\Sigma}\boldsymbol{d}}{\boldsymbol{d}'\boldsymbol{\Sigma}\boldsymbol{d}'}, \quad \boldsymbol{\Sigma} = \boldsymbol{W} + \boldsymbol{F}\left(\boldsymbol{\Sigma} - \frac{\boldsymbol{\Sigma}\boldsymbol{d}\boldsymbol{d}'\boldsymbol{\Sigma}'}{\boldsymbol{d}'\boldsymbol{\Sigma}\boldsymbol{d}}\right)\boldsymbol{F}'. \tag{49}$$

3.2.3 MV LQG Control System Assessment

Output and control variances for systems with continuous-time filters can be expressed by following formulae:

$$\sigma_y^2 = \mathrm{var}\{y_i\} = \boldsymbol{d}_0'\boldsymbol{V}^o\boldsymbol{d}_0, \tag{50}$$

$$\sigma_u^2 = \mathrm{var}\{u_i\} = \boldsymbol{k}_x'\boldsymbol{V}^f\boldsymbol{k}_x, \tag{51}$$

where $\boldsymbol{V}^o$, $\boldsymbol{V}^f$, end $\boldsymbol{V}^{fo}$ are submatrices of matrix $\boldsymbol{V}$

$$\boldsymbol{V} = \mathrm{E}\left\{\begin{bmatrix}\boldsymbol{x}_i \\ \hat{\boldsymbol{x}}_{i|i}\end{bmatrix}\begin{bmatrix}\boldsymbol{x}_i' & \hat{\boldsymbol{x}}_{i|i}'\end{bmatrix}\right\} = \begin{bmatrix}\boldsymbol{V}^o & \boldsymbol{V}^{of} \\ \boldsymbol{V}^{fo} & \boldsymbol{V}^f\end{bmatrix} \tag{52}$$

which is a solution of the following matrix Lyapunov equation:

$$\boldsymbol{V} = \boldsymbol{\Phi}\boldsymbol{V}\boldsymbol{\Phi}' + \boldsymbol{\Omega}\boldsymbol{W}\boldsymbol{\Omega}', \tag{53}$$

with:

$$\boldsymbol{\Lambda} = (\boldsymbol{I} - \boldsymbol{k}^f \boldsymbol{d}')(\boldsymbol{F} + \boldsymbol{g}\boldsymbol{k}_x'), \qquad \boldsymbol{\Psi} = (\boldsymbol{\Lambda} + \boldsymbol{k}^f \boldsymbol{d}' \boldsymbol{g}\boldsymbol{k}_x'),$$

$$\boldsymbol{\Phi} = \begin{bmatrix} \boldsymbol{F} & \boldsymbol{g}\boldsymbol{k}_x' \\ \boldsymbol{k}^f \boldsymbol{d}' \boldsymbol{F} & \boldsymbol{\Psi} \end{bmatrix}, \qquad \boldsymbol{\Omega} = \begin{bmatrix} \boldsymbol{I} \\ \boldsymbol{k}^f \boldsymbol{d}' \end{bmatrix}.$$

4. Examples

We will study the properties of control systems for a plant having control path

$$K_c(s) = \frac{1}{(1+0.5s)^2}, \tag{54}$$

with disturbance modeled by:

$$K_d(s) = \frac{k_d}{(1+T_d s)^2}, \tag{55}$$

with $T_d = 2$ and k_d chosen such, that $\operatorname{var} d(t) = 1$. For the noise model in Fig.2 we use three different transfer functions

$$K_n^1(s) = \frac{k_n^1}{T_n^2 s^2 + 2\zeta_n T_n s + 1}, T_n = 0.05, \zeta_n = 1 \tag{56}$$

$$K_n^2(s) = \frac{k_n^2}{T_n^2 s^2 + 2\zeta_n T_n s + 1}, T_n = 0.05, \zeta_n = 0.05 \tag{57}$$

$$K_n^3(s) = k_n^3 \cdot (K_n^1(s) + K_n^2(s)) \tag{58}$$

with $k_n^i, i = 1,2,3$ chosen such that $\operatorname{var} n(t) = \sigma_n^2$. The model in eq. (56) produces a wide-band noise, the one in eq. (57) a narrow band, while the model in eq. (58) a mixed character one. Spectral density characteristics of $K_n(s)$ and $K_d(s)$) are presented in Fig. 3.

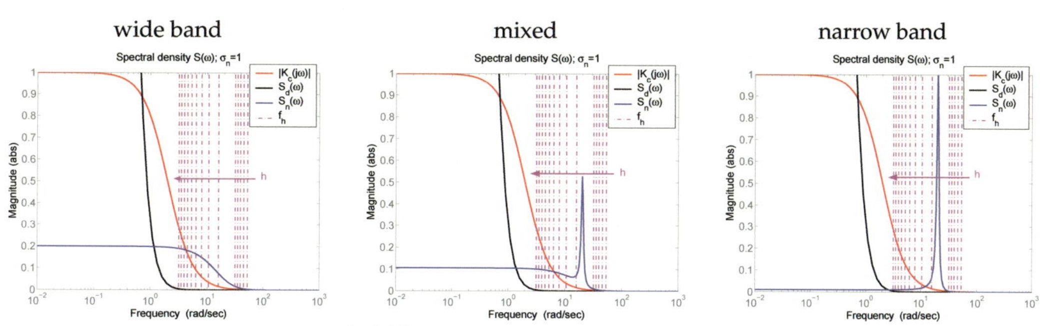

Fig. 3. Spectral density for std $\{n(t)\} = 1.0$

4.1 Open-loop results

The effect of Butterworth filter compared with continuous-time Kalman filter in the pure signal processing context is presented in Fig. 4a - b for a wide-band noise. In Fig. 4a it is clearly seen, that for small level of noise the only result is that filtration error increases with increasing sampling period h. This is due to the signal deformation caused by filtering. At high noise levels there are two effects: decreasing influence of noise with increasing sampling period accompanied by increasing deformation of the useful signal. This situation becomes greatly improved when Butterworth filter is followed by a discrete-time Kalman filter of (47)-(48), see Fig. 4b. In this figure we have $\text{std}\,(\Delta d^*) = \lim_{i\to\infty} \text{std}\,\{\Delta d^*(i)\}$, where $\Delta d^*(i)$ is the difference between actual value d_i and a sample s_i, and $\text{std}\,(\Delta s) = \lim_{i\to\infty} \text{std}\,\{\Delta s(i)\}$, where $\Delta d(i) = d_i - \hat{d}_{i|i}$ is the difference between d_i and its estimate $\hat{d}_{i|i}$ produced by the discrete-time Kalman filter These phenomena will play important role in the control context in closed loop.

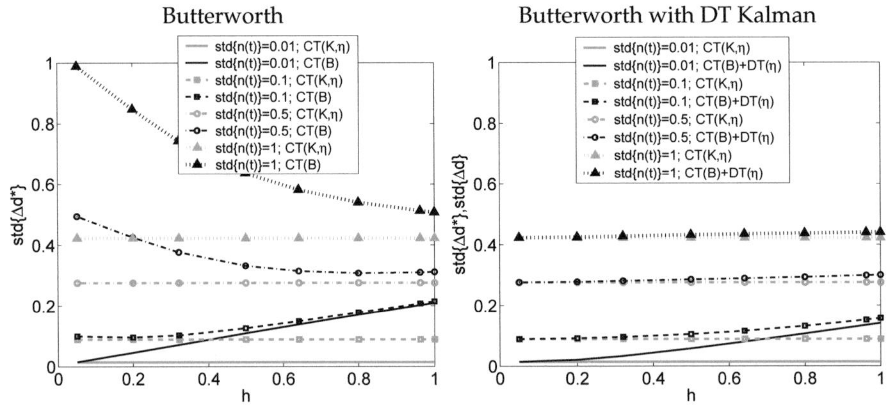

Fig. 4. Wide-band noise filtering results: CT Butterworth filter and CT Butterworth with DT Kalman compared with CT Kalman filter

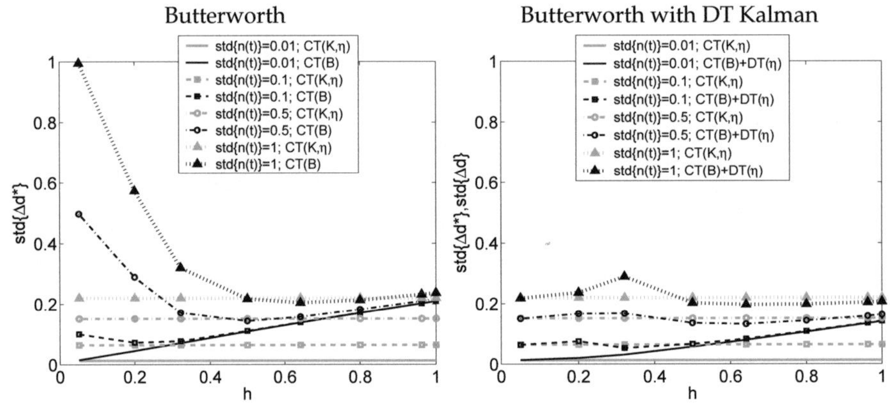

Fig. 5. Narrow-band noise filtering results: CT Butterworth filter and CT Butterworth with DT Kalman compared with CT Kalman filter

4.2 Closed-loop results

The results for PID QDR, optimal PID and LQG controlled systems are presented in figure Fig. 6 as functions of the sampling period h. The main conclusion is that all control systems

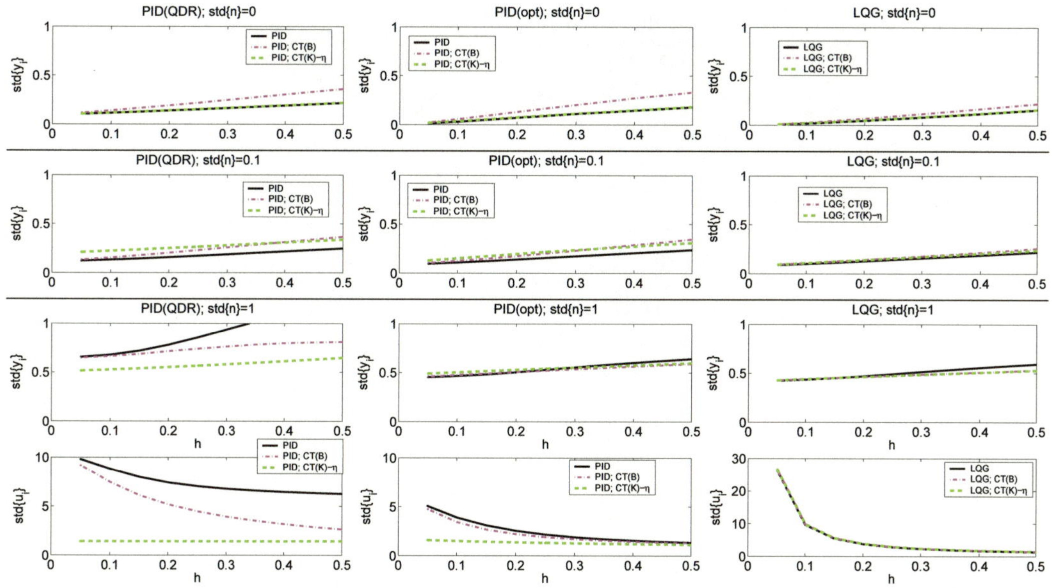

Fig. 6. Control errors and control efforts as functions of h for various noise magnitudes

behave worse when the anti-aliasing filter is used in the noiseless case. This is also true in the case of small noise level and PID controllers.

In contrast to the LQG control, the continuous-time Kalman filter does not help either. Very small improvement is attained in MV LQG system at very high noise level and longer sampling periods. The characteristic feature of MV LQG is that the control magnitudes do not depend on the type of filter used.

The improvement in terms of output variance is better visible in the case of PID controllers. Systems with Kalman filter behave then better in wide range of sampling instants.

Rather large improvement is seen, however, in terms of control signal magnitudes. It does not depend practically on sampling period in the case of CT Kalman filter, and tends to it with increasing sampling period in the case of Butterworth filter.

Selected results for PID and LQG controllers with parameters collected in Table 2 are illustrated in Fig.7 on the plane std$\{u\}$–std$\{y\}$ for $h = 0.2$. It is readily seen that analog filtering makes restricted sense only for PID controllers with QDR tuning and high noise level. Unfortunately the quality of control remains then very poor, even if the continuous-time Kalman filter is applied as analog filter. Application of optimally tuned PID controllers leads to an even more surprising result: from figure Fig.7 it is seen that even at large noise level very good results close to the LQG benchmark can be obtained without any analog filter.

In Fig.7the results are plotted on the plane std$\{u\}$–std$\{y\}$ for various values of h, showing again that the use of anti-aliasing filter makes no sense, and that the quality of disturbance attenuation of optimally tuned PID controllers is very similar to that of MV LQG controller. Unfortunately, Nyquist plots of a series connection of the plant and the controller depicted in Fig.8 show that PID systems are less robust than the MV LQG ones. Moreover, the usage of anti-aliasing filters makes this even worse.

	QDR	std $\{y_i\}$	OPTIMAL	std $\{y_i\}$
PID	$k_P = 2.8146$ $T_I = 0.7045$ $T_D = 0.1761$	0.78	$k_P = 0.9383$ $T_I = 0.9647$ $T_D = 0.2199$	0.50
PID;B	$k_P = 2.2328$ $T_I = 0.8843$ $T_D = 0.2211$	0.71	$k_P = 0.9293$ $T_I = 0.9486$ $T_D = 0.2427$	0.50
PID;K	$k_P = 1.8621$ $T_I = 1.6319$ $T_D = 0.4080$	0.55	$k_P = 1.4118$ $T_I = 1.5648$ $T_D = 0.6619$	0.53

Table 2. QDR PID & Optimal PID controller settings for std $\{n\} = 1$ and $h = 0.2$

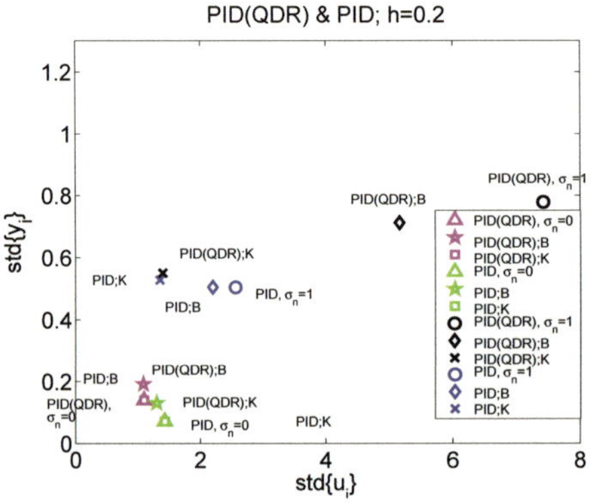

Fig. 7. PID QDR & optimal PID controller results, for $h = 0.2$ with std $\{n(t)\} = 0$ and std $\{n(t)\} = 1$

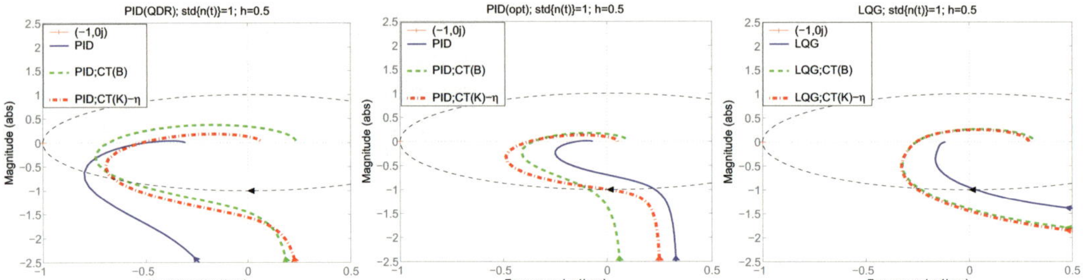

Fig. 8. Nyquist plots and robustness of various control systems

Influence of sampling period and noise character is further studied in figures Fig.9 - Fig.14

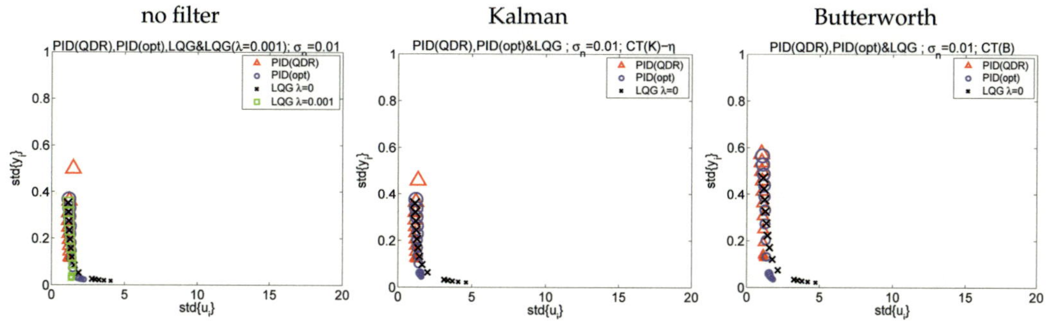

Fig. 9. Negligible noise level results as functions of h, std $\{n_i\} = 0.01$

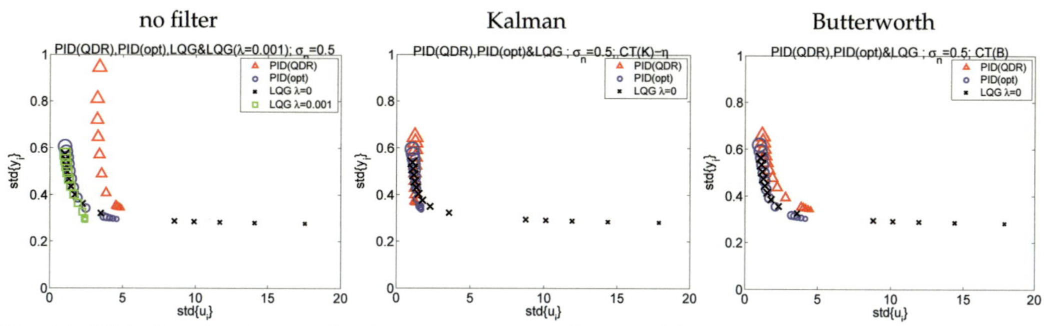

Fig. 10. Wide-band noise results for various controllers and filters as functions of h

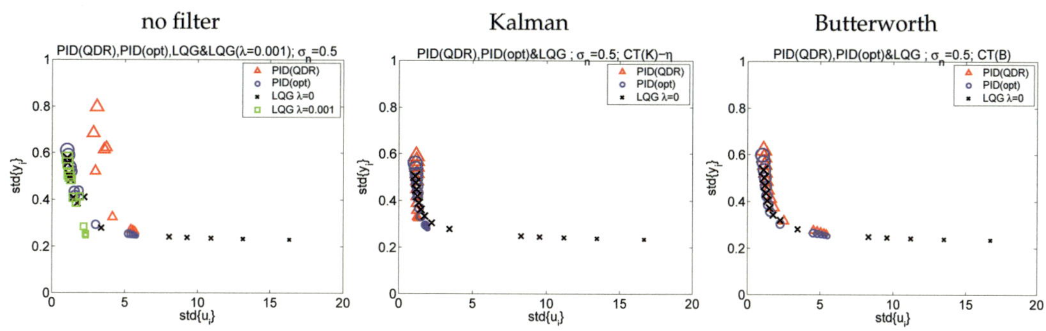

Fig. 11. Mixed-band noise results for various controllers and filters as functions of h

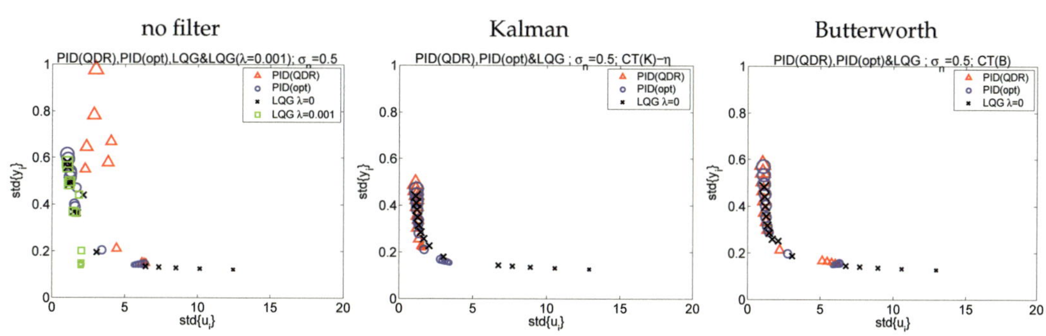

Fig. 12. Narrow-band noise results for various controllers and filters as functions of h

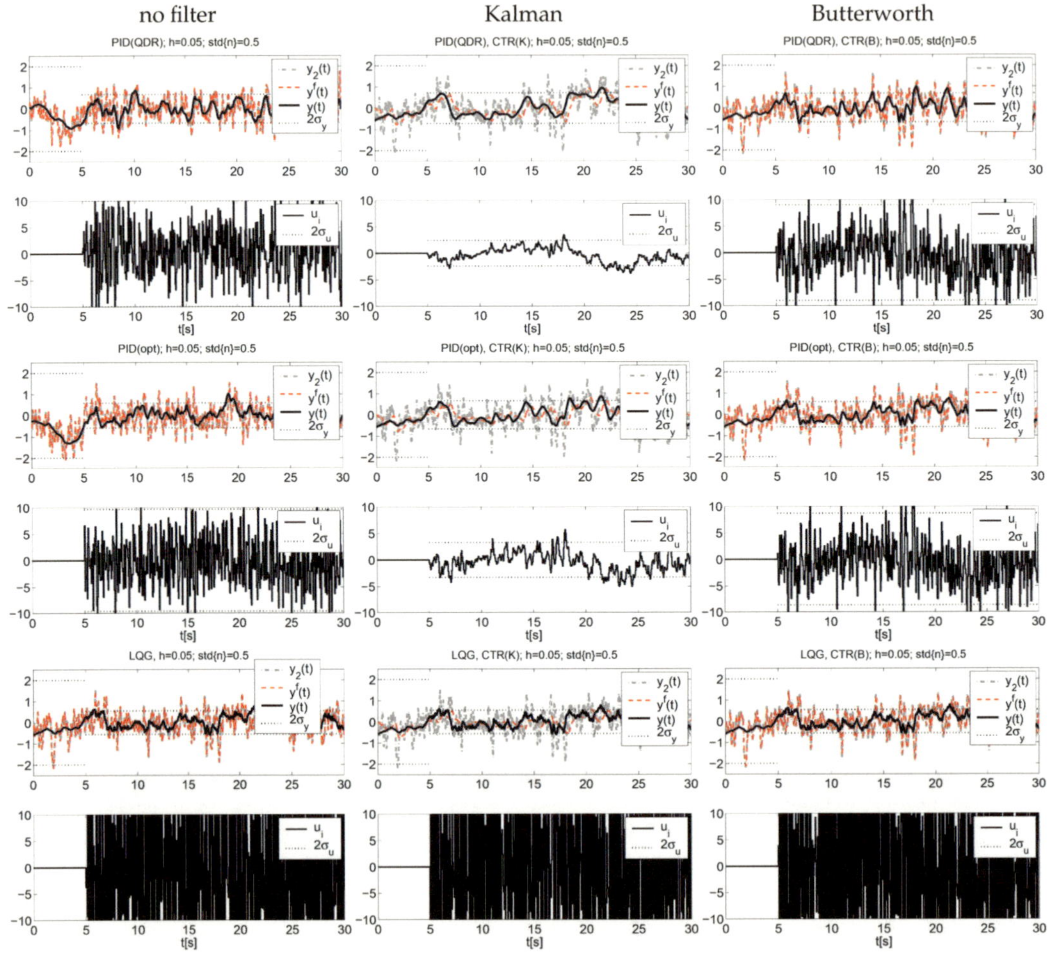

Fig. 13. Wide-band noise: realizations of output and control signals

5. Conclusion

It has been shown that the use of anti-aliasing filters is not justified in sampled-data MV LQG and PID control systems with noiseless measurements, or when the level of noise is small. Certain improvement can be made in the case of PID control systems with QDR and optimal settings in terms of both, output signal and control signal variance, in the case of large level of noise. However, continuous-time Kalman filter is then much better in the wide range of sampling periods, while the effect of Butterworth filter becomes better with increasing sampling period. Unfortunately the usage of any analog filters deteriorates the robustness of control systems. This makes the claim of uselessness of anti-aliasing filters even stronger.

Optimal tuning of PID controllers that takes the disturbance and noise parameters into account leads to the results comparable with those of LQG controllers without any analog prefilters. (Goodwin et al., 2001)

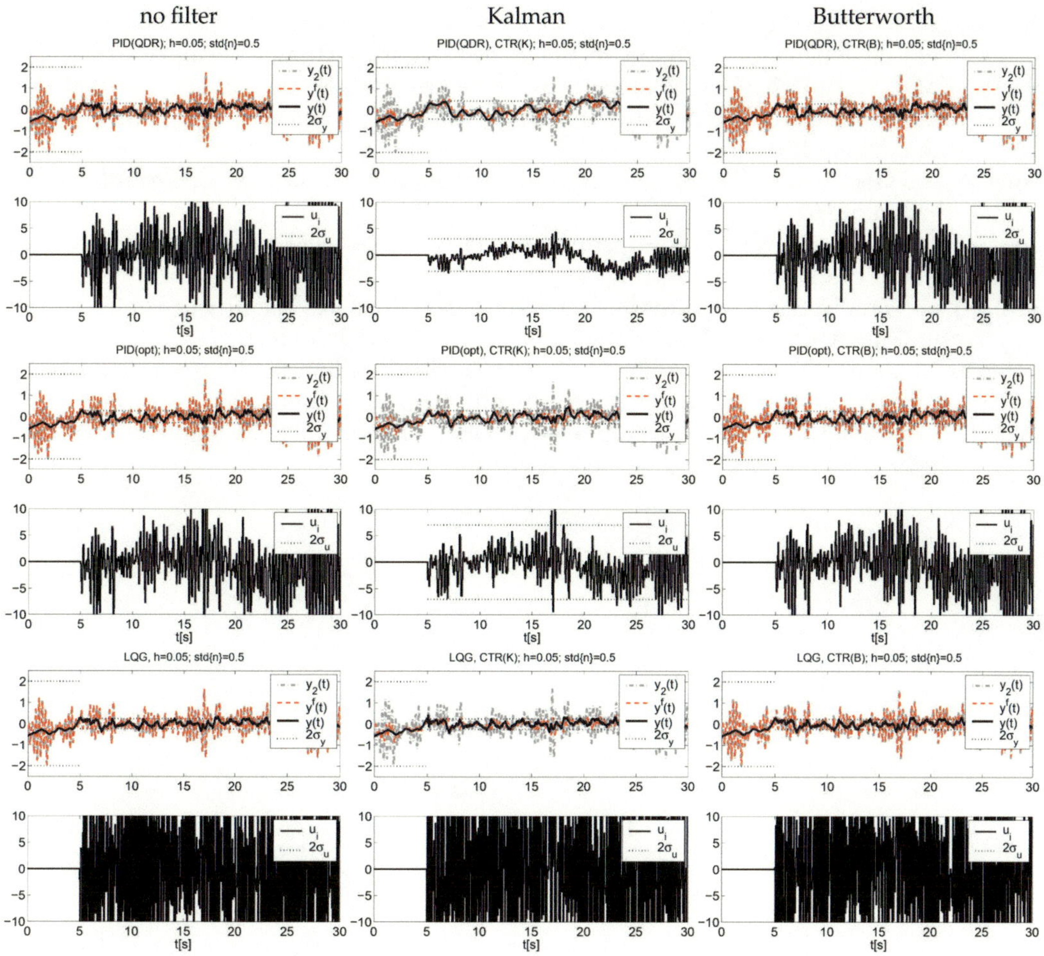

Fig. 14. Narrow-band noise: realizations of output and control signals

6. References

Anderson, B.D.O. and Moore, J.B. (1979). *Optimal Filtering*, Prentice Hall, Inc., Englewood Cliffs, New Jersey, .

Åström, K. and Wittenmark, B. (1997). *Computer–Controlled Systems*, Prentice Hall, 1997.

Blachuta, M. J., Grygiel, R. T. (2008a). Averaging sampling: models and properties. *Proc. of the 2008 American Control Conference*, pp. 3554-3559, Seattle USA, June 2008.

Blachuta, M. J., Grygiel, R. T. (2008b). Sampling of noisy signals: spectral vs anti-aliasing filters, *Proc. of the 2008 IFAC World Congress*, pp. 7576-7581, Seul Korea, July 2008.

Blachuta, M. J., Grygiel, R. T. (2009a). On the Effect of Antialiasing Filters on Sampled-Data PID Control, *Proc. of 21th Chinese Conference on Decision and Control*, Guilin China, June 2009.

Blachuta, M. J., Grygiel, R. T. (2009b). Are anti-aliasing filters really necessary for sampled-data control? *Proc. of the 2009 American Control Coference*, pp. 3200-3205, St Louis USA, June 2009.

Blachuta, M. J., Grygiel, R. T. (2009c). Are anti-aliasing filters necessary for PID sampled-data control? *Proc. of European Control Conference*, Budapest Hungary, August 2009.

Blachuta, M. J., Grygiel, R. T. (2010). Impact of Anti-aliasing Filters on Optimal Sampled-Data PID Control. *Proc. of 8th IEEE International Conference on Control & Automation*, Xiamen China, June 2010.

Feuer, A. and Goodwin, G. (1996). *Sampling in Digital Signal Processing and Control*. Birkhäuser Boston, 1996.

Goodwin, G.C.; Graebe S.F.; and Salgado M.F. (2001). *Control System Design*. Prentice Hall, 2001.

Jerri, A.J. (1977). The Shannon sampling theorem - its variuos extensions and applications: a tutorial review. *Proc. IEEE*, Vol.(65), 1977, pp. 1656-1596

Steinway, W.J. and Melsa, J.L. (1971). Discrete Linear Estimation for Previous Stage Noise Correlation. *Automatica*, Vol. 7, pp. 389-391, Pergamin Press, 1971.

Shats, S. and Shaked U. (1989). Exact discrete-time modelling of linear analogue system. *Int. J. Control*, Vol. 49, No. 1, pp 145-160, 1989.

Part 2

PID Tuning

7

Multi-Loop PID Control Design by Data-Driven Loop-Shaping Method

Masami Saeki and Ryoyu Kishi
Department of Mechanical System Engineering, Hiroshima University,
1-4-1 Higashi-Hiroshima, 739-8527
Japan

1. Introduction

In the analysis and synthesis of control systems, model-based design methods are standard and powerful. However, the plant property is wide-ranging, and the identification of the mathematical model requires much effort and expert knowledge. Since the purpose of control design is to find a controller that optimizes a performance index using plant responses and other preliminary knowledge, a mathematical model is not necessarily required for design though it is very useful. We consider that essential merit of this data-driven design approach lies in the fact that the controller structure is known completely, whereas it is impossible to identify the plant structure without uncertainties.

Design methods that satisfy the following conditions are considered to be more user friendly. a) Plant responses used for design can be obtained in the normal plant operation. b) Not so many plant responses are required for design. For example, a few step responses may be desirable, preferably in the closed-loop operation. c) The design method is applicable to various plants by tuning one or two design parameters. d) The parameter value of the design specification has absolute meaning for control performance. Namely, it is desirable to be plant independent.

Recently, there have been two major data-driven approaches proposed. One is the iterative feedback tuning (IFT) (Hjalmarsson et al. (1999); Lequin et al. (2003)). Since IFT requires special experiments to get the plant responses iteratively, it does not satisfy the requirements a) and b). The other is the virtual reference feedback tuning (VRFT) (Campi et al. (2002)). VRFT is based on model matching, and the controller that gives a desired closed-loop transfer function is sought. We consider that VRFT almost satisfies a), b), c). Since preliminary knowledge is necessary to give an adequate and realizable target closed-loop transfer function, the requirement d) is not satisfied. VRFT is suitable for those problems where the target closed-loop transfer function can be given or easily found from some preliminary knowledge.

In the classical control and robust control, loop-shaping is recognized as a very practical and useful design specification (Skogestad & Postlethwaite (2007)). PID controller is widely used for the industrial plants and the tuning of the PID gains is easier compared with other controllers (Åström & Hägglund (1995)). Therefore, we have developed a data driven method for the mixed sensitivity control problem of PID control (Saeki (2004a), Saeki et al. (2006)) based on unfalsified control (Safonov & Tsao (1997)). After this, we found a simpler problem setting for PI control in the reference (Åström et al. (1998)), where the integral gain of PI controller is maximized subject to the maximum sensitivity condition and this problem is treated on the frequency domain. Since this problem setting and the solutions satisfy c) and d), we have studied a data-driven method for this problem in order to develop a method that satisfies all the requirements. This problem can be considered as a loop shaping problem, which will be explained in Section 2.

The basic idea of unfalsified control is to remove the controllers from the candidate controllers if they do not satisfy the design specification for given plant responses, and to apply an unfalsified controller to the plant. We have examined application of this idea to robust control design. Since we found by simulation that the falsification condition of an L_2 gain performance index cannot efficiently falsifies the controllers by a single plant response, we proposed a method of generating many virtual responses by filtering the measured data with many bandpass filters (Saeki et al. (2006)). We have obtained a data-driven method that almost satisfies a) and b) for a single-input single-output plant (Saeki (2008)). We refer to this method as the data driven loop shaping method (DDLS).

In this paper, we will study an extension of DDLS to multi-loop PID control, and we will examine the possibility of this approach because the design of multi-loop PID control systems is much harder than that of single-input single-output plants (Johnson & Moradi (2005)). A design problem is formulated in Section 2, the constraints on PID gains are derived in Section 3, and a method of generating plant response data and the design procedure are explained in Section 4. A numerical example for a two-input two-output time-delay plant is shown in Section 5, and an experimental result for a two-rotor hovering system is shown in Section 6.

For signals $w(t) \in R^n, v(t) \in R^n, t \in [0, \infty)$, we use the following notations. $\|w\|_2 = \sqrt{\int_0^\infty w(\tau)^T w(\tau) d\tau}$, $\|w\|_{2T} = \sqrt{\int_0^T w(\tau)^T w(\tau) d\tau}$, $\langle w, v \rangle = \int_0^\infty w(\tau)^T v(\tau) d\tau$, $\langle w, v \rangle_T = \int_0^T w(\tau)^T v(\tau) d\tau$. Denote the (i, j)-element of a matrix A as $[A]_{ij}$ and the ith-element of a vector b as $[b]_i$.

2. Problem setting

Let us consider the feedback system described by

$$y = Pe \tag{1}$$

$$e = w - u \tag{2}$$

$$u = Ky \tag{3}$$

where $y, e, u, w \in R^m$. The plant P is linear time-invariant and m-input and m-output. K is a multi-loop PID controller given by

$$K(s) = K_P + K_I \frac{1}{s} + K_D s \tag{4}$$

where K_P, K_I, K_D are constant diagonal matrices. We will use the notation $\hat{K} = [K_P, K_I, K_D]$. Since we are considering a data-driven method, we assume that a few input-output responses of the plant, $e(t), y(t)$, are given in the finite interval $t \in [0, T]$, where the plant is at the steady state at $t = 0$, i.e., $e(t) = 0, y(t) = 0, t < 0$. If $e(t) = e(0) \neq 0, y(t) = y(0) \neq 0$ for $t < 0$, the bias must be eliminated by $e(t) - e(0), y(t) - y(0)$. These data will be used for design.

The sensitivity and complementary sensitivity functions at the plant input are denoted by

$$S_I = (I + KP)^{-1} \tag{5}$$

$$T_I = (I + KP)^{-1} KP \tag{6}$$

For this system, following properties are known.

a) The maximum sensitivity defined by

$$M_s = \max_{0 \leq \omega < \infty} \sigma_{max} \{S_I(j\omega)\} \tag{7}$$

is a useful measure for stability margin, and the typical values of M_s are in the range of 1.2 to 2. This condition is represented by

$$\sigma_{max} \{S_I(j\omega)\} < \gamma_1, \ \ \omega \in R, \gamma_1 \in [1.2, 2] \tag{8}$$

In the time domain, this is equivalent to the L_2-gain condition;

$$\|e\|_2 < \gamma_1 \|w\|_2 \tag{9}$$

for all $w \in L_2$ and $e = S_I w$.

b) A robust stability condition is given by

$$\sigma_{max} \{T_I(j\omega)\} < \gamma_2, \ \ \omega \in R \tag{10}$$

, which is equivalent to the L_2-gain condition;

$$\|u\|_2 < \gamma_2 \|w\|_2 \tag{11}$$

for all $w \in L_2$ and $u = T_I w$.

c) Let $y_i(t)$ be the response for a step disturbance $w_i(t) = 1$ and $w_j(t) = 0, \ j \neq i$. Then the intergal of $y_i(t)$ satisfies

$$\int_0^\infty y_i(\tau) d\tau = \frac{1}{[K_I]_{ii}} \tag{12}$$

From this property, disturbance attenuation is attained by making $|[K_I]_{ii}|$ larger for $i = 1, 2, \cdots, m$. We formulate the plant description so that $[K_I]_{ii} > 0, \ i = 1, 2, \cdots, m$ can be a necessary condition for the closed-loop stability, and, for this system, we adopt the next performance index to measure the largeness of K_I.

$$J = \sum_{i=1}^{m} [K_I]_{ii} \tag{13}$$

d) When $\sigma_{min} \{K_I P(0)\} \neq 0$, the next approximation is satisfied at low frequencies.

$$S_I(j\omega) \approx j\omega (K_I P(0))^{-1} \tag{14}$$

In this paper, we will study a maximization problem of the integral gain of the PID controller under the maximum sensitivity condition and, if necessary, the robust stability condition. From the above properties a), b), and c), this problem is considered as a disturbance attenuation problem with adequate stability margin. This is also considered as a loop shaping problem. Namely, from the properties a) and d), if $\sigma_{min} \{K_I P(0)\} \neq 0$, the system has a loop shape illustrated in Fig. 1. By substituting (14) into $\sigma_{max} \{S_I(j\omega)\} < 1, \ \omega < \sigma_{min} \{K_I P(0)\}$. Therefore, the control bandwidth is estimated by $\sigma_{min} \{K_I P(0)\}$, which can be made larger by making K_I larger.

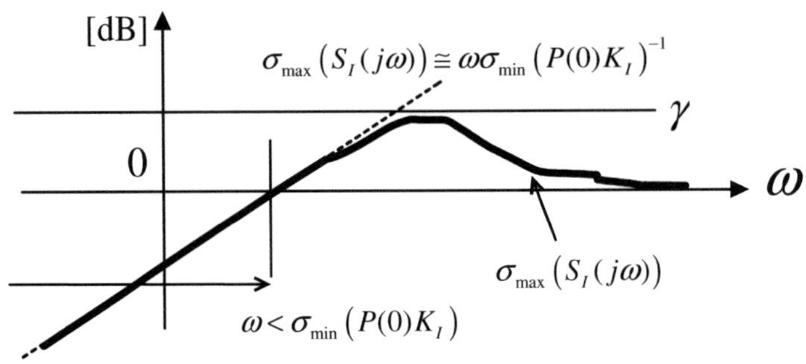

Fig. 1. Loop shaping for the sensitivity function

Lemma 1(Vidyasagar (1993)) Suppose that the system satisfies causality and it is in the steady state at $t = 0$. Then, if (9) is satisfied,

$$\| e \|_{2T} < \gamma_1 \| w \|_{2T} \tag{15}$$

for $T > 0$. Similarly, if (11) is satisfied,

$$\| u \|_{2T} < \gamma_2 \| w \|_{2T} \tag{16}$$

for $T > 0$.
Compared with (9), the merit of the condition (15) is that it can be calculated for the finite length data $e(t), y(t), t \in [0, T]$, and the demerit is that (15) is only a necessary condition for (9). Since we can only use the finite length data, we will use the condition (15) instead of (9). The same idea is applied to (11) and (16). In this paper, we will examine the next design problem.
Loop-shaping problem: For the feedback system (1)-(3), find a PID controller that maximizes J defined by (13) subject to

$$\|e\|_{2T} < \gamma_1 \|w\|_{2T} \tag{17}$$

$$\|u\|_{2T} < \gamma_2 \|w\|_{2T} \tag{18}$$

for sufficiently many disturbances $w = w_i \in L_{2e},\ i = 1, 2, \cdots, N$.
In this problem setting, it is ideal to test the constraints for all $w \in L_{2e}$, but practically we can only generate a finite number of disturbances from the measured data $e(t), y(t)$ in the following discussion. Therefore, the number of w is finite in the above problem setting.

3. Derivation of convex constraints on PID gains

3.1 Derivation of a constraint from (17)

From (17),

$$\langle w, w \rangle_T > \frac{1}{\gamma_1^2} \langle e, e \rangle_T \tag{19}$$

The disturbance w that gives the plant response e, y is given by

$$w(t) = e(t) + u(t) \quad (20)$$
$$u(t) = K_P y(t) + K_I y_I(t) + K_D y_D(t) \quad (21)$$

where $y_I(t) = \int_0^t y(\tau)d\tau$ and $y_D(t) = \frac{dy}{dt}(t)$. Substitution of (20) into (19) gives

$$\langle e+u, e+u\rangle_T > \frac{1}{\gamma_1^2}\langle e, e\rangle_T \quad (22)$$

This is a concave constraint on the PID gains. Next, we will derive a linear constraint from (22) as a sufficient condition. From

$$\langle u-u_0, u-u_0\rangle_T \geq 0 \quad (23)$$

for any $u_0(t)$, a sufficient condition for (22) is given by

$$\langle e+u, e+u\rangle_T > \frac{1}{\gamma_1^2}\langle e, e\rangle_T + \langle u-u_0, u-u_0\rangle_T \quad (24)$$

By expanding this,

$$\langle e+u_0, u\rangle_T > b \quad (25)$$

where

$$b = \frac{1}{2}\left\{\left(\frac{1}{\gamma_1^2} - 1\right)\langle e, e\rangle_T + \langle u_0, u_0\rangle_T\right\} \quad (26)$$

This is a linear constraint on the PID gains. Thus, we have the next lemma by substituting (21) into (25).

Lemma 2 If the next linear constraint on the PID gains is satisfied for a data $e(t), y(t)$, (17) is also satisfied for the same data.

$$\sum_{i=1}^{m}\{a_{Pi}[K_P]_{ii} + a_{Ii}[K_I]_{ii} + a_{Di}[K_D]_{ii}\} > b \quad (27)$$

where $a_{Pi} = \langle [e+u_0]_i, [y]_i\rangle_T$, $a_{Ii} = \langle [e+u_0]_i, [y_I]_i\rangle_T$, and $a_{Di} = \langle [e+u_0]_i, [y_D]_i\rangle_T$.

The linear constraint (27) is satisfied for any u_0, but u_0 should be chosen so that the gain set defined by the constraint may contain the set of stabilizing PID gains. We assume that a stabilizing PID gain $\hat{K} = \hat{K}_a$ is given. Denote $u(t)$ of (21) as $u_a(t)$ for $\hat{K}_a$ and further assume that (22), which is equivalent to (17), is satisfied for $u(t) = u_a(t)$.

The set of u that satisfies (22) corresponds to the outside region of the sphere with center $-e$ and radius $\|e\|_2/\gamma_1$ as illustrated in Fig. 2. This set is concave and u_a lies outside the sphere by assumption. Let u_0 be the intersection of the segment that connects $-e$ and u_a and the sphere. We consider that the sphere is approximated by the plane which touches the sphere at the point u_0 as illustrated shown in Fig. 2. Note that this convex set determined by the plane is described by (27) with this u_0.

Let us calculate u_0. The segment is described by

$$u = qu_a + (1-q)(-e), \ \ 0 \leq q \leq 1. \quad (28)$$

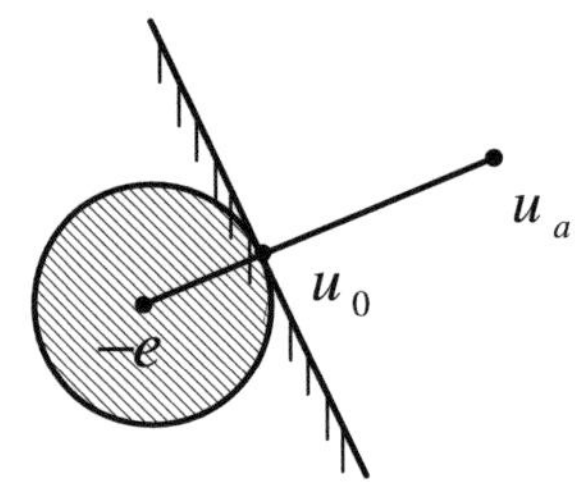

Fig. 2. Approximation of the concave region by plane

By substituting this into (22),

$$q^2 \langle e+u_a, e+u_a \rangle_T > \frac{1}{\gamma_1^2} \langle e, e \rangle_T . \tag{29}$$

From this, the minimum value of q is found to be

$$q_0 = \frac{1}{\gamma_1} \frac{\|e\|_{2T}}{\|u_a + e\|_{2T}}, \tag{30}$$

and

$$u_0 = q_0 u_a - (1 - q_0)e. \tag{31}$$

From the above derivation, we have the next lemma.
Lemma 3 The stabilizing gain K_a satisfies the linear constraint (27) for u_0 that is given by (30) and (31).
The above discussions are summarized as the next theorem.
Theorem 1 Suppose that a data $e(t), y(t), t \in [0, T]$ and a stabilizing PID gain K_a that satisfies (17) are given. The linear constraint (27) with u_0 given by (30) and (31) is a sufficient condition for (17), and the linear constraint is satisfied for the stabilizing PID gain.

3.2 Derivation of a constraint from (18)

By substituting (20) into (18),

$$\frac{1}{\gamma_2^2} \langle u, u \rangle_T > \langle e+u, e+u \rangle_T \tag{32}$$

By expanding this,

$$(1 - \frac{1}{\gamma_2^2}) \langle u, u \rangle_T + 2\langle e, u \rangle_T + \langle e, e \rangle_T > 0 \tag{33}$$

We will derive convex constraints from (33), where three cases are considered depending on the value of γ_2.
If $\gamma_2 = 1$, (33) becomes

$$2\langle e, u \rangle_T + \langle e, e \rangle_T > 0 \tag{34}$$

From this inequality, a linear constrains on PID gains can be derived immediately. Namely,

$$2\sum_{i=1}^{m} P_i x_i + c > 0 \tag{35}$$

where

$$x_i = \left[\ [K_P]_{ii} \quad [K_I]_{ii} \quad [K_D]_{ii}\ \right]^T,$$
$$P_i = [\langle [e]_i, [y]_i\rangle_T, \langle [e]_i, [y_I]_i\rangle_T, \langle [e]_i, [y_D]_i\rangle_T],$$
$$c = \langle e, e\rangle_T$$

If $\gamma_2 > 1, 1 - 1/\gamma_2^2 > 0$ and (33) can be represented as

$$\langle u, u\rangle_T + (\frac{2\gamma_2^2}{\gamma_2^2 - 1})\langle e, u\rangle_T + (\frac{\gamma_2^2}{\gamma_2^2 - 1})\langle e, e\rangle_T > 0 \tag{36}$$

Further, by denoting $\tilde{e} = (\gamma_2^2/(\gamma_2^2 - 1))e$, this inequality can be represented as

$$\langle u, u\rangle_T + 2\langle \tilde{e}, u\rangle_T + \langle \tilde{e}, \tilde{e}\rangle_T > \frac{1}{\gamma_2^2}\langle \tilde{e}, \tilde{e}\rangle_T \tag{37}$$

Since this condition has the same form as (22), Theorem 1 with e replaced with $\tilde{e}$ is satisfied. If $\gamma_2 < 1$, (33) is a convex constraint and represented as

$$\left(1 - \frac{1}{\gamma_2^2}\right)\sum_{i=1}^{m} x_i^T Q_i x_i + 2\sum_{i=1}^{m} P_i x_i + c > 0 \tag{38}$$

where $Q_i = Q_i^T$ and

$$Q_i = \begin{bmatrix} \langle [y]_i, [y]_i\rangle_T & \langle [y]_i, [y_I]_i\rangle_T & \langle [y]_i, [y_D]_i\rangle_T \\ & \langle [y_I]_i, [y_I]_i\rangle_T & \langle [y_I]_i, [y_D]_i\rangle_T \\ * & & \langle [y_D]_i, [y_D]_i\rangle_T \end{bmatrix}$$

By representing Q_i by the singular value decomposition form $Q_i = U_{i1}^T \Sigma_i U_{i1}$ where $\Sigma_i > 0$ and applying Schur complement, the next LMI (linear matrix inequality) with respect to x_i's is obtained.

$$\sum_{i=1}^{m} \begin{bmatrix} P_i x_i + x_i^T P_i^T + \frac{1}{m}c & x_i^T U_{i1}^T \\ U_{i1} x_i & \frac{\gamma_2^2}{1-\gamma_2^2}\Sigma_i^{-1} \end{bmatrix} > 0 \tag{39}$$

The above discussions are summarized as the following two theorems.

Theorem 2 Suppose that a data $e(t), y(t), t \in [0, T]$ is given. If $\gamma_2 = 1$, (18) is equivalent to the linear constraint (36). If $\gamma_2 < 1$, (18) is equivalent to the LMI constraint (39).

Theorem 3 Suppose that a data $e(t), y(t), t \in [0, T]$ and a stabilizing PID gain that satisfies (18) are given. If $\gamma_2 > 1$, the linear constraint (27) with u_0 given by (30) and (31), where e is replaced with $\tilde{e}$, is a sufficient condition for (18), and the linear constraint is satisfied for the stabilizing PID gain.

4. Data generation and design procedure

4.1 Data generation by filtering

Since the multi-loop PID controller contains many variables to be determined, many linear constraints are necessary for the determination. Since one linear constraint (27) is derived from one input-output response $e(t), y(t), t \in [0, T]$, many input output responses would be necessary.

In order to obtain the plant response $e(t)$ and $y(t)$, we may give the test input to $w(t)$ of the system (1)-(3) at the steady state, or to the reference $r(t)$ of the system described by

$$y = Pe \quad (40)$$

$$e = K(r - y). \quad (41)$$

Since the plant is m-input and m-output, m sets of responses $e(t)$ and $y(t)$ may be necessary at least. Therefore, we give a test input for the j th input $[w]_j$ or $[r]_j$ and measure the input-output response $\{e(t), y(t)\}$, which are denoted by e^j, y^j. By iterating this experiment m times, m sets of data $e^j, y^j,\ j = 1, 2, \ldots, m$ are obtained.

Next, we will generate many fictitious data $e^{ij}(t), y^{ij}(t), i = 1, 2, \ldots, n_F, j = 1, 2, \ldots, m$ by

$$e^{ij}(t) = F_i(s)e^j(t) \quad (42)$$

$$y^{ij}(t) = F_i(s)y^j(t), t \in [0, T] \quad (43)$$

where the filter $F_i(s)$ is a stable transfer function. Note that the notation $F_i(s)e^j(t)$ means that $F_i(s)$ filters each element of the m-dimensional vector $e^j(t)$.

From the assumptions that P is linear time-invariant and that the system is in the steady state at $t = 0$,

$$y^{ij}(t) = P(s)e^{ij}(t) \quad (44)$$

is satisfied. Namely, the data $e^{ij}(t), y^{ij}(t)$ can be considered as the input-output response of the plant.

Remark 1 Even if the condition that P is linear time-invariant is not assumed, the above loop shaping problem can be interpreted for a nonlinear plant as a problem with the weighted L_2 gain criterion given by

$$\|F_i(s)e\|_2 < \gamma_1 \|F_i(s)w\|_2, i = 1, 2, \ldots, n_F. \quad (45)$$

Namely, if a controller is falsified by the condition (17) for the filtered responses of a nonlinear plant, we can say that the controller is falsified by the criterion (45).

Remark 2 From the previous discussions, the L_2 gain constraint (17) is evaluated for the fictitious disturbances $w(t)$ given by (20), i.e. $w(t) = e(t) + Ky(t)$ for the data $e(t) = e^{ij}(t)$, $y(t) = y^{ij}(t), i = 1, 2, \ldots, n_F, j = 1, 2, \ldots, m$ and the number of disturbances is $N = n_F m$.

4.2 Filter selection

We use the next bandpass filters $F_i(s)$ for the sample frequencies $\omega_i, i = 1,\ 2,\ \cdots, n_F$.

$$F_i(s) = \hat{\psi}(s/\omega_i) \quad (46)$$

$$\hat{\psi}(s) = \left(\frac{s}{(s+\alpha)^2 + 1}\right)^4 \quad (47)$$

The gain plot of $\hat{\psi}(s)$ is shown in Fig. 3. Since the peak gain is taken at $\omega = \omega_i(1+\alpha^2)^{0.5}$, this filter can be used for extracting this frequency component.
Let us consider the filtering from the viewpoint of the wavelet transform (Addison (2002)). In the last decade, wavelet transform has become popular as a time-frequency analysis tool. Wavelet transform is useful to get important information regarding the frequency properties lies locally in the time-domain from the non-stationary signals e, y .
If we denote the impulse response of $F_i(s) = \hat{\psi}(s/\omega_i)$ as $L^{-1}\{F_i(s)\} = \omega_i\psi(\omega_i t)$, then the correspondence

$$a \leftrightarrow \frac{1}{\omega_i}, \ b \leftrightarrow t, \ -\phi(-t) \leftrightarrow \psi(t). \tag{48}$$

is satisfied between the filtering;

$$y_i(t) = F_i(s)y(t) \tag{49}$$

$$= \omega_i \int_0^t \psi(\omega_i(t-\tau))y(\tau)d\tau. \tag{50}$$

and the integral wavelet transform;

$$\left(W_\phi y\right)(b,a) = |a|^{-1}\int_{-\infty}^{\infty}\phi\left(\frac{\tau-b}{a}\right)y(\tau)d\tau. \tag{51}$$

The impulse response $\psi(t)$ of $\hat{\psi}(s)$ with $\alpha = 0.5$ is shown in Fig. 4, and the graph of $-\phi_{db10}(-t)$ is shown in Fig. 5 for the Daubechies wavelet "db10"$\phi_{db10}(t)$. From the uncertainty principle in the wavelet analysis, there is a trade-off between the time window and the frequency window. The time-frequency window can be tuned by the parameter α. $\alpha = 0.5$ is the value with which $\psi(t)$ can be close to $-\phi_{db10}(-t)$.
By the way, since $F_i(s)$ has four zeros at $s=0$, $F_i(s)e(t) = 0$ for $e(t) = a_0 + a_1 t + a_2 t^2 + a_3 t^3$. Namely, the output becomes zero for this class of smooth inputs. For step or ramp inputs, their time-derivatives have discontinuity and so we have nonzero outputs. For the response $e(t), y(t)$ shown in Fig. 6, the responses filtered by $F_i(s)$ are shown in Fig. 7.

4.3 Design procedure

Step 1 Measure the input output responses $e^j(t), y^j(t), t \in [0,T], j = 1,2,\ldots,m$ by exciting the system at the steady state. If the response has bias, eliminate it.

Step 2 Set $\omega_i, i = 1,2,\ldots,n_F$ as logarithmically equally spaced n_F points in the important frequency range for control. Generate the fictitious responses

$$e^{ij}(t), y^{ij}(t), t \in [0,T], i = 1,2,\ldots,n_F. \tag{52}$$

from $e^j(t), y^j(t), t \in [0,T], j = 1,2,\ldots,m$ by (42) and (43). Set the value of γ_1. Set the value of γ_2 if necessary.

Step 3 Give a stabilizing PID gain $\hat{K}_a$ that satisfies (17) and (18) for γ_1 and γ_2. Then, compute the constraints on the PID gains for the n_F set of responses $e^{ij}(t), y^{ij}(t)$ following Theorems 1, 2, 3.

Step 4 If (17) is only considered as the constraints, solve a linear programming problem of maximizing J subject to (13) and the linear constrains on the PID gains. Otherwise, if both (17) and (18) are considered, solve an LMI problem of maximizing J defined by (13) and the linear constrains on the PID gains.

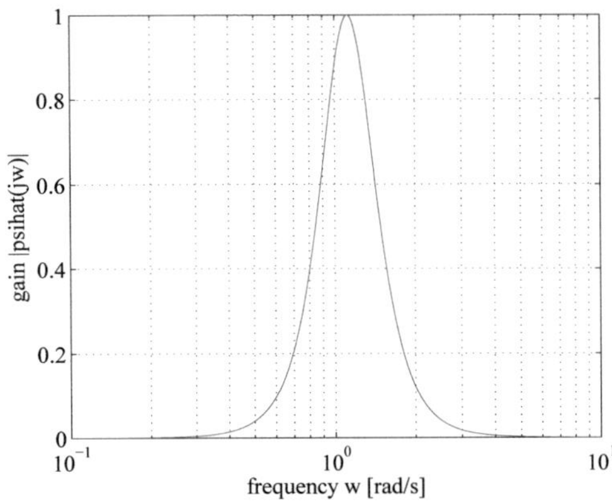

Fig. 3. Gain plot of $\hat{\psi}(j\omega)$

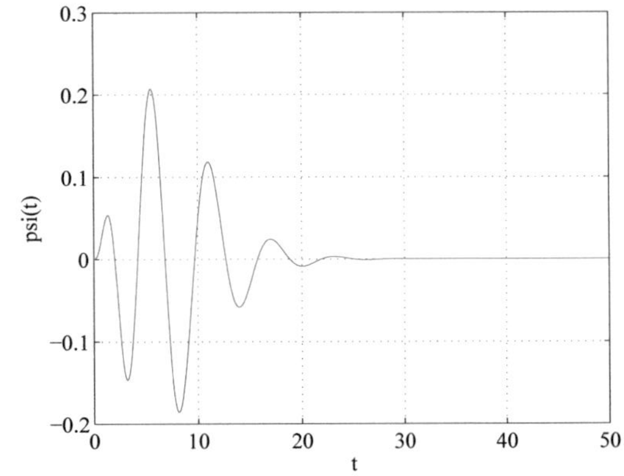

Fig. 4. Impulse response $\psi(t)$ for $\sigma = 0.5$

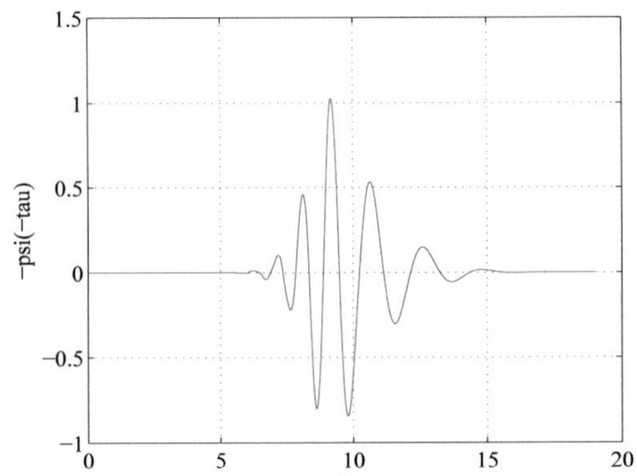

Fig. 5. Mother wavelet db10 $y = -\phi_{db10}(-\tau)$

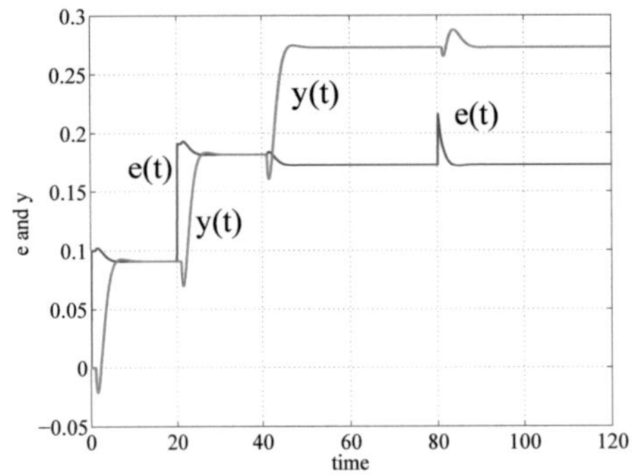

Fig. 6. $e(t)$ and $y(t)$

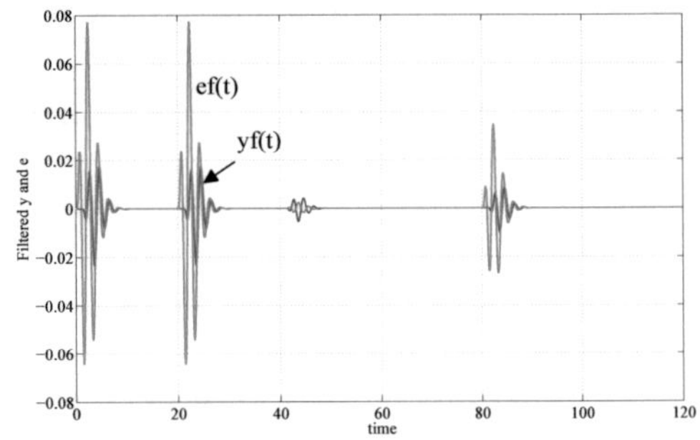

Fig. 7. $e_f(t) = F_i(s)e(t), y_f(t) = F_i(s)y(t)$

Step 5 Implement the PID controller.

If the plant is stable, a low gain P or PD controller is usually a stabilizing PID gain $\hat{K}_a$ that satisfies (17) and (18) in Step 3. However, if the plant is marginally stable or unstable, it may be not so easy to find such a stabilizing gain.

5. A numerical examples for a plant with time-delay

Let us consider the feedback system described by (40)(41), where the plant transfer function is given by

$$P(s) = \begin{bmatrix} \frac{12.8}{1+16.7s}e^{-s} & \frac{18.9}{1+21s}e^{-3s} \\ \frac{6.6}{1+10.9s}e^{-7s} & \frac{19.4}{1+14.4s}e^{-3s} \end{bmatrix}. \tag{53}$$

This transfer function is obtained from that of the Wood and Berry's binary distillation column process (Wood & Berry (1973)) by changing the sign of the $(1, 2)$ and $(2, 2)$ elements so that the plant may be stabilized by positive $K_I(1)$ and $K_I(2)$. Therefore, a solution for the Wood and Berry's binary distillation column process can be obtained by changing the sign of the second PI controller designed by our method.
First, we will get the plant responses with a stabilizing controller $K(s) = 0.1I_2$. Measurement noises with zero mean values and variances 0.0001 are given at the output y_1 and y_2 in the closed-loop operation, respectively. Fig. 8 shows the response $e(t)$ and $y(t)$ for the reference input $r_1(t) = 1, r_2(t) = 0$, and Fig. 9 for $r_1(t) = 0, r_2(t) = 1$.

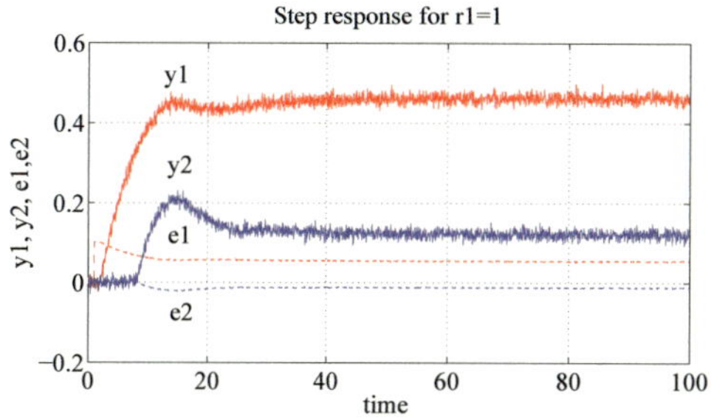

Fig. 8. Inputs and outputs of the plant for $r_1(t) = 1$ with $K = 0.1I_2$

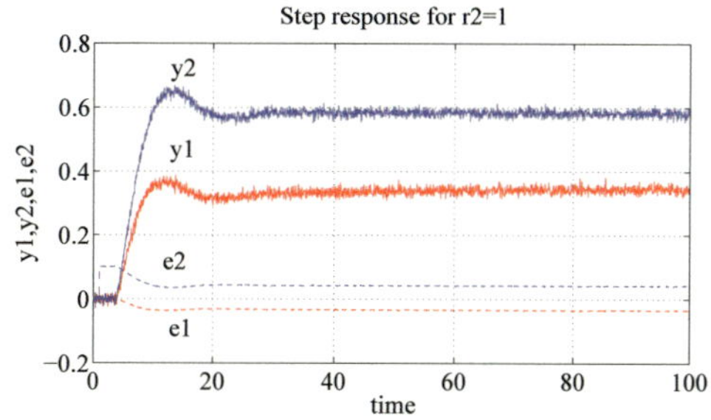

Fig. 9. Inputs and outputs of the plant for $r_2(t) = 1$ with $K = 0.1I_2$

Now, design a diagonal PI controller using these step response data. We will only consider the main constraint (17), and hence a solution can be obtained by applying linear programming. We set $\gamma_1 = 1.5$ and $\omega_i, i = 1, 2, \ldots, 40$ logarithmically equally spaced frequencies between $0.1[rad/s]$ and $10[rad/s]$, and give the bandpass filters by (46). The derivative and integral calculations in the continuous time are executed approximately in the discrete time, where the sampling interval is $\Delta T = 0.05[s]$. A solution that maximizes $J = [K_I]_{11} + [K_I]_{22}$ is given by

$$K(s) = \begin{bmatrix} 0.279 + \frac{0.0368}{s} & 0 \\ 0 & 0.0698 + \frac{0.00834}{s} \end{bmatrix}. \tag{54}$$

Fig. 10 shows the singular value plots of $S_I(s)$ and $T_I(s)$. In this figure, the horizontal line shows the bound $\gamma_1 = 1.5$. Note that since the condition (17) is a necessary condition for the L_2 gain constraint (9), the maximum singular value tends to become larger than γ_1. Fig. 11 shows the step response $y(t)$ for the reference input $r_1(t) = 1, r_2(t) = 0$, and Fig. 12 for $r_1(t) = 0, r_2(t) = 1$.

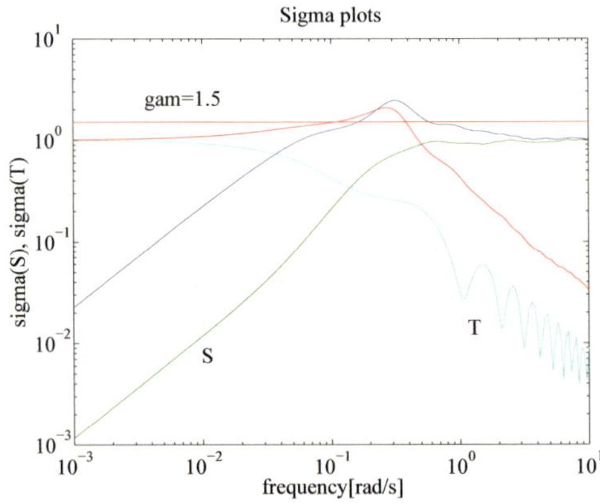

Fig. 10. Singular value plots of S_I and T_I with PI control

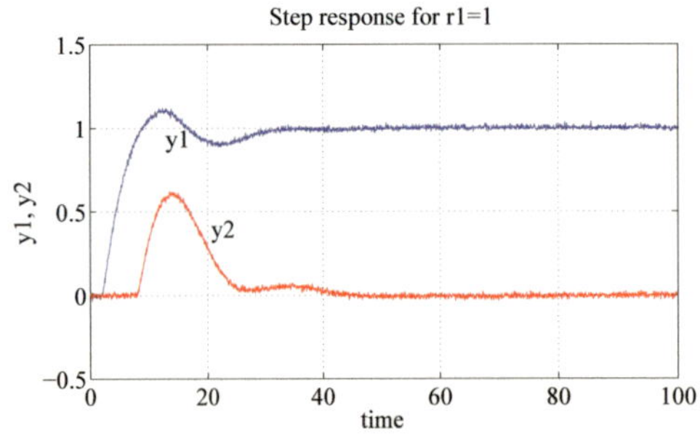

Fig. 11. Output response of the plant for $r_1(t) = 1$ with PI control

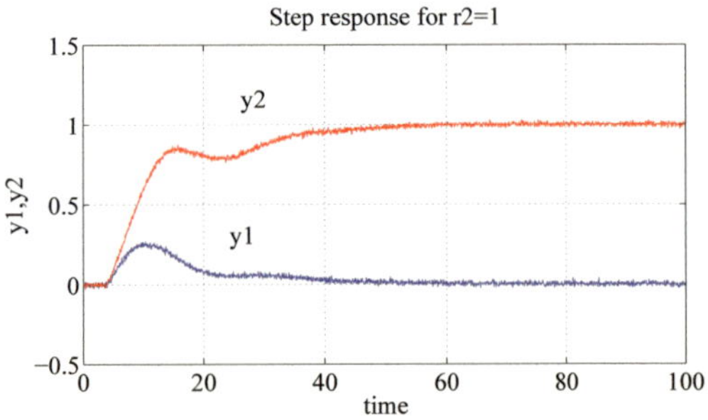

Fig. 12. Output response of the plant for $r_2(t) = 1$ with PI control

Next, design a diagonal PID controller with a first order lowpass filter of the next form using the above plant responses. Note that our method can be directly applied to this design problem by considering the plant as $P(s)/(0.1s + 1)$. This filter is used for the attenuation of the loop gain at high frequencies.

$$K(s) = \frac{1}{0.1s + 1}\left(K_P + K_I \frac{1}{s} + K_D s\right) \tag{55}$$

Then, we obtain the next controller.

$$K(s) = \begin{bmatrix} \frac{0.383s + 0.0798s + 0.477s^2}{(0.1s+1)s} & 0 \\ 0 & \frac{0.118s + 0.0246 + 0.247s^2}{(0.1s+1)s} \end{bmatrix}.$$

Fig. 13 shows the singular value plots, and Fig. 14 and Fig. 15 show the responses of the closed-loop system for the reference inputs.

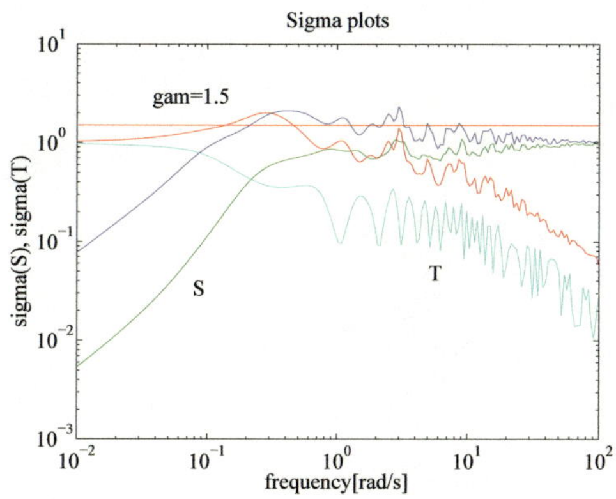

Fig. 13. Singular value plots of S_I and T_I with PID control

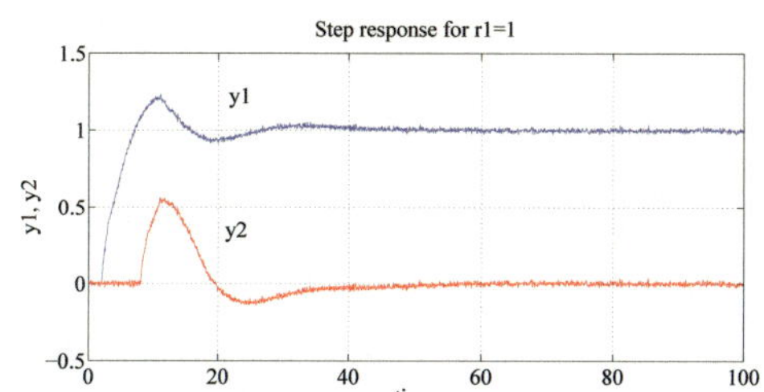

Fig. 14. Output response of the plant for $r_1(t) = 1$ with PID control

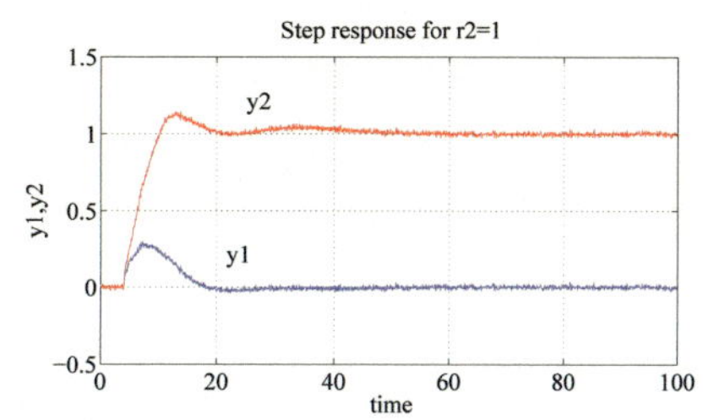

Fig. 15. Output response of the plant for $r_2(t) = 1$ with PID control

6. Experiment using a two-rotor hovering system

We will design a multi-loop PID controller for a two-rotor hovering system. The general view of our experimental apparatus is shown in Fig.16. The arm AB can rotate around the center O freely, and y_1 and y_2 are the yaw and the roll angles, respectively. The airframe CD can also rotate freely on the axis AB, and θ is the pitch angle. Thus, this system has three degrees of freedom. The rotors are driven separately by two DC motors. The rotary encoders are mounted on the joint O to measure the angles y_1 and y_2[rad], respectively. The encoder for θ is mounted on the position A. The actuator part is illustrated in Fig. 17. The control inputs u_1 and u_2 are the thrust and the rolling moment, and $\tilde{f}_1$ and $\tilde{f}_2$ are the lift forces of the two rotors, respectively. In our previous study , we designed a nonlinear controller for a mathematical model (Saeki & Sakaue (2001)). Those who are interested in the plant property, please see the reference.

The feedback control system is illustrated in Fig. 18. PID controller K will be designed to track the references r_1, r_2 [rad]. We use a PD controller $0.4 + 0.2s/(1 + 0.01s)$ in order to control θ, and this gain is determined by trail and error. Then, we treat the plant as a two-input two-output system. The element denoted by K_{uv} is a constant matrix that transforms the control inputs u to the input voltages u_v to the motors. The input voltages are limited to be less than ± 5[V]. We consider the subsystem shown by the dotted line as the plant P to be controlled.

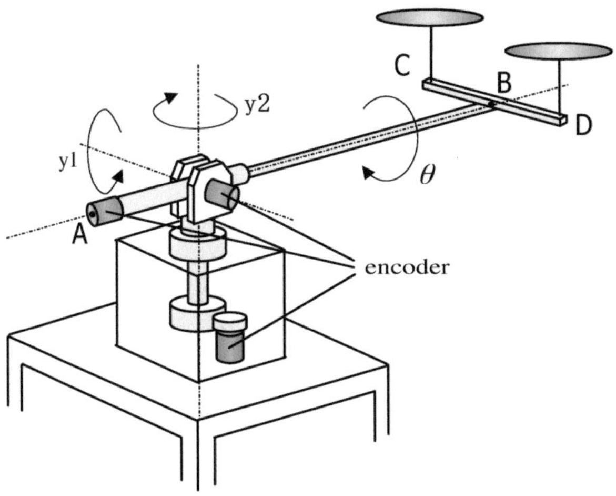

Fig. 16. Experimental setup

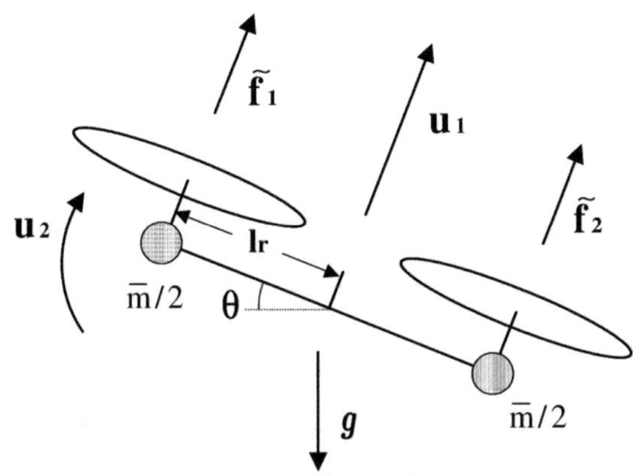

Fig. 17. Illustration of the actuator part

Thus, the feedback system is described by

$$y = P(s)e \tag{56}$$

$$e = K(s)(r - y) \tag{57}$$

The plant responses shown in Fig. 19 - Fig. 22 are obtained by experiment in the closed-loop operation for the controller

$$K(s) = \begin{bmatrix} 0.5 & 0 \\ 0 & 0.1 \end{bmatrix} + \begin{bmatrix} 1 & 0 \\ 0 & 0 \end{bmatrix} \frac{1}{s} + \begin{bmatrix} 1 & 0 \\ 0 & 0.5 \end{bmatrix} \frac{s}{0.01s + 1} \tag{58}$$

Now, let us design a PID controller by using the responses. Since this plant is marginally stable, it is not so easy to give a stabilizing PID controller compared with stable plants. It is

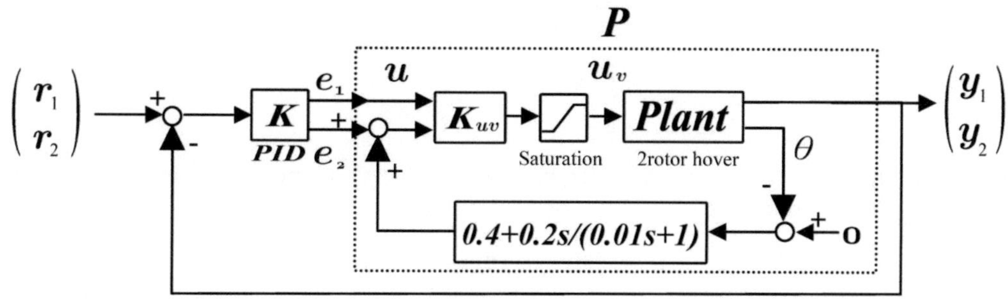

Fig. 18. Feedback control system

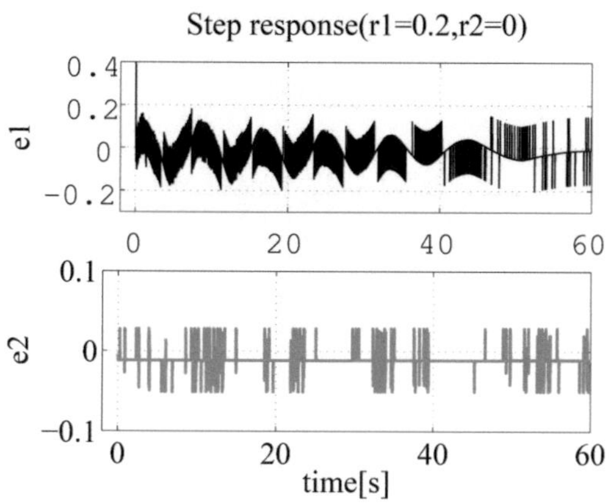

Fig. 19. Input response used for design

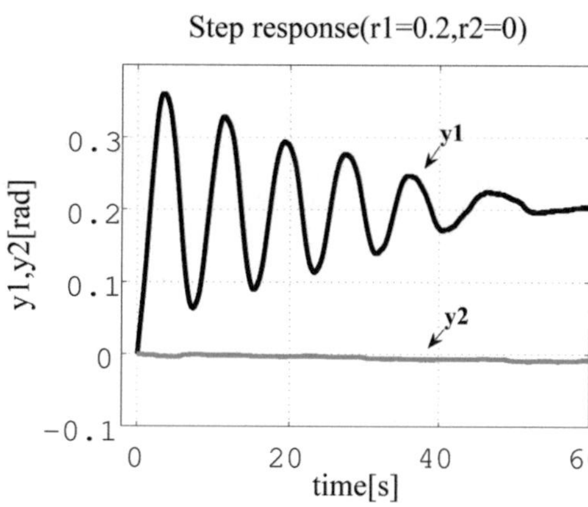

Fig. 20. Output response used for design

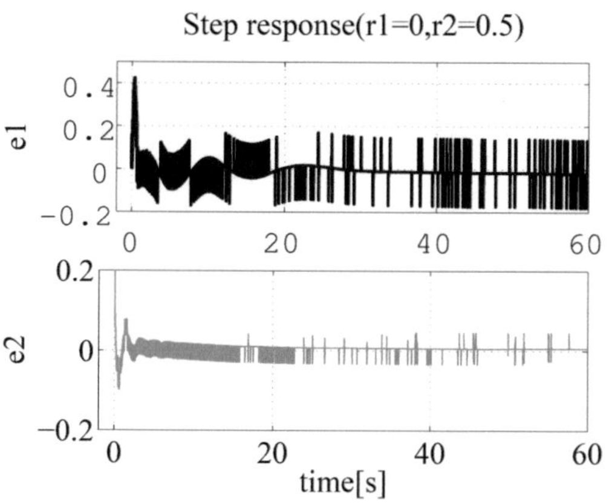

Fig. 21. Input response used for design

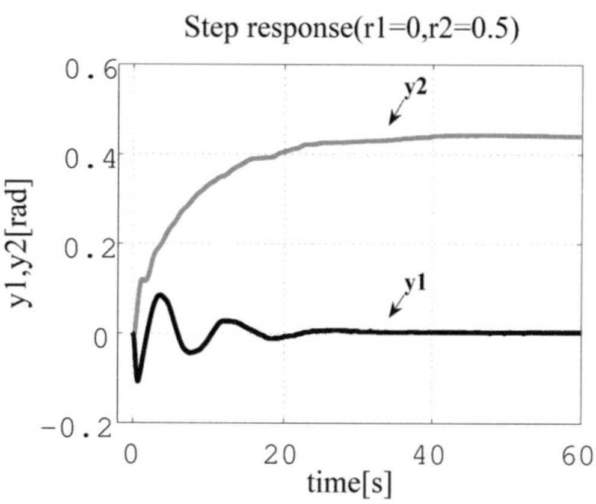

Fig. 22. Output response used for design

easier to find a stabilizing PD controller than PID controller. Therefore, we give the next PD controller, which is found by trial and error.

$$K_a = \begin{bmatrix} 0.4 & 0 \\ 0 & 0.4 \end{bmatrix} + \begin{bmatrix} 1 & 0 \\ 0 & 0.5 \end{bmatrix} s \tag{59}$$

Sample frequencies ω_i are logarithmically equally spaced 100 points between 10^{-2} and 10^{2}. By solving an LMI once, we obtain the next controller.

$$K(s) = \begin{bmatrix} 1.4549 & 0 \\ 0 & 1.0624 \end{bmatrix} + \begin{bmatrix} 0.0980 & 0 \\ 0 & 0.1309 \end{bmatrix} \frac{1}{s} + \begin{bmatrix} 1.4914 & 0 \\ 0 & 1.2581 \end{bmatrix} \frac{s}{0.01s+1} \tag{60}$$

The step responses are shown in Fig. 23 - Fig. 26. It is necessary to develop an efficient method of finding a stabilizing controller that satisfies (17)(18) for marginally stable or unstable plants. This is our future work.

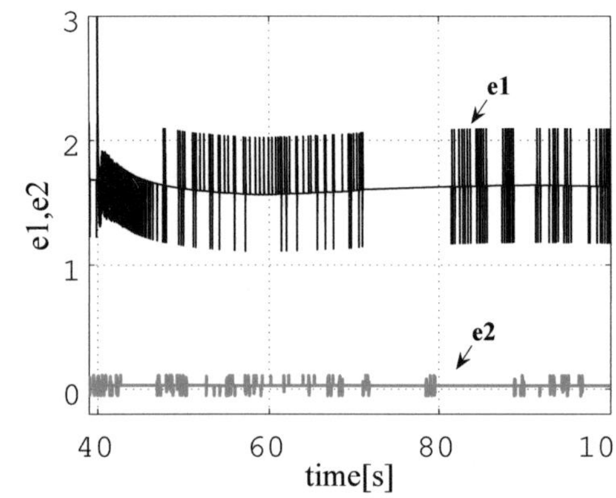

Fig. 23. Input response(r1=0.2,r2=0)

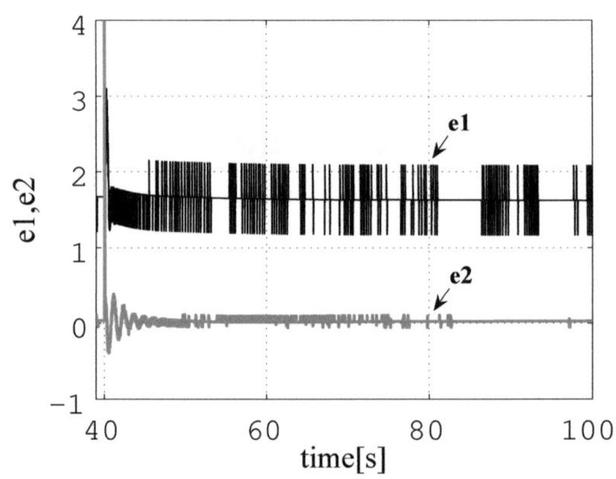

Fig. 24. Input response(r1=0,r2=0.5)

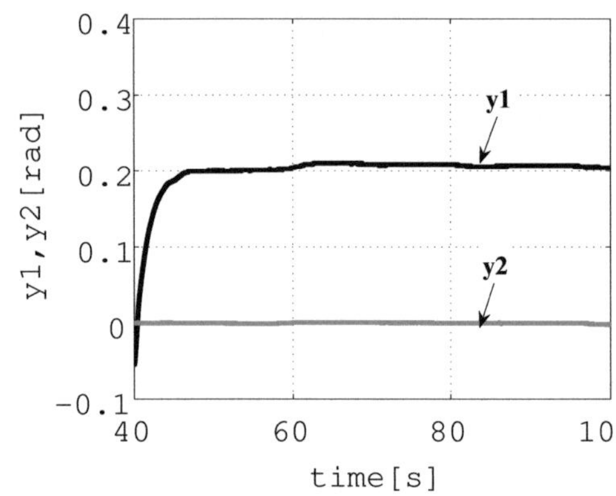

Fig. 25. Output response 1 (r1=0.2,r2=0)

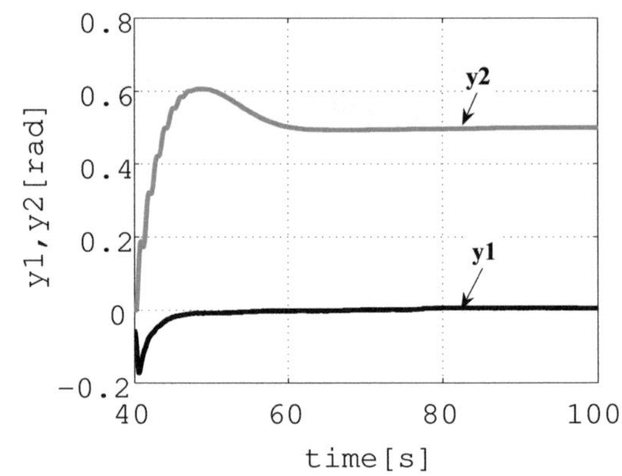

Fig. 26. Output response 2 (r1=0,r2=0.5)

7. Conclusion

DDLS (data driven loop shaping method) has been developed for the multi-loop PID control tuning. The constraints on the PID gains are directly derived from a few input-output responses based on falsification conditions without explicitly identifying the plant model. The design problem is reduced to a linear programming or a linear matrix inequality problem, and the solution is obtained by solving it only once.

We have applied our method to the Wood and Berry's binary distillation column process, and our method gives good loop shapes where only two step responses of the closed-loop system are used for design. However, it is difficult to specify the transient response property such as overshoot by our method, because our method treats the optimization problem of disturbance attenuation. Two-degree of freedom control systems may be suitable for the improvement of the transient response. Further, we have applied our method to the control problem of a two-rotor hovering system. From our experience including these examples, our method seems considerably robust against noises of the plant input output signals obtained in the closed-loop operation. Our design method can be extended to the PID controllers whose gains are full square matrices.

8. References

Addison, P.S. (2002). *The Illustrated Wavelet Transform Handbook*, IOP Publishing Ltd., England.

Åström, K & Hägglund, T (1995). *PID Controllers: Theory, Design, and Tuning*, ISA, Research Triangle Park, North Carolina.

Åström, K.; Panagopoulous, H.; Hägglund, T.(1998). Design of PI controllers based on non-convex optimization, *Automatica*, pp. 585-601.

Åström, K & Hägglund, T (2006). *Advanced PID Control*, ISA.

Campi,M.C.; Lecchini, A.; Savaresi, S.M.(2002). Virtual reference feedback tuning: a direct method for the design of feedback controllers, *Automatica*,Vol. 38, pp. 1337-1346.

Hjalmarsson, H.; Gevers, M.; Gunnarsson, S.;Lequin, O.(1999). Iterative feedback tuning: Theory and application, *IEEE Control Systems Magazine*, Vol. 42, No. 6, pp. 843-847.

Johnson, M.A. & Moradi, M.H. (Editors)(2005). *PID Control; New identification and design methods*, Springer-Verlag London Limited.

Lequin O.; Gevers M.; Mossberg M.; Bosmans E.; Triest L. (2003). Iterative feedback tuning of PID parameters: comparison with classical tuning rules, *Control Engineering Practice*, Vol. 11, pp. 1023-1033.

Saeki M. & Sakaue, Y. (2001). Flight control design for a nonlinear non-minimum phase VTOL aircraft via two-step linearization, *Proceedings of the 40th IEEE Conf. on Decision and Control*, pp. 217-222, Orland, Florida USA.

Saeki, M.(2004a). Unfalsified control approach to parameter space design of PID controllers, *Trans. of the Society of Instrument and Control Engineers*, Vol. 40, No. 4, pp. 398-404.

Saeki,M.; Hamada, O.; Wada,N.; Masubuchi, I. (2006). PID gain tuning based on falsification using bandpass filters, *Proc. of SICE-ICCAS*, Busan, Korea, pp. 4032–4037.

Saeki, M.(2008). Model-free PID controller optimization for loop shaping, *Proc. of the 17th IFAC World Congress*, pp. 4958-4963.

Safonov, M.G. & Tsao, T.C. (1997). The unfalsified control concept and learning, *IEEE Trans. on Automatic Control*, Vol. AC-42, No. 6, pp. 843-847.

Skogestad, S. & Postlethwaite, I. (2007), *Multivariable Feedback Control*, John Wiley & Sons, Ltd.

Vidyasagar, M. (1993). *Nonlinear systems analysis*, second edition, PRENTICE HALL, Englewood Cliffs.
Wood, R.K. & Berry, M.W. (1973). Terminal composition control of a binary distillation column, *Chem., Eng., Sci.*, Vo. 28, pp. 1707-1717, 1973.

8

Neural Network Based Tuning Algorithm for MPID Control

Tamer Mansour, Atsushi Konno and Masaru Uchiyama
Tohoku University
Japan

1. Introduction

One of the biggest problems for space manipulators are to cope with flexibility. If manipulator links undergo deflection during the course of operation, it may prove difficult to reach a desired position or to avoid obstacles. Furthermore, once the manipulator has reached a set point, the residual vibration may degrade positioning accuracy and may cause a delay in task execution. At the same time the flexible manipulators has the advantages of high payload to weight ratio, which make them superior in the space exploration and orbital operation. The high payload to weight ration is not the only merits of using flexible manipulators in space application. Lower power consumption, smaller actuators and speedy operation make the flexible manipulators the optimum choice for space manipulators. Since Cannon et al. (Cannon & Schmitz, 1984) started initial experiments using the linear quadratic approach methods to control flexible link manipulators, much researches on the usage of flexible manipulator had been developed.

Using the approach of enhancement the measurements of the vibration variables was studied by (Ge et al., 1999; Sun et al., 2005) while Etxebarria et al. (Etxebarria et al., 2005) gives attention to the algorithms used in controlling the flexible manipulators. The enhancement of the traditional PD controller by adding a vibration control term is one of the most effective methods for the flexible manipulators. Lee et al. proposed PDS (proportional-derivative strain) control for vibration suppression of multi-flexible-link manipulators and analysed the Liapunov stability of the PDS control (Lee et al., 1988). Maruyama et al. (Maruyama et al., 2006) developed a golf robot whose swing simulates human motion. They presented model accounting for golf club flexibility with all parameters identified in experiments and generated and implemented trajectories for different criterion such as minimizing total consumed work, minimizing summation of the squared derivative of active torque and maximizing impact speed. Matsuno and Hayashi applied the PDS control to a cooperative task of two one-link flexible arms (Matsuno & Hayashi, 2000). They aimed to accomplish the desired grasping force for a common rigid object and the vibration absorption of the flexible arms.

A neural network is a data modelling tool that is able to capture and represent complex input/output relationships. The motivation for the development of neural network technology stemmed from the desire to develop an artificial system that could perform "intelligent" tasks similar to those performed by the human brain. Neural Networks resemble the human brain in the following two ways:

- A neural network acquires knowledge through learning.
- A neural network knowledge is stored within inter neuron connection strengths known as weights.

The true power and advantages of neural networks lies in their ability to represent both linear and non-linear relationships and in their ability to learn these relationships directly from the data being modelled. Traditional linear models are simply inadequate when it comes to modelling data that contains non-linear characteristics. Some researchers tried to use the neural network (herein after abbreviated as NN) as a main controller like (Talebi et al., 1998). In their research the controllers are designed by utilizing the modified output re-definition approach. The modified output re-definition approach requires only a priori knowledge about the linear model of the system but does not require a priori knowledge about the payload mass. Various NN schemes have been proposed so far such as a modified version of the "feedback error-learning" approach to learn the inverse dynamics of the flexible manipulator (Kawato et al., 1987). On the other proposed NN structure the controller is designed based on tracking the reference joint angle while controlling the elastic deflection at the tip. Isogai et al. (Isogi et al., 1999) proposed a fault-tolerant system using inverse dynamics constructed by NN for sensor fault detection and NN adaptive control for the actuator fault to reconfigure control to compensate for parameter changes due to actuator faults.

Other researches like Lianfang (Lianfang et al., 2004) use the neural network as a correction for the main control signal coming from the main feed-back controller. In his research the neural network approach is presented for the motion control of constrained flexible manipulators where both the contact forces exerted by the flexible manipulator and the position of the end-effectors contacting with a surface are controlled. Cheng and Patel in (Cheng & Patel, 2003) tried to made stable tracking control of a flexible macro-micro manipulator utilizing two layer neural network to approximate the non-linear robot dynamic behavior. A learning algorithm for the neural network using Lyapunov stability is derived. Yazdizadeh et al. proposed two neuro-dynamic identifiers to identify the input-output relationship of two-link flexible manipulator. They provided in (Yazdizadeh et al., 2000) a selection criterion for specifying the fixed structural parameters as well as the adaptation laws for updating the adjustable parameters of the networks.

A Modified PID control (MPID) is proposed to control a single-link flexible manipulator by Mansour et al. (Mansour et al., 2008). The MPID control depends mainly on vibration feedback to improve the response of the flexible arm without the massive need of measurements. The advantage of the MPID is that it is not affected by residual strain due to material defect and/or static deformation. The residual strain and material defect may lead to inaccurate movement. The difficulty with the MPID is that it includes nonlinear terms and so the standard gain tuning method can not be used for the controller. The motivation for this research is to find a fast and simple way to tune the MPID controller, which is able to achieve final accurate tip position for the flexible arm and at the same time reduce resulting vibration. The NN is used to solve this problem. In this research a NN is used to find an optimum vibration gain of MPID controller. The main advantage of the NN approach to tune the vibration control gain of the MPID control is the considerable low computational cost to find an optimal tuned gain with different tip payload.

The neural network is used to estimate a result of the dynamic simulation when the simulation condition is given. As a result of the dynamic simulation, integral of the squared tip deflection weighted by exponential function

$$\text{Criterion function} = \int_0^{t_s} \delta^2 e^t dt, \qquad t_s : \text{settling time} \tag{1}$$

is considered in this work. Therefore, the input to the neural network is the simulation condition, while the output is the criterion function defined in 1. The mapping from the input to the output is many-to-one. In order to train the neural network, the results of a dynamic simulator for a given condition are used as teacher signals. In this shadow the feed-forward neural network can be used as a mapping between the simulation conditions and the output response all over the time span which is represented by $\int_0^{t_s} \delta^2 e^t dt$.
The powerful ability of the neural network to model nonlinear model is utilized to map the relation between the vibration control gain of the MPID and the output response represented by the criterion function. Once this relation is drawn, the optimum value of the vibration control gain is corresponding to the minimum value of the criterion function.
The sequence of finding the optimum value for the vibration control gain for the single link flexible manipulator is summarized in Fig. 1. This chapter is organized as follows: An introduction to the control of flexible manipulator and using NN in the control process is highlighted in section 1. The mathematical model of the flexible manipulator is shown in section 2. The detailed of the controller structure and the simulatiom model are presented in sections 3 and 4. In section 5 the NN algorithm used in this research is explained, the structure of the NN is also shown and the optimal vibration control gain finding procedure are highlighted. The learning and training process of the NN is shown in section 6. The response results for the flexible manipulator with the tuned gain using the NN is shown in section 7. Finally, section 8 concludes this chapter with some remarks.

2. Mathematical Model

Before discussing the NN based gain tuning method, the MPID controller (Mansour et al., 2008) is briefly introduced in sections 2 and 3. From the analysis of the single-link flexible arm shown in Fig. 2, the flexible link is approximated by a continuous clamped-free beam. The flexible arm is rotating in the horizontal plane with a rotational angle $\theta(t)$ and the effect of gravity is not taken into consideration. Frame *O-XY* is the fixed base frame and frame *O-xy* is the local frame rotating with the hub. The tip deflection $\delta(L,t)$ is the difference between the actual tip position and the rotating frame *O-xy*. The deflection $\delta(x,t)$ is assumed to be small compared to the length of the arm. Let $p(x,t)$ represents the tangential position of a point on the flexible arm with respect to frame *O-xy*. From the assumption of the deflection of the flexible arm, the tangential position is expressed as:

$$p(x,t) = x\theta(t) + \delta(x,t). \tag{2}$$

The flexible arm is treated as Euler-Bernoulli beam with uniform cross-sectional area and constant characteristics. Then, the Euler-Bernoulli equation for the link is given as follows :

$$EI\frac{\partial^4 p(x,t)}{\partial x^4} + \rho\frac{\partial^2 p(x,t)}{\partial t^2} = 0, \tag{3}$$

where ρ is the sectional density, E is the Young (elastic) modulus, and I is the second moment of area. Substituting (2) into (3) the following equation is obtained :

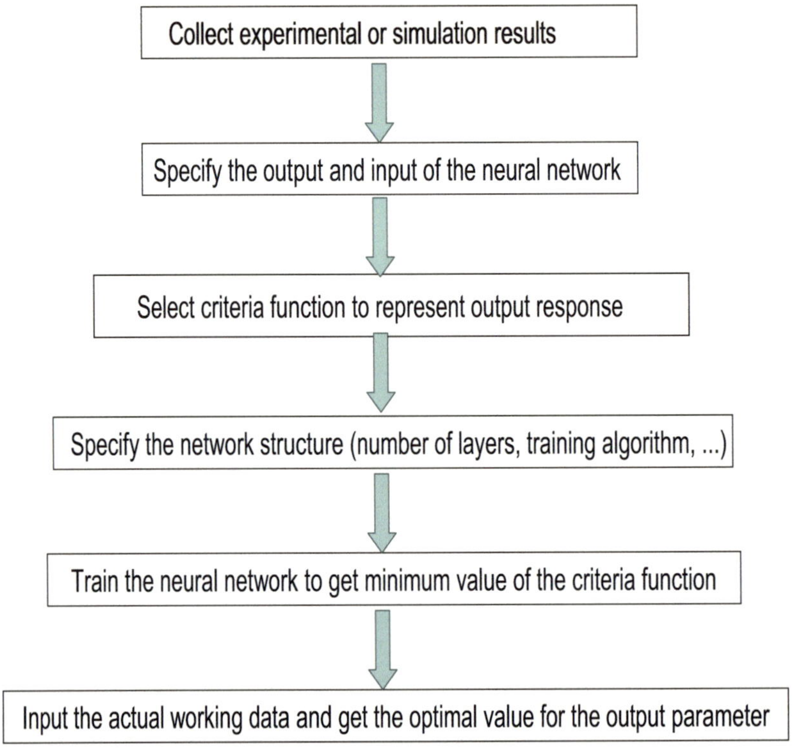

Fig. 1. Neural network using sequence.

$$EI\frac{\partial^4\delta(x,t)}{\partial x^4}+\rho\frac{\partial^2\delta(x,t)}{\partial t^2}=-\rho x\ddot{\theta}(t). \tag{4}$$

The flexible arm is clamped at its base, so both the deflection and slope of the deflection curve must be zero at the clamped end. Bending moment at the free end also equals zero. Making force balance at the tip obtains the following boundary conditions:

$$\delta(x,t)|_{x=0} = 0, \tag{5}$$

$$\left.\frac{\partial\delta(x,t)}{\partial x}\right|_{x=0} = 0, \tag{6}$$

$$EI\left.\frac{\partial^2\delta(x,t)}{\partial x^2}\right|_{x=L} = 0, \tag{7}$$

$$EI\left.\frac{\partial^3\delta(x,t)}{\partial x^3}\right|_{x=L} = m_t\left[x\ddot{\theta}(t)+\frac{\partial^2\delta(x,t)}{\partial t^2}\right]_{x=L}, \tag{8}$$

where L is the arm length. The dynamic equation describing the system presented in (?) is written as follows:

$$\begin{aligned} T(t) &= \left(I_h+\frac{1}{3}\rho L^3\right)\ddot{\theta}(t)+\rho\int_0^L x\ddot{\delta}(x,t)dx \\ &\quad +m_tL\left(L\ddot{\theta}(t)+\ddot{\delta}(L,t)\right). \end{aligned} \tag{9}$$

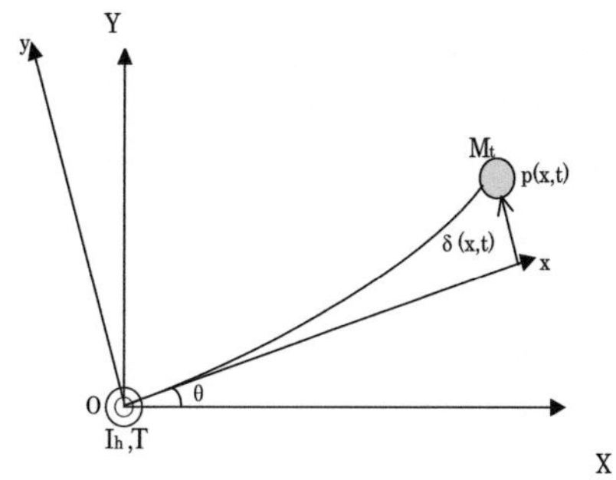

Fig. 2. Single-link flexible manipulator.

A flexible manipulator simulator is built in MATLAB Simulink software using the mathematical model shown before to study the performance of the MPID control with different loading and gains conditions.

3. Controller

A Modified PID controller (MPID) is proposed for controlling the tip position of the single-link flexible manipulator (Mansour et al., 2008). This controller used three measurements to generate the control signal, the hub rotational angle $\theta(t)$, the tip deflection $\delta(L,t)$, and the velocity of the hub $\dot{\theta}(t)$. If we choose the tip position as the output from the system then the error includes two components. The first component $e_j(t)$ is a result of the joint motion and is equal to $L(\theta_{ref} - \theta(t))$ which is identical with the rigid arm error where θ_{ref} is the reference joint angle. The second one is much more important and is due to the flexibility of the arm and equals $\delta(L,t)$. These two error components are coupled to each other. The Modified PID (MPID) controller replaces the classical integral term of a PID controller with a vibration feedback term to affect the flexible modes of the beam in the generated control signal. The MPID controller is formed as follows (Mansour et al., 2008):

$$\begin{aligned} u(t) &= K_p e_j(t) + K_d \dot{e}_j(t) \\ &\quad + K_{vc} g(t) \, \mathrm{sgn}(\dot{e}_j(t)) \int_0^t g(t) dt, \end{aligned} \tag{10}$$

where $u(t)$ is the control signal, K_p, K_d are the proportional and derivative gains for the joint control,respectively, K_{vc} is the vibration control feedback gain, $e_j(t)$ is the tangential position error and $g(t)$ is a vibration variable such as strain, deflection, shear force or acceleration under a single condition that the vibration variable value equal zero when the flexible manipulator is static and under goes no deformation. The stability of the proposed controller had been studied previously in (Mansour et al., 2008). It was proved that the system is stable as long as $K_d \geq 0$. The flexible manipulator simulator is used to validate the MPID controller given by (10) and the results are shown in Figs. 3 and 4. We found from the simulation results that the response of the flexible manipulator is sensitive to the change of the controller gains. In addition to that, the change in the tip payload have a noticeable influence on the response of the flexible manipulator end effector. If the controller gain is not tuned well, the response with the new loading condition will suffer a performance degradation. As shown in Fig. 3, a

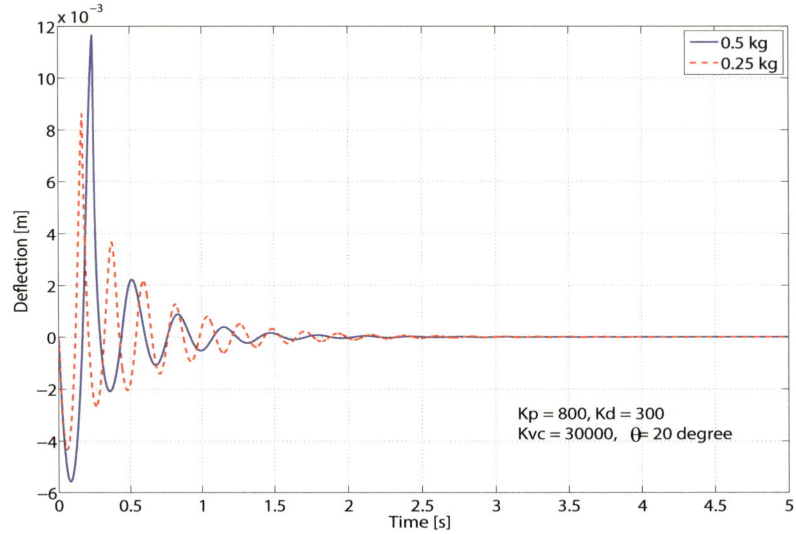

Fig. 3. Effect of changing tip load.

change in the tip load of the flexible manipulator has an undesirable effect for the vibration of the end effector.

Not only the change of the environment parameters like the tip payload causes an undesirable effect on the response as shown in Fig. 3, but also changing the system configuration like joint angle causes the same effect. Unlike industrial manipulators, both the environment parameter (i.e. tip payload) and the system configuration (i.e. joint angle) are always changeable in the case of space manipulators. This highlights the importance of optimization the gain used with this controller.

Another important point is that the change of the vibration control gain K_{vc} has seriously affects on the response of the single-link flexible manipulator. This is completely noticeable from the results in Fig. 4. This fact is the main motivation to find out a good way for tuning K_{vc} that brings the minimum vibration for the tip as well as a fast response for the joint position. It is predicted from Fig. 4 that the damping effect becomes stronger as the vibration control gain K_{vc} increases to a certain limit. However if the gain K_{vc} exceeds the limits it start to create an overshoot in the joint response.

The most difficulty of using the MPID controller is the adjustment of the vibration control gain. Ge et al. tried to use the genetic algorithm optimization process to find the suitable gain for the controller (Ge et al., 1996). In their research they consider the fixed tip payload of the flexible manipulator and generate a set of gains for this configuration using the genetic algorithm. However in general, the tip payloads and the joint angle are not the same in each operation but it varies from one task to another. Hence the tuning of the vibration control gain K_{vc} becomes the most importance issue to achieve the required position with a minimum vibration. To overcome the lake of consideration with the changing of tip payload and joint angle in the tuning of the MPID we proposed to use the NN in the tuning of the MPID.

In this research the vibration control gain K_{vc} for the MPID controller given by equation (10) is tuned using the NN for the environment parameter (i.e. tip payload), the system configuration

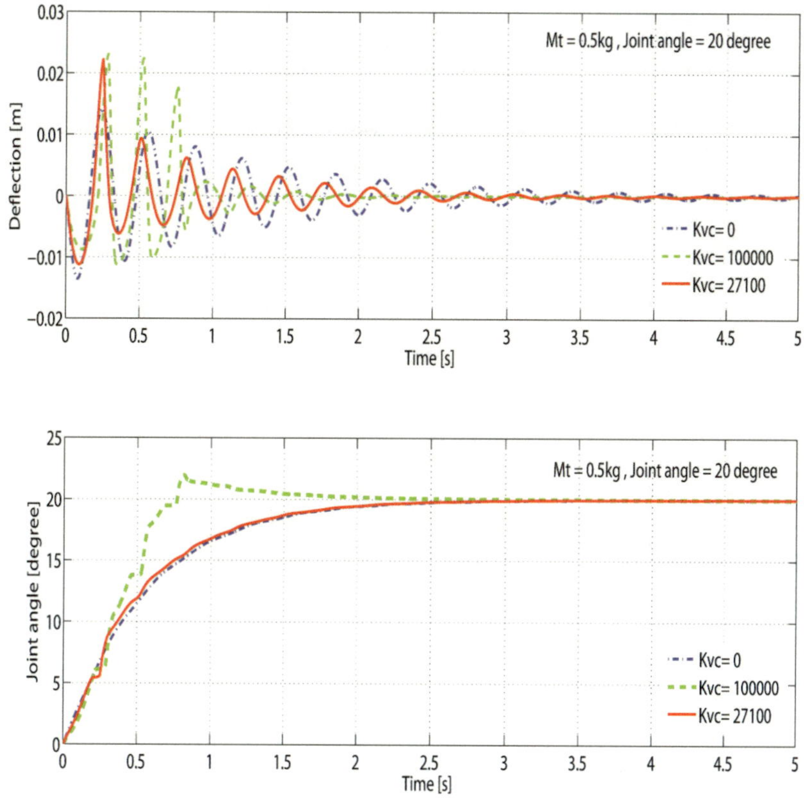

Fig. 4. Effect of changing vibration control gain.

(i.e. joint angle) and for both the other controller gains (i.e. K_p, K_d). By this way the controller gives the best response with respect to all the parameters related to the flexible manipulator.

4. Simulation Model

In this section, a highlight to the simulation which is used to simulate the flexible manipulator is given. A model of the flexible robot control system is simulated in Matlab-Simulink software. The development of the Matlab-Simulink model allows control algorithms to be evaluated before we use the neural network. Also, the simulation program is used to provide information about the behave of the system. The result we get from the simulation will be used in selecting the criterion function, which we will train the neural network on it. In this simulation, we used the mathematical equation derived on section (2) for the flexible manipulator. The block of the simulation model which had been used is shown in Fig. 5. We wish to give attention to some point we consider in make the simulation and important simulation parameters. In the simulation, we use the variable step solvers not the fixed step solvers. The Variable step solvers vary the step size during the simulation, reducing the step size to

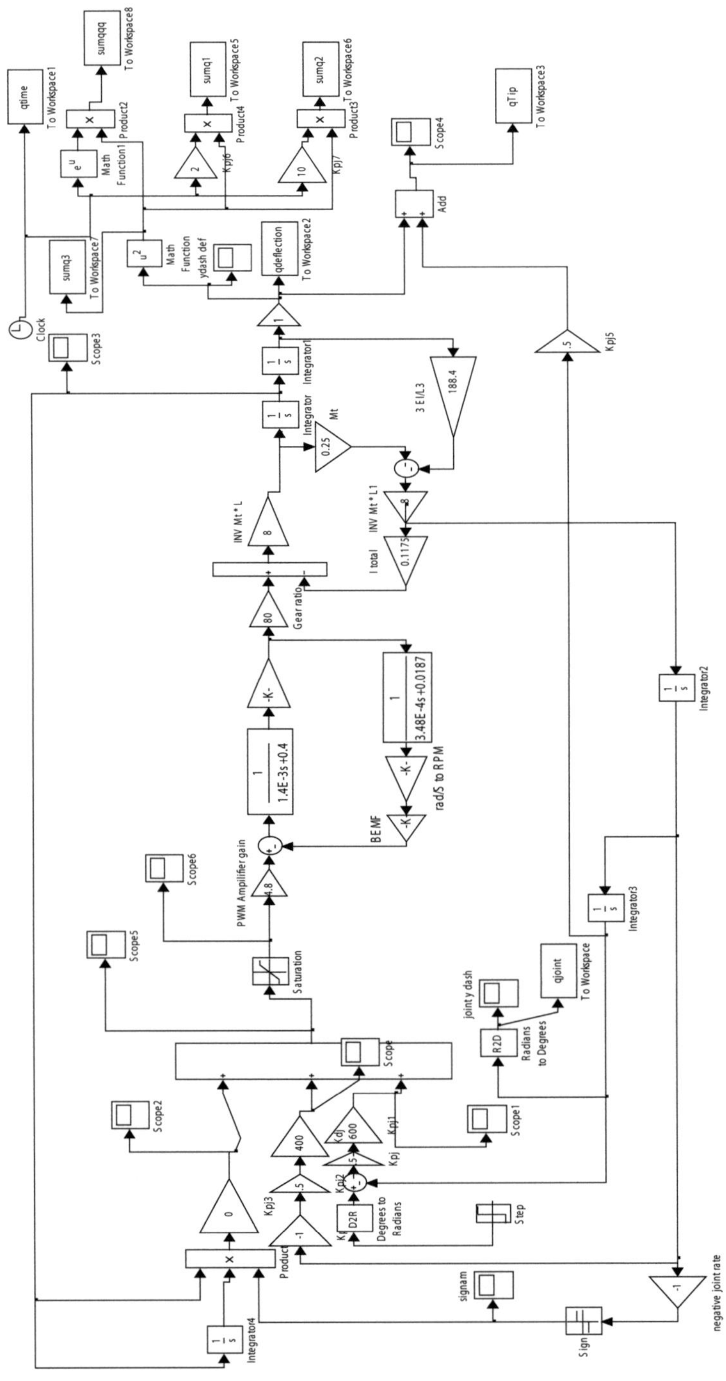

Fig. 5. Simulation model for the flexible manipulator.

increase accuracy when a model's states are changing rapidly and increasing the step size to avoid taking unnecessary steps when the model's states are changing slowly. Computing the step size adds to the computational overhead at each step but can reduce the total number of steps, and hence simulation time, required to maintain a specified level of accuracy for models with rapidly changing or piecewise continuous states. Also for the numerical integration techniques for solving the ordinary differential equations (ODEŠs) that represent the continuous states of dynamic systems. Simulink provides an extensive set of fixed-step and variable-step continuous solvers, each implementing a specific ODE solution method. There are many solver types in the solution methods. First we identifying the optimal solver for the model, optimal means acceptable accuracy with the shortest simulation. We use in the solution of the numerical integration for the ordinary differential equations the ode45 (Dormand-Prince). ODE45 is based on an explicit Runge-Kutta (4,5) formula. It is a one-step solver; that is, in computing $y(t_n)$, it needs only the solution at the immediately preceding time point, $y(t_{n-1})$. In general, ode45 is the best solver to apply if the problem is not stiff. It is also the default solver used by Simulink for models with continuous states.

5. Neural Network

There is no precise agreed definition among researchers as to what a neural network is, but most would agree that it involves a network of simple processing elements (neurons), which can exhibit complex global behavior, determined by the connections between the processing elements and element parameters. The original inspiration for the technique was from examination of the central nervous system and the neurons (and their axons, dendrites and synapses) which constitute one of its most significant information processing elements. In a neural network model, simple nodes (called variously "neurons," "neurodes," or "PEs- processing elements") are connected together to form a network of nodes Ů hence the term "neural network." While a neural network does not have to be adaptive itself, its practical use comes with algorithms designed to alter the strength (weights) of the connections in the network to produce a desired signal flow. These networks are also similar to the biological neural networks in the sense that functions are performed collectively and in parallel by the units, rather than there being a clear delineation of subtasks to which various units are assigned. Currently, the term Artificial Neural Network (ANN) tends to refer mostly to neural network models employed in statistics, cognitive psychology and artificial intelligence.
In modern software implementations of artificial neural networks, the approach inspired by biology has more or less been abandoned for a more practical approach based on statistics and signal processing. In some of these systems, neural networks or parts of neural networks (such as artificial neurons) are used as components in larger systems that combine both adaptive and non-adaptive elements. While the more general approach of such adaptive systems is more suitable for real-world problem solving, it has far less to do with the traditional artificial intelligence connection models. What they do however have in common is the principle of non-linear, distributed, parallel and local processing and adaptation. Neural networks, with their remarkable ability to derive meaning from complicated or imprecise data, can be used to extract patterns and detect trends that are too complex to be noticed by either humans or other computer techniques. A trained neural network can be thought of as an "expert" in the category of information it has been given to analyse. This expert can then be used to provide projections given new situations of interest and answer "what if" questions. Other advantages include:

- Adaptive learning: An ability to learn how to do tasks based on the data given for training or initial experience.
- Self-Organization: An ANN can create its own organization or representation of the information it receives during learning time.
- Real Time Operation: ANN computations may be carried out in parallel, and special hardware devices are being designed and manufactured which take advantage of this capability.
- Fault Tolerance via Redundant Information Coding: Partial destruction of a network leads to the corresponding degradation of performance. However, some network capabilities may be retained even with major network damage.

A simple representation of neural network is shown in Fig. 6. The Input to the neural network is presented by $X_1, X_2, ..., X_R$ where R is the number of inputs in the input layer, S is the number of neuron in the hidden layer and w is the weight. The output from the neural network Y is given by

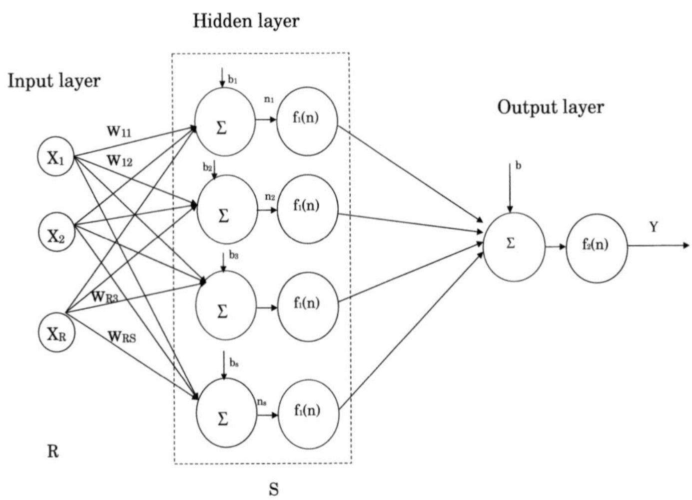

Fig. 6. Simple presentation of neural network.

$$Y = f_2(\sum_{j=1}^{j=S} f_1(n_j) + b) \tag{11}$$

$$n_j = \sum_{j=1}^{j=S} \sum_{i=1}^{i=R} X_i w_{ij} + b_j \tag{12}$$

where $i = 1, 2, \ldots, R$, $j = 1, 2, \ldots, S$,
f_1 and f_2 represents transfer functions.

To overcome the problem of tuning the vibration control gain K_{vc} due to the changing in the manipulator configuration, environment parameter or the other controller gains the neural network is proposed. The main task of the neural network is to get the optimum vibration control gain which can achieve the vibration suppression while reaching the desired position for the flexible manipulator.

So the function of the neural network is to receive the desired position θ_{ref} and the manipulator tip payload M_t with the classical PD controller gains K_p, K_d. The neural network will give out the relation between the vibration control gain K_{vc} and the criterion function at a certain inputs $\theta_{ref}, M_t, K_p, K_d$. From this relation the value of the value of optimum vibration control gain K_{vc} is corresponding to the minimum criterion function.
A flow chart for the training process of the neural network with the parameters of the manipulator and gains of the controller is shown in Fig. 7. The details of the learning algorithm and how is the weight in changed will be discussed later in the training of the neural network.

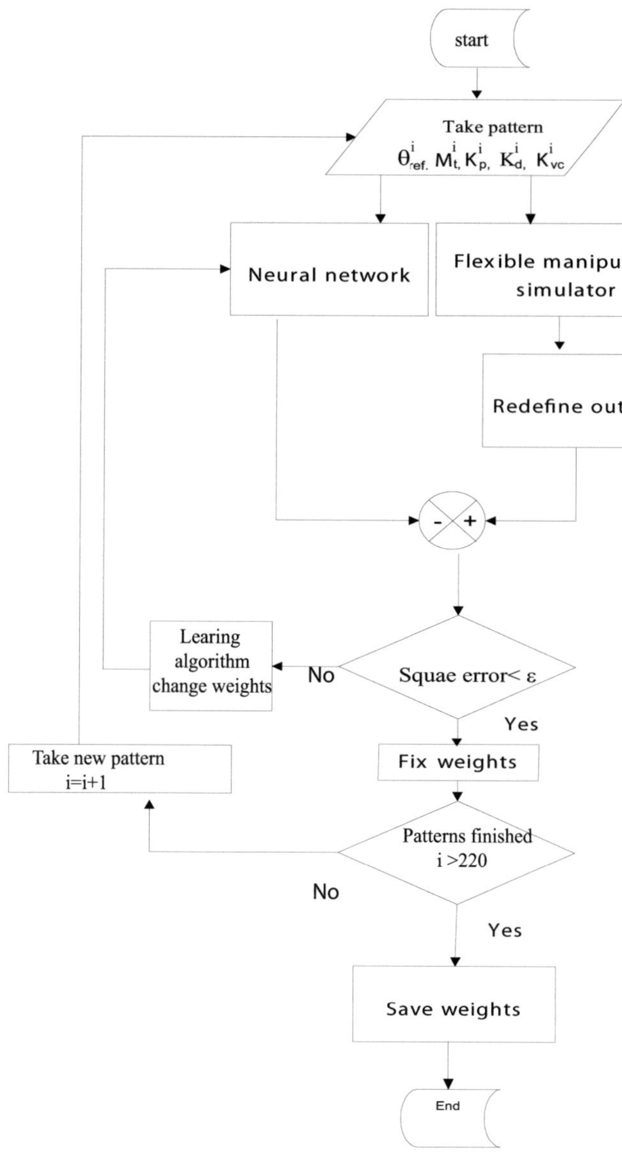

Fig. 7. Flow chart for the training of the neural network.

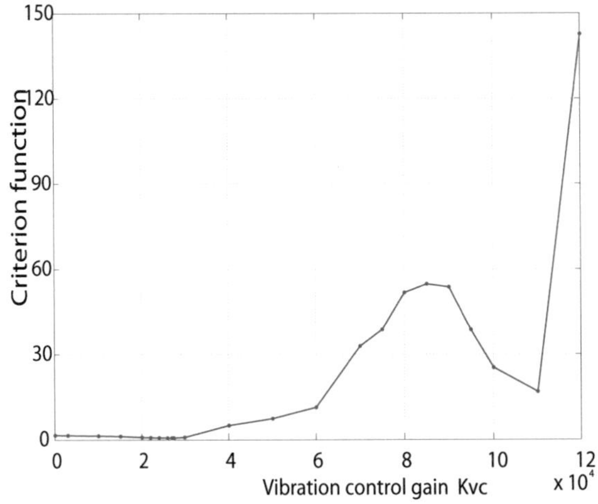

Fig. 8. Relation between vibration control gain and criterion function.

By trying many criterion function to select one of them as a measurement for the output response from the simulation. We put in mind when selecting the criterion function to include two parameters. The first one is the amplitude of the defection of the end effector and the second one is the corresponding time. A set of criterion function like $\int_0^{t_s} t\delta^2 dt$, $\int_0^{t_s} 10t\delta^2 dt$, $\int_0^{t_s} \delta^2 e^t dt$ is tried and a comparison between the behave for all of them and the vibration control gain K_{vc} is done. The value of t_s here represent the time for simulation and on this research we take it as 10 seconds. The criterion function $\int_0^{t_s} \delta^2 e^t dt$ is selected as its value is always minimal when the optimum vibration control gain is used. The term optimum vibration control gain K_{vc} pointed here to the value of K_{vc} which give a minimum criterion function $\int_0^{t_s} \delta^2 e^t dt$ and on the same time keep stability of the system.
The neural network is trained on the results from the simulation with different $\theta_{ref}, M_t, K_p, K_d, K_{vc}$. The neural network is trying to find how the error in the response from the system (represented by the criterion function $\int_0^{t_s} \delta^2 e^t dt$ is changed with the manipulator parameter (tip payload, joint angle) i.e. M_t, θ_{ref} and also how it changes with the other controller parameters K_p, K_d, K_{vc}. The relation between the vibration control gain of the controller, K_{vc} which will be optimized using the neural network and the criterion function, $\int_0^{t_s} \delta^2 e^t dt$ which represent a measurement for the output response from the simulation is shown in Fig. 8. After the input and output of the neural network is specified, the structure of the neural network have to been built. In the next section the structure of the neural network used to optimize the vibration control gain K_{vc} will be explained.

5.1 Design

The neural network structure mainly consists of input layer, output layer and it also may contain a hidden layer or layers. Depending on the application whether it is a classification, prediction or modelling and the complexity of the problem the number of hidden layer is decided. One of the most important characteristics of the neural network is the number of neurons in the hidden layer(s). If an inadequate number of neurons are used, the network will be unable to model complex data, and the resulting fit will be poor. If too many neurons

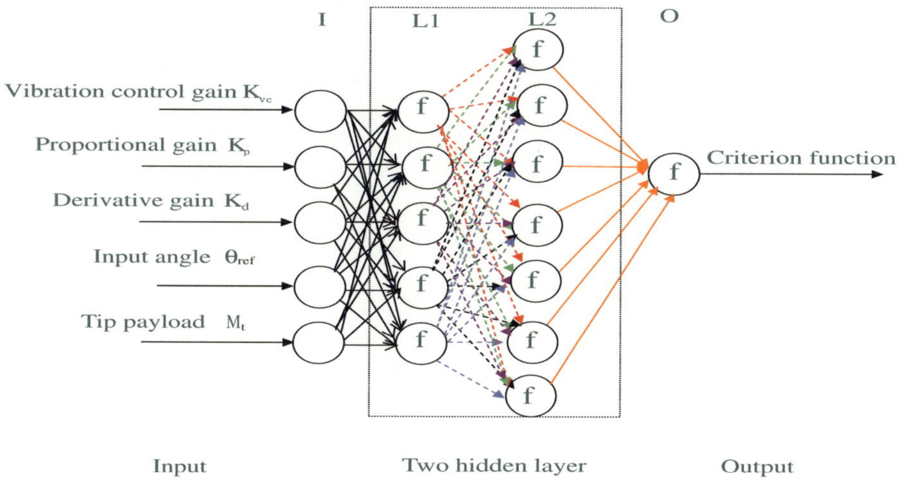

Fig. 9. NN structure.

are used, the training time may become excessively long, and, worse, the network may over fit the data. When over fitting occurs, the network will begin to model random noise in the data. The result is that the model fits the training data extremely well, but it generalizes poorly to new, unseen data.

Validation must be used to test for this. There are no reliable guidelines for deciding the number of neurons in a hidden layer or how many hidden layers to use. As a result, the number of hidden neurons and hidden layers were decided by a trial and error method based on the system itself (Principe et al., 2000). Networks with more than two hidden layers are rare, mainly due to the difficulty and time of training them. The best architecture to be used is problem specific.

A proposed neural network structure is shown in Fig. 9. A neural network with one input layer and one output layer and two hidden layers is proposed. In the proposed neural network the input layer contains five inputs, $\theta_{ref}, M_t, K_p, K_d, K_{vc}$. Those inputs represent the manipulator configuration, environment variable and controller gains. The output layer is consists of one output which is the criterion function and a bias transfer function on the neuron of this layer. The first one of the two hidden layers is consists of 5 neuron and the second one is consists of 7 neurons. For the transfer function used in the neuron of the two hidden layer first we use the sigmoid function described by 13 to train the neural network.

$$f(x_i, w_i) = \frac{1}{1 + \exp(-x_i^{bias})}, \tag{13}$$

where $x_i^{bias} = x_i + w_i$.

The progress of the training of the neural network is shown when using sigmoid transfer function in Fig. 10. As we notice that no good progress in the training we propose to use the tanh as a transfer function for the neuron for both of the two layers. Tanh applies a biased tanh function to each neuron/processing element in the layer. This will squash the range of each neuron in the layer to between -1 and 1. Such non-linear elements provide a network with the ability to make soft decisions. The mathematical equation of the tanh function is give

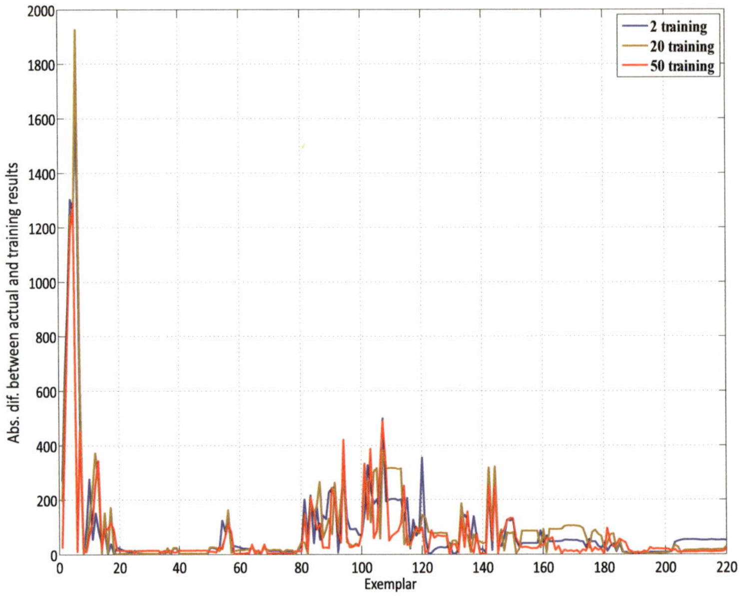

Fig. 10. Progress in training using sigmoid function.

by 14.

$$f(x_i, w_i) = \frac{2}{1 + \exp(-2x_i^{bias})} - 1, \tag{14}$$

where $x_i^{bias} = x_i + w_i$. Also the progress in the training of the neural network using the tanh function is shown in Fig. 11.

5.2 Optimal Vibration Control Gain Finding Procedure

The MPID controller includes non-linear terms such as $\text{sgn}(\dot{e}_j(t))$, therefore standard gain tuning method like Ziegler-Nichols method can not be used for the controller. For the optimal control methods like pole placement, it involves specifying closed loop performance in terms of the closed-loop poles positions.

However such theory assumes a linear model and a controller. Therefore it can not be directly applied to the MPID controller.

In this research we propose a NN based gain tuning method for the MPID controller to control flexible manipulators. The true power and advantages of NN lies in its ability to represent both linear and non-linear relationships and in their ability to learn these relationships directly from the data being modelled. Traditional linear models are simply inadequate when it comes to modelling data that contains non-linear characteristics. The basic idea to find the optimal gain K_{vc} is illustrated in Fig. 12 (a). The procedure is summarized as follows.

1. A task, i.e. the tip payload M_t and reference angle θ_{ref}, is given.
2. The joint angle control gains K_p and K_d are appropriately tuned without considering the flexibility of the manipulator.
3. Initial K_{vc} is given.

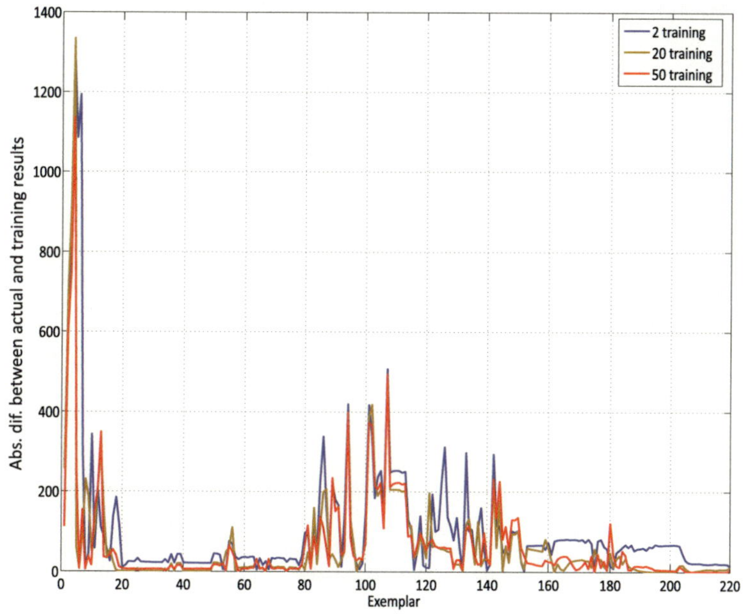

Fig. 11. Progress in training using tanh function.

4. The control input $u(t)$ is calculated with given $K_p, K_d, K_{vc}, \theta_{ref}$ and θ_t using (10).
5. Dynamic simulation is performed with given tip payload M_t and the control input $u(t)$
6. 4 and 5 are iterated when $t \leq t_s$ (t_s: given settling time).
7. Criterion function is calculated using (15).
8. 4 $\sim$ 7 are iterated for another K_{vc}.
9. Based on the obtained criterion function for various K_{vc}, an optimal gain K_{vc} is found

As the criterion function $C(M_t, \theta_{ref}, K_p, K_d, K_{vc})$, the integral of the squared tip deflection weighted by exponential function is considered as:

$$C(M_t, \theta_{ref}, K_p, K_d, K_{vc}) = \int_0^{t_s} \delta^2(t) e^t dt, \tag{15}$$

where t_s is a given settling time and $\delta(t)$ is one of the output of the dynamic simulator (see Fig. 12 (a)).
The NN replaces the MPID control and dynamic simulator and bring out the relation between the input to the simulator, control gains and the criterion function. Based on this relation we can get the optimal vibration gain K_{vc} for any combination of simulator input and PD joint gains K_p, K_d.
However the procedure 5 (dynamic simulation) requires high computational cost and procedure 5 is iterated plenty of times. Consequently it is difficult to find an optimal gain K_{vc} on-line.
Therefore we propose to replace the blocks enclosed by a dashed rectangle in Fig. 12 (a) by a NN model illustrated in Fig. 12 (b). By this way the input to the NN is the simulation

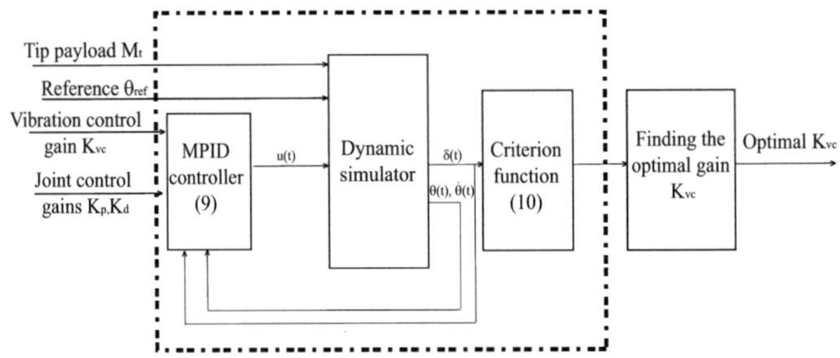

(a) Concept behind finding optimal gain K_{vc}.

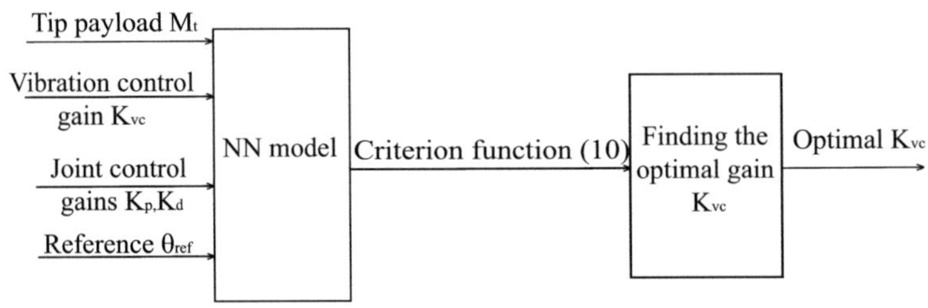

(b) Finding optimal gain K_{vc} using a NN model.

Fig. 12. Finding optimal gain K_{vc}.

condition, $\theta_{ref}, M_t, K_p, K_d, K_{vc}$ while the output is the criterion function defined in (15). The mapping from the input to the output is many-to-one.

5.3 A NN Model to Simulate Dynamic of A Flexible Manipulator

The NN structure generally consists of input layer, output layer and hidden layer(s). The number of hidden layer is depending on the application such as classification, prediction or modelling and on the complexity of the problem. One of the most important problems of the NN is the determination of the number of neurons in the hidden layer(s). If an inadequate number of neurons are used, the network will be unable to model complex function, and the resulting fit will not be satisfactory. If too many neurons are used, the training time may become excessively long, and, if the worst comes, the network may over fit the data. When over fitting occurs, the network will begin to model random noise in the data. The result of the over fitting is that the model fits the training data well, but it is failed to be generalized for new and untrained data. The over fitting should be examined (Principe et al., 2000). The proposed NN structure is shown in Fig. 9. The NN includes one input layer, one output layer and two hidden layers. In the designed NN the input layer contains five inputs: $\theta_{ref}, M_t, K_p, K_d, K_{vc}$ (see also Fig. 12). Those inputs represent the manipulator configuration, environment variable and controller gains. The output layer consists of one output which is the criterion function, $\Sigma\delta^2 e^t$ and a bias transfer function on the neuron of this layer. The first hidden layer consists of five neurons and the second hidden layer consists of seven neurons. For the transfer function used in the neurons of the two hidden layers a tanh function is used.
The mathematical equation of the tanh function is give by:

$$f(x_i, w_i) = \frac{2}{1 + \exp(-2x_i^{bias})} - 1, \tag{16}$$

where x_i is the ith input to the neuron, w_i is the weight for the input x_i and $x_i^{bias} = x_i + w_i$. After the NN is structured, it is trained using a various examples to generate the correct weights to be used in producing the data in the operating stage.
The main task of the NN is to represent the relation between the input parameters to the simulator, MPID gains and the criterion function.

6. Learning and Training

The training for the NN is analogous to the learning process of the human. As human starts in the learning process to find the relationship between the input and outputs. The NN does the same activity in the training phase.
The block diagram which represents the system during the training process is shown in Fig. 13.

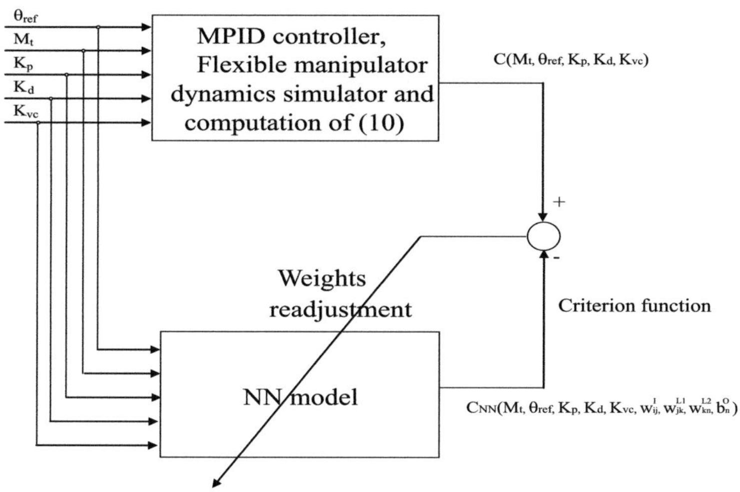

Fig. 13. Block diagram for the training the NN.

After the NN is constructed by choosing the number of layers, the number of neurons in each layer and the shape of transfer function in each neuron, the actual learning of NN starts by giving the NN teacher signals. In order to train the NN, the results of the dynamic simulator for given conditions are used as teacher signals. In this shadow the feed-forward NN can be used as a mapping between $\theta_{ref}, M_t, K_p, K_d, K_{vc}$ and the output response all over the time span which is calculated by (15).
For the NN illustrated in Fig. 9, the output can be written as

$$\text{Output} = C_{NN}(M_t, \theta_{ref}, K_p, K_d, K_{vc}, w^I_{ij}, w^{L1}_{jk}, w^{L2}_{k1}, b^O_1), \tag{17}$$

where w^I_{ij} is the weight from element i $(i = 1 \sim 5)$ in input layer (I) to element j $(j = 1 \sim 5)$in next layer $(L1)$. w^{L1}_{jk} is the weight from element j $(j = 1 \sim 5)$ in first hidden layer $(L1)$ to element k $(k = 1 \sim 7)$ in next layer $(L2)$. w^{L2}_{k1} is the weight from element k $(k = 1 \sim 7)$ in second hidden layer $(L2)$ to element n in output layer (O). b^O_1 is the bias of the output layer. The NN begins to adjust the weights is each layer to achieve the desired output.

Herein, the performance surface $E(w)$ is defined as follows:

$$E(w) = (C(M_t, \theta_{ref}, K_p, K_d, K_{vc}) - C_{NN}(M_t, \theta_{ref}, K_p, K_d, K_{vc}))^2. \tag{18}$$

The conjugate gradient method is applied to readjustment of the weights in NN. The principle of the conjugate gradient method is shown in Fig. 14.

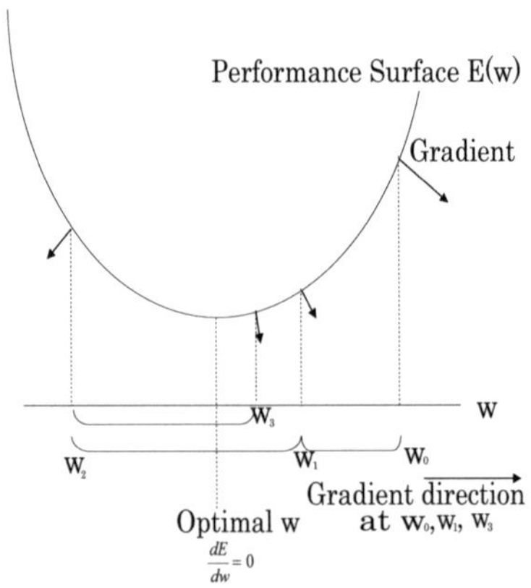

Fig. 14. Conjugate gradient for minimizing error.

By always updating the weights in a direction that is conjugate to all past movements in the gradient, all of the zigzagging of 1st order gradient descent methods can be avoided. At each step, a new conjugate direction is determined and then move to the minimum error along this direction. Then a new conjugate direction is computed and so on. If the performance surface is quadratic, information from the Hessian can determine the exact position of the minimum along each direction, but for non quadratic surfaces, a line search is typically used. The equations which represent the conjugate gradient method are:

$$\Delta \boldsymbol{w} = \alpha(n)\boldsymbol{p}(n), \tag{19}$$

$$\boldsymbol{p}(n+1) = -\boldsymbol{G}(n+1) + \beta(n)\boldsymbol{p}(n), \tag{20}$$

$$\beta(n) = \frac{\boldsymbol{G}^T(n+1)\boldsymbol{G}(n+1)}{\boldsymbol{G}^T(n)\boldsymbol{G}(n)}, \tag{21}$$

where $\boldsymbol{w}$ is a weight, $\boldsymbol{p}$ is the current direction of weight movement, α is the step size, $\boldsymbol{G}$ is the gradient (back propagation information) and β is a parameter that determines how much of the past direction is mixed with the gradient to form the new conjugate direction. And as a start for the searching we put $\boldsymbol{p}(0) = -\boldsymbol{G}(0)$. The equation for α in case of line search to find the minimum mean squared error (MSE) along the direction $\boldsymbol{p}$ is given by:

$$\alpha = \frac{-\boldsymbol{G}^T(n)\boldsymbol{p}(n)}{\boldsymbol{p}^T(n)\boldsymbol{H}(n)\boldsymbol{p}(n)}, \tag{22}$$

where $\boldsymbol{H}$ is the Hessian matrix. The line search in the conjugate gradient method is critical for finding the right direction to move next. If the line search is inaccurate, then the algorithm

may become brittle. This means that we may have to spend up to 30 iterations to find the appropriate step size.
The scaled conjugate is more appropriate for NN implementations. One of the main advantages of the scaled conjugate gradient (SCG) algorithm is that it has no real parameters. The algorithm is based on computing $\boldsymbol{Hd}$ where $\boldsymbol{d}$ is a vector. It uses equation (22) and avoids the problem of non-quadratic surfaces by manipulating the Hessian so as to guarantee positive definiteness, which is accomplished by $\boldsymbol{H} + \lambda \boldsymbol{I}$, where $\boldsymbol{I}$ is the identity matrix. In this case α is computed by:

$$\alpha = \frac{-\boldsymbol{G}^T(n)\boldsymbol{p}(n)}{\boldsymbol{p}^T(n)\boldsymbol{H}(n)\boldsymbol{p}(n) + \lambda \mid \boldsymbol{p}(n) \mid^2}, \tag{23}$$

instead of using (22). The optimization function in the NN learning process is used in the mapping between the input to the simulator and the output criterion function not in the optimization of the vibration gain.

6.1 Training result

The SCG is chosen as the learning algorithm for the NN. Once the algorithm for the learning process is selected, the NN is trained on the patterns. The result of the learning process is shown in this subsection. The teacher signals (training data set) are generated by the simulation system illustrated in Fig. 12 (a). The examples of the training data set are listed in Table 1. 220 data sets are used for the training. The data is put in a scattered order to allow the NN to get the relation in a correct manner.

Pattern	θ_{ref}	M_t	K_p	K_d	K_{vc}	$\Sigma\delta^2 e^t$
1	5	0.5	300	100	20000	0.0129
2	15	0.25	800	300	80000	7.242
3	10	0.25	600	200	0	1.21
4	25	0.5	600	200	10000	0.1825
5	25	0.5	600	200	10000	0.1825
6	15	0.25	600	150	70000	4.56
...	...	...	...	...	...	...

Table 1. Sample of NN training patterns.

As shown in Fig. 15, two curves are drawn relating the value of the normalized criterion for each example used in the training. The normalized the criterion function $C(M_t, \theta_{ref}, K_p, K_d, K_{vc}$ obtained from the simulation is plotted in circles while the normalized criterion function $C_{NN}(M_t, \theta_{ref}, K_p, K_d, K_{vc})$ generated by the NN in the training process is plotted in cross marks. The results of Fig. 15 show that training of the NN enhance its ability to follow up the output from the simulation. A performance measure is used to evaluate whether the training of the NN is completed. In this measurement, the normalized mean squared error (NMSE) between the two datasets (i. e. the dataset the NN trained on and the dataset the NN generate) is calculated. For this case NMSE is 0.0054. Another performance index is also used which is the correlation coefficient r between the two datasets. The correlation coefficient r is 0.9973. When a test is done for the trained NN upon a complete new set of data the NMSE is 0.0956 and r is 0.9664.

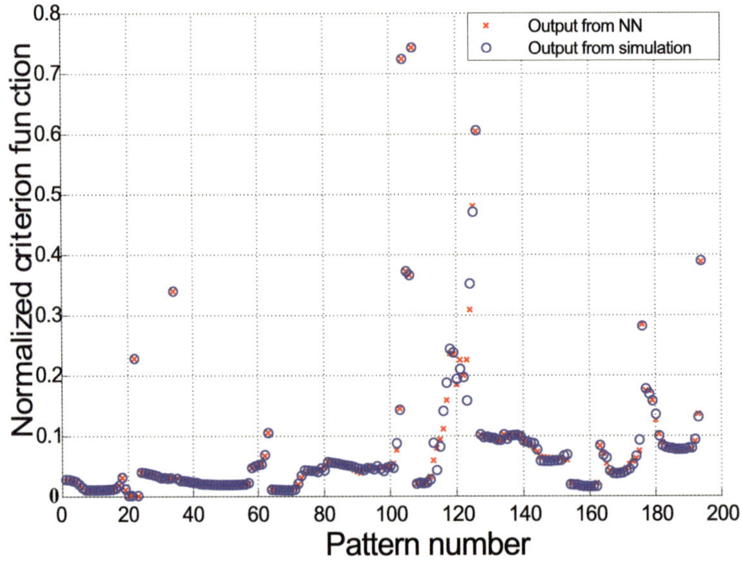

Fig. 15. NN training.

7. Optimization result

In this section, the results obtained using the simulation are compared with the results obtained using the NN. The criterion function C computed by (15) and the output of NN, C_{NN}, for the vibration control gain K_{vc} are plotted in Fig. 16. Comparing the results obtaind using the NN for the criterion function with the results obtained using dynamic simulator in Fig. 16. shows good coincidence. This means that the NN network can successfully replace the dynamic simulator to find how the criterion function changes with the changing of the system parameters.

Form Fig. 16 the optimum gain K_{vc} can be easily found. One of the main advantages of using the NN to find the optimal gain for the MPID control is the computional speed. To generate the data of the simulation curve, which is indicated by the triangles in Fig. 16, 1738 seconds is needed while only 6 seconds are needed to generate the data using the NN, which is indicated by the circles. The minimum values of the criterion function occurs when the value of the vibration control gain K_{vc} equals 22500 V s/m^2.

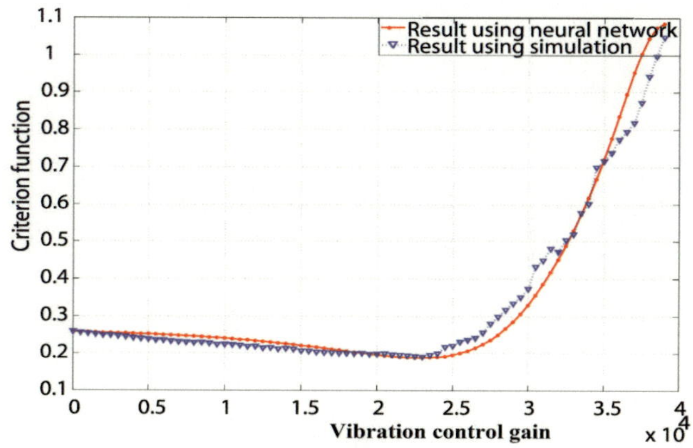

Fig. 16. Vibration control gain vs. criterion function.

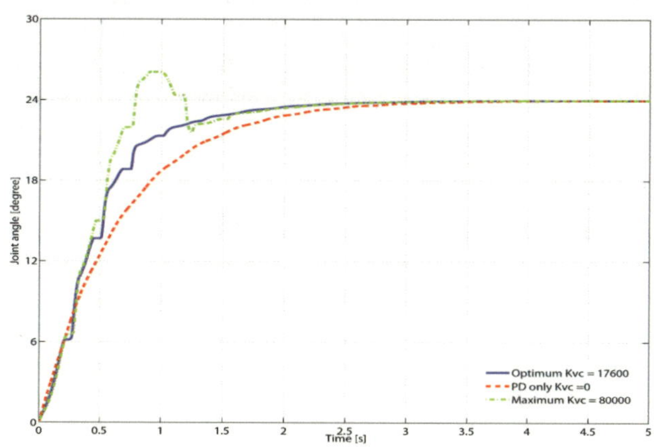

Fig. 17. Response using optimum gain.

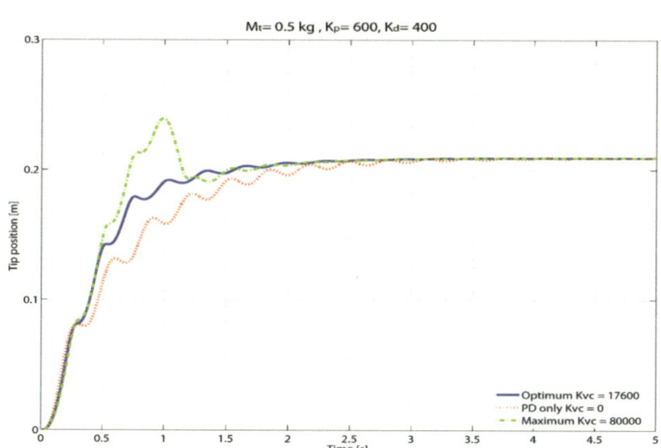

Fig. 18. Response using optimum gain.

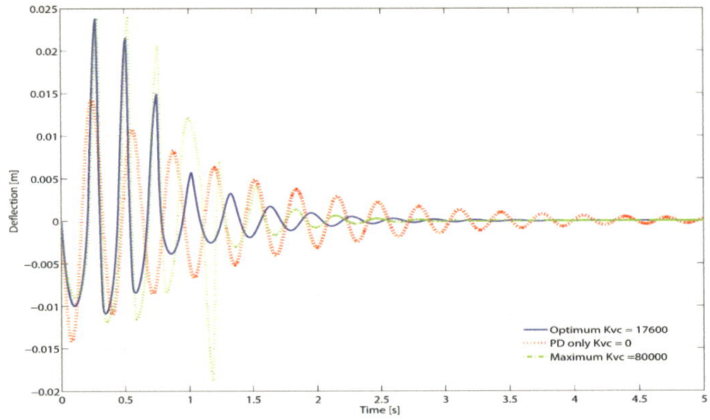

Fig. 19. Response using optimum gain.

The response of the flexible manipulator using the optimal gain K_{vc} is shown in Fig. 17, Fig. 18 and Fig. 19. 0.5 kg is used as a tip payload M_t with 24 degree for the joint reference angle θ_{ref}. For the controller described by equation (10), the values of K_p and K_d are set at 600 V rad/m and 400 V s rad/m respectively. The response with different vibration gains K_{vc} is plotted. In the beginning the response with PD control only (i.e. $K_{vc} = 0$) is plotted in dash line while the response with the maximum K_{vc} which is 80000 V s/m^2 is plotted in a dash-dot line. The response with the optimum K_{vc} -which was tuned using NN- appears in a continous line. The value of the optimum vibration control gain K_{vc} is 17600 V s/m^2. Increasing the vibration control gain K_{vc} leads the system to have fast response for the joint position as shown in Fig. 17 but more increasing in the value of the vibration control gain leads to an undesirable overshoot as shown in Fig. 18 with a dash-dot line. To focus on the effect of the vibration gain on the end-effector vibration Fig. 19 is plotted. It is clear from the figure that the optimum vibration control gain for the MPID succeed to suppress the vibration at the end of the flexible manipulator.

8. Conclusions

This chapter discusses a NN based gain tuning method for the vibration control PID (MPID) controller of a single-link flexible manipulator. The NN is trained to simulate the dynamics of the single-link flexible manipulator and to produce the integral of the squared tip deflection weighted by exponential function. A dynamic simulator is used to produce the teacher signals.

The main advantage of using NN to find an optimal gain is the computational speed. The NN based method is approximately 290 times faster than the dynamic simulation based method. Simulation results with the obtained optimal gain validate the proposed method.

9. References

Cannon, R. H. Jr. & Schmitz, E. (1984). Initial Experiments on the End Point Control of a Flexible One-Link Robot. *International Journal of Robotics Research*, Vol. 3, No. 4, pp. 62–75.

Ge, S. S.; Lee, T. H. & Gong, J. Q. (1999). A Robust Distributed Controller of a Single-Link SCARA /Cartesian Smart Materials Robot, *Mechatronics*, Vol. 9, No. 1, 1999, pp. 65–93.

Sun, D.; Shan J.; Su, Y.; Liu H. & Lam, C. (2005). Hybrid Control of a Rotational Flexible Beam Using Enhanced PD Feedback with a Non-Linear Differentiator and PZT Actuators, *Smart Mater. Struct.*, Vol. 14, pp 69–78.

Etxebarria, V.; Sanz, A. & Lizarraga, I. (2005). Control of a Lightweight Flexible Robotic Arm Using Sliding Modes, *International Journal of Advanced Robotic Systems*, Vol. 2, No. 2, pp. 103–110.

Lee,H. G.; Arimoto, S., & Miyazaki, F. (1988). Liapunov Stability Analysis for PDS Control of Flexible Multi-link Manipulators, *Proceeding of the Conference on Decision and Control, Austin*, pp. 75–80.

Maruyama, T.; Xu, C.; Ming, A. & Shimojo, M. (2006). Motion Control of Ultra-High-Speed Manipulator with a Flexible Link Based on Dynamically Coupled Driving, *Joural of Robotics and Mechatronics*, Vol. 18, No. 5, pp. 598–607.

Matsuno, F. & Hayashi, A. (2000). PDS Cooperative Control of Two One-link Flexible Arms, *Proceeding of the 2000 IEEE International Conference on Robotics and Automation, San Francisco*, pp. 1490–1495.

Talebi, H. A.; Khorasani, K.,& Patel, R. V. (1998).Neural Network Based Control Schemes for Flexible Link Manipulators: Simulations and Experiments, *Neural Networks*, Vol. 11, pp. 1357–1377.

Kawato, M.; Furukawa, K. & Suzuki, R. (1987). A Hierarchical Neural Network Model for Control and Learning of Voluntary Movement, *Biological Cybernetics*, Vol. 57, pp. 169–185.

Isogai, M.; Arai, F. & Fukuda, T. (1999). Intelligent Sensor Fault Detection of Vibration Control for Flexible Structures, *Joural of Robotics and Mechatronics*, Vol. 11, No. 6, pp. 524–530.

Lianfang, T.; Wang, J., & Mao, Z. (2004). Constrained Motion Control of Flexible Robot Manipulators Based on Recurrent Neural Networks, *IEEE Transactions On Systems, Man, And Cybernetics Part B: Cybernetics*, Vol. 34, No. 3, pp. 1541–1552.

Cheng, X. P. & Patel, R. V. (2003). Neural Network Based Tracking Control of a Flexible MacroŰMicro Manipulator System, *Neural Networks*, Vol. 16, pp. 271–286.

Yazdizadeh, A.; Khorasani, K. & Patel, R.V. (2000). Identification of a Two-Link Flexible Manipulator Using Adaptive Time Delay Neural Networks, *IEEE Transactions On Systems, Man, And Cybernetics Part B: Cybernetics*, Vol. 30, No. 1, pp. 165–172.

Ge, S. S.; Lee, T. H. & Zhu, G. (1996). Genetic Algorithm Tuning of Lyapunov-Based Controllers" An Application to a Single-Link Flexible Robot System, *IEEE Transactions On Industrial Electronics*, Vol. 43, No. 5, pp. 567–573.

Principe, J.; Euliano, N. & Lefebvre, W. (2000). Neural and Adaptive Systems: Fundamentals Through Simulations, John Wiley and Sons, New York, pp. 100–172.

Mansour, T.; Konno, A. & Uchiyama, M. (2008). Modified PID Control of a Single-Link Flexible Robot, *Advanced Robotics*, Vol. 22, pp. 433–449.

Adaptive PID Control for Asymptotic Tracking Problem of MIMO Systems

Kenichi Tamura[1] and Hiromitsu Ohmori[2]
[1]*Tokyo Metropolitan University*
[2]*Keio University*
JAPAN

1. Introduction

PID control, which is usually known as a classical output feedback control for SISO systems, has been widely used in the industrial world(Åström & Hägglund, 1995; Suda, 1992). The tuning methods of PID control are adjusting the proportional, the integral and the derivative gains to make an output of a controlled system track a target value properly. There exist much more researches on tuning methods of PID control for SISO systems than MIMO systems although more MIMO systems actually exist than SISO systems. The tuning methods for SISO systems are difficult to apply to PID control for MIMO systems since the gains usually become matrices in such case.

MIMO systems usually tend to have more complexities and uncertainties than SISO systems. Several tuning methods of PID control for such MIMO system are investigated as follows. From off-line approach, there are progressed classical loop shaping based methods (Ho et al., 2000; Hara et al., 2006) and H_∞ control theory based methods (Mattei, 2001; Saeki, 2006; Zheng et al., 2002). From on-line approach, there are methods from self-tuning control such as the generalized predictive control based method (Gomma, 2004), the generalized minimum variance control based method (Yusof et al., 1994), the model matching based method (Yamamoto et al., 1992) and the method using neural network (Chang et al., 2003).

These conventional methods often require that the MIMO system is stable and are usually used for a regulator problem for a constant target value but a tracking problem for a time-varying target value, which restrictions narrow their application. So trying these problems is significant from a scientific standpoint how there is possibility of PID control and from a practical standpoint of expanding applications. In MIMO case, there is possibility to solve these problems because PID control has more freedoms in tuning of PID gain matrices.

On the other hand, adaptive servo control is known for a problem of the asymptotic output tracking and/or disturbances rejection to unknown systems under guaranteeing stability. There are researches for SISO systems (Hu & Tomizuka, 1993; Miyasato, 1998; Ortega & Kelly, 1985) and for MIMO systems (Chang & Davison, 1995; Dang & Owens, 2006; Johansson, 1987). Their controllers generally depend on structures of the controlled system and the reference system, which features are undesirable from standpoint of utility (Saeki, 2006; Miyamoto, 1999). So it is important to develop the fixed controller like PID controller to solve the servo problem and to show that conditions. But they are difficult to apply to the tuning of PID controller because of differences of their construction.

In this paper, we consider adaptive PID control for the asymptotic output tracking problem of MIMO systems with unknown system parameters under existence of unknown disturbances.

The proposed PID controller has constant gain matrices and adjustable gain matrices. The proposed adaptive tuning laws of the gain matrices are derived by using Lyapunov theorem. That is a Lyapunov function based on characteristics of the proposed PID controller is constructed. This method guarantees the asymptotic output tracking even if the controlled MIMO system is unstable and has uncertainties and unknown constant disturbances. Finally, the effectiveness of the proposed method is confirmed with simulation results for the 8-state, 2-input and 2-output missile control system and the 4-state, 2-input and 2-output unstable system.

2. Problem statement

Consider the MIMO system:

$$\dot{x}(t) = Ax(t) + Bu(t) + d_i, \tag{1}$$

$$y(t) = Cx(t) + d_o, \tag{2}$$

where $x(t) \in \mathcal{R}^n, u(t) \in \mathcal{R}^m, y(t) \in \mathcal{R}^m$ are the state vector, the input vector and the output vector respectively, $d_i \in \mathcal{R}^n, d_o \in \mathcal{R}^m$ are unknown constant disturbances, and A, B, C are unknown system matrices.
The target signal of the output is $y_M(t) \in \mathcal{R}^m$ generated by the reference system:

$$\dot{x}_M(t) = A_M x_M(t) + B_M u_M, \tag{3}$$

$$y_M(t) = C_M x_M(t), \tag{4}$$

where $x_M(t) \in \mathcal{R}^{n_M}$ and $u_M \in \mathcal{R}^{r_M}$ are the state vector and the constant input vector, respectively. Note that A_M, B_M, C_M are allowed to be unknown matrices.
In this article, we propose the new adaptive PID controller:

$$u(t) = K_{I0}\int_0^t e_y(\tau)d\tau + (K_{P0} + K_{P1}(t))e_y(t) + K_{D1}(t)\dot{e}_y(t) + K_{P2}(t)y_M(t) + K_{D2}(t)\dot{y}_M(t) \tag{5}$$

which has the adjustable gain matrices $K_{P1}(t),\ K_{P2}(t), K_{D1}(t), K_{D2}(t) \in \mathcal{R}^{m\times m}$ and the constant gain matrices $K_{I0}, K_{P0} \in \mathcal{R}^{m\times m}$, and

$$e_y(t) = y_M(t) - y(t) \tag{6}$$

denotes the error of the output from the target signal $y_M(t)$. The diagram of the proposed PID controller is shown in Fig. 1.
The objective is to design the constant gain matrices K_{I0}, K_{P0} and the adaptive tuning laws of the adjustable gain matrices $K_{P1}(t), K_{P2}(t), K_{D1}(t), K_{D2}(t)$ to solve the asymptotic output tracking, i.e. $e_y(t) \to 0$ as $t \to \infty$.
Here we assume the following conditions:
Assumption 1: rank $\begin{bmatrix} A & B \\ C & 0 \end{bmatrix} = n + m$, and $\lambda_i(M_{11})\lambda_j(A_M) \neq 1, i = 1, 2, \cdots, n, j = 1, 2, \cdots, n_M$,
where $\begin{bmatrix} M_{11} & M_{12} \\ M_{21} & M_{22} \end{bmatrix} := \begin{bmatrix} A & B \\ C & 0 \end{bmatrix}^{-1}$, $M_{11} \in \mathcal{R}^{n\times n}$ and $\lambda(\cdot)$ denotes eigenvalues of a matrix.
Assumption 2: rank $\begin{bmatrix} C_M \\ C_M A_M \end{bmatrix} = n_M$.
Assumption 3: The zero-dynamics of $\{A, B, C\}$ is asymptotically stable.
Assumption 4: (a) $CB = 0,\ CAB > 0$ or (b) $CB > 0$.

Let us explain these assumptions. Assumption 1 is well known condition for a servo problem. Assumption 2 means the output of the reference system and its derivative contain the information of its state. Assumption 3 equals to the minimum phase property of the controlled system. Assumption 4 contains the condition that the relative degrees are ≤ 2. It is inevitable that these conditions seem a little severe because these are conditions for the PID controller that has the structural constraint. But also there is an advantage that the controlled system's stability property, which is often assumed in other PID control's methods, is not assumed.

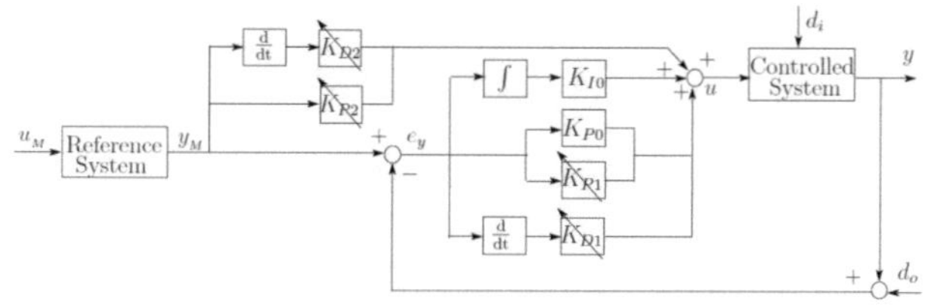

Fig. 1. Proposed Adaptive PID Controller

3. Error system with proposed adaptive PID controller

In this section, we derive the error system with the adaptive PID controller. When the perfect output tracking occurs (i.e. $y(t) = y_M(t),\ \forall t \geq 0$), we can define the corresponding state and input trajectories as $x^*(t), u^*(t)$, respectively. That is $x^*(t), u^*(t)$ are trajectories satisfying the following relation:

$$\dot{x}^*(t) = Ax^*(t) + Bu^*(t) + d_i, \tag{7}$$

$$y_M(t) = Cx^*(t) + d_o,\ \forall t \geq 0. \tag{8}$$

From Appendix A inspired by (Kaufman et al., 1994), there exist matrices $M_{ij}, T_{ij}, i, j = 1, 2$ under Assumption 1, and the ideal trajectories $x^*(t), u^*(t)$ satisfying relations (7), (8) can be expressed as

$$x^*(t) = T_{11}x_M(t) + T_{12}u_M - M_{11}d_i - M_{12}d_o, \tag{9}$$

$$u^*(t) = T_{21}x_M(t) + T_{22}u_M - M_{21}d_i - M_{22}d_o. \tag{10}$$

Introducing these ideal trajectories, we can define the following state error

$$e_x(t) = x^*(t) - x(t). \tag{11}$$

Then, the output tracking error (6) can be described as

$$e_y(t) = y_M(t) - y(t) = (Cx^*(t) + d_o) - (Cx(t) + d_o) = Ce_x(t), \tag{12}$$

which means that if the error system obtained by differentiating (11):

$$\dot{e}_x(t) = Ae_x(t) + B(u^*(t) - u(t)) \tag{13}$$

can be asymptotically stabilized i.e. $e_x(t) \to 0$, then the asymptotic output tracking can be achieved i.e. $e_y(t) \to 0$.
Now, substituting (5) and (10) into (13), we get the following closed loop error system:

$$\begin{aligned}\dot{e}_x(t) = Ae_x(t) - B\Big[&- T_{21}x_M(t) - T_{22}u_M + M_{21}d_i + M_{22}d_o + K_{I0}\int_0^t e_y(\tau)\mathrm{d}\tau \\ &+ K_{P0}e_y(t) + K_{P1}(t)e_y(t) + K_{D1}(t)\dot{e}_y(t) + K_{P2}(t)y_M(t) + K_{D2}(t)\dot{y}_M(t)\Big].\end{aligned} \tag{14}$$

From Appendix B, there exist matrices $S_1, S_2 \in \mathcal{R}^{m\times m}$ under Assumption 2, and $T_{21}x_M(t)$ in (14) can be decomposed as

$$T_{21}x_M(t) = S_1 y_M(t) + S_2(\dot{y}_M(t) - C_M B_M u_M). \tag{15}$$

Hence, (14) can be expressed as

$$\begin{aligned}\dot{e}_x(t) = Ae_x(t) - B\Big[&(S_2C_MB_M - T_{22})u_M + M_{21}d_i + M_{22}d_o + K_{I0}\int_0^t e_y(\tau)\mathrm{d}\tau + K_{P0}e_y(t)\\ &+ K_{P1}(t)e_y(t) + K_{D1}(t)\dot{e}_y(t) + (K_{P2}(t) - S_1)y_M(t) + (K_{D2}(t) - S_2)\dot{y}_M(t)\Big].\end{aligned} \tag{16}$$

Here put the constant term of the above equation as

$$\widetilde{d} := (S_2C_MB_M - T_{22})u_M + M_{21}d_i + M_{22}d_o$$

to represent (16) simply as

$$\begin{aligned}\dot{e}_x(t) = Ae_x(t) - B\Big[&\widetilde{d} + K_{I0}\int_0^t e_y(\tau)\mathrm{d}\tau + K_{P0}e_y(t) + K_{P1}(t)e_y(t) + K_{D1}(t)\dot{e}_y(t)\\ &+ (K_{P2}(t) - S_1)y_M(t) + (K_{D2}(t) - S_2)\dot{y}_M(t)\Big].\end{aligned} \tag{17}$$

Therefore, if the origin of this close-loop error system is asymptotically stabilized i.e. $e_x(t) \to 0$, the asymptotic output tracking i.e. $e_y(t) \to 0$ is achieved. We derive the constant gain matrices and the adaptive tuning laws of adjustable gain matrices to accomplish $e_x(t) \to 0$ in the next section.

4. Adaptive tuning laws of PID gain matrices

In this section, we show the constant gain matrices K_{I0}, K_{P0} and the adaptive tuning law of the adjustable gain matrices $K_{P1}(t), K_{P2}(t), K_{D1}(t), K_{D2}(t)$ to asymptotically stabilize the error dynamics (17) (i.e. $e_x \to 0$ as $t \to \infty$) at *Case A* when Assumption 4(a) is hold or at *Case B* when Assumption 4(b) is hold.

4.1 Case A

Theorem 1: Suppose Assumption 3 and Assumption 4(a). Give the constant gain matrices K_{I0}, K_{P0} as

$$K_{I0} = \gamma_I H_1, \quad K_{P0} = \gamma_I H_2, \tag{18}$$

and the adaptive tuning laws of the adjustable gain matrices $K_{Pi}(t), K_{Di}(t), i = 1, 2$ as

$$\dot{K}_{P1}(t) = \Gamma_{P1}\big(H_1e_y(t) + H_2\dot{e}_y(t)\big)e_y(t)^{\mathrm{T}}, \tag{19a}$$
$$\dot{K}_{D1}(t) = \Gamma_{D1}\big(H_1e_y(t) + H_2\dot{e}_y(t)\big)\dot{e}_y(t)^{\mathrm{T}}, \tag{19b}$$
$$\dot{K}_{P2}(t) = \Gamma_{P2}\big(H_1e_y(t) + H_2\dot{e}_y(t)\big)y_M(t)^{\mathrm{T}}, \tag{19c}$$
$$\dot{K}_{D2}(t) = \Gamma_{D2}\big(H_1e_y(t) + H_2\dot{e}_y(t)\big)\dot{y}_M(t)^{\mathrm{T}} \tag{19d}$$

where

$$H_1 = \mathrm{diag}\{h_{11}, \cdots, h_{1m}\},\ H_2 = \mathrm{diag}\{h_{21}, \cdots, h_{2m}\},\ h_{1j}, h_{2j} > 0,\ j = 1, \cdots, m, \tag{20}$$

then the origin of (17) is asymptotically stable ($e_x(t) \to 0$ as $t \to \infty$) and the adjustable gain matrices are bounded. Here $\Gamma_{P1}, \Gamma_{P2}, \Gamma_{D1}, \Gamma_{D2} \in \mathcal{R}^{m\times m}$ are arbitrary positive definite matrices and γ_I is arbitrary positive scalar.

Proof: From Assumption 4(a), the error dynamics (17) is transformed into the normal form (see e.g. (Isidori, 1995)):

$$\begin{bmatrix}\dot{\xi}_1(t)\\ \dot{\xi}_2(t)\\ \dot{\eta}(t)\end{bmatrix} = \begin{bmatrix}0 & I_m & 0\\ Q_{21} & Q_{22} & Q_{23}\\ Q_{31} & Q_{32} & A_\eta\end{bmatrix}\begin{bmatrix}\xi_1(t)\\ \xi_2(t)\\ \eta(t)\end{bmatrix} - \begin{bmatrix}0\\ CAB\\ 0\end{bmatrix}\Big[K_{I0}\int_0^t \xi_1(\tau)\mathrm{d}\tau + K_{P0}\xi_1(t) + \tilde{d}$$
$$+ K_{P1}(t)\xi_1(t) + K_{D1}(t)\xi_2(t) + (K_{P2}(t) - S_1)y_M(t) + (K_{D2}(t) - S_2)\dot{y}_M(t)\Big], \quad (21)$$

which Q_{ij} are unknown matrices, by transformation

$$\begin{bmatrix}\xi_1(t)\\ \xi_2(t)\\ \eta(t)\end{bmatrix} = \begin{bmatrix}C\\ CA\\ T\end{bmatrix} e_x(t) \quad (22)$$

where $TB = 0$, $T \in \mathcal{R}^{(n-2m)\times n}$ and

$$\xi_1(t) = Ce_x(t) = e_y(t),\ \xi_2(t) = CAe_x(t) = \dot{e}_y(t). \quad (23)$$

Note that when $\xi_1(t), \xi_2(t) \equiv 0$,

$$\dot{\eta}(t) = A_\eta \eta(t), \quad (24)$$

which is called zero-dynamics, is asymptotic stable from Assumption 3.
Thus (21) is can be rewritten as

$$\begin{bmatrix}\dot{\xi}_1(t)\\ \dot{\xi}_2(t)\\ \dot{\eta}(t)\end{bmatrix} = \begin{bmatrix}0 & I_m & 0\\ -K_{\xi 1} & -K_{\xi 2} & Q_{23}\\ Q_{31} & Q_{32} & A_\eta\end{bmatrix}\begin{bmatrix}\xi_1(t)\\ \xi_2(t)\\ \eta(t)\end{bmatrix} - \begin{bmatrix}0\\ I_m\\ 0\end{bmatrix}\Big[CAB(K_{I0}\int_0^t \xi_1(\tau)\mathrm{d}\tau + K_{P0}\xi_1(t) + \tilde{d})$$
$$+ (CABK_{P1}(t) - Q_{21} - K_{\xi 1})\xi_1(t) + (CABK_{D1}(t) - Q_{22} - K_{\xi 2})\xi_2(t)$$
$$+ CAB(K_{P2}(t) - S_1)y_M(t) + CAB(K_{D2}(t) - S_2)\dot{y}_M(t)\Big] \quad (25)$$

where $K_{\xi 1}, K_{\xi 2} \in \mathcal{R}^{m\times m}$ are the constant matrices only used in the proof.

For simplicity, put

$$\xi(t) := \begin{bmatrix}\xi_1(t)\\ \xi_2(t)\end{bmatrix}, A_\xi := \begin{bmatrix}0 & I_m\\ 0 & 0\end{bmatrix}, B_\xi := \begin{bmatrix}0\\ I_m\end{bmatrix}, \quad (26)$$

$$K_\xi := \begin{bmatrix}K_{\xi 1} & K_{\xi 2}\end{bmatrix}, \overline{Q}_1 := \begin{bmatrix}0\\ Q_{23}\end{bmatrix}, \overline{Q}_2 := \begin{bmatrix}Q_{31} & Q_{32}\end{bmatrix}, \quad (27)$$

$$\psi_I(t) := CAB(K_{I0}\int_0^t \xi_1(\tau)\mathrm{d}\tau + K_{P0}\xi_1(t) + \tilde{d}), \quad (28)$$

$$\Psi_{P1}(t) := CABK_{P1}(t) - Q_{21} - K_{\xi 1}, \quad (29a)$$
$$\Psi_{D1}(t) := CABK_{D1}(t) - Q_{22} - K_{\xi 2}, \quad (29b)$$
$$\Psi_{P2}(t) := CAB(K_{P2}(t) - S_1), \quad (29c)$$
$$\Psi_{D2}(t) := CAB(K_{D2}(t) - S_2) \quad (29d)$$

to describe (25) as

$$\begin{bmatrix}\dot{\xi}(t)\\ \dot{\eta}(t)\end{bmatrix} = \begin{bmatrix}A_\xi - B_\xi K_\xi & \overline{Q}_1\\ \overline{Q}_2 & A_\eta\end{bmatrix}\begin{bmatrix}\xi(t)\\ \eta(t)\end{bmatrix} - \begin{bmatrix}B_\xi\\ 0\end{bmatrix}\Big[\psi_I(t) + \Psi_{P1}(t)\xi_1(t)$$
$$+ \Psi_{D1}(t)\xi_2(t) + \Psi_{P2}(t)y_M(t) + \Psi_{D2}(t)\dot{y}_M(t)\Big], \tag{30}$$

where

$$\dot{\psi}_I(t) = CAB(K_{I0}\xi_1(t) + K_{P0}\xi_2(t)), \tag{31}$$
$$\dot{\Psi}_{P1}(t) = CAB\dot{K}_{P1}(t), \tag{32a}$$
$$\dot{\Psi}_{D1}(t) = CAB\dot{K}_{D1}(t), \tag{32b}$$
$$\dot{\Psi}_{P2}(t) = CAB\dot{K}_{P2}(t), \tag{32c}$$
$$\dot{\Psi}_{D2}(t) = CAB\dot{K}_{D2}(t). \tag{32d}$$

Meanwhile because $\{A_\xi, B_\xi\}$ is controllable pair from (26), there exist K_ξ such that Lyapunov equation

$$P_\xi(A_\xi - B_\xi K_\xi) + (A_\xi - B_\xi K_\xi)^{\mathrm{T}} P_\xi = -Q, \;\; Q > 0$$

has an unique positive solution $P_\xi > 0$. So here we set $Q = 2\varepsilon I_{2m}$, $\varepsilon > 0$ and select K_ξ as

$$K_{\xi_1} = \varepsilon H_1^{-1}, \;\; K_{\xi_2} = \varepsilon H_2^{-1}(I_m + (1/\varepsilon)H_1), \tag{33}$$
$$H_i = \mathrm{diag}\{h_{i1}, \cdots, h_{im}\}, h_{ij} > 0, i = 1, 2, \; j = 1, \cdots, m,$$

such that

$$P_\xi(A_\xi - B_\xi K_\xi) + (A_\xi - B_\xi K_\xi)^{\mathrm{T}} P_\xi = -2\varepsilon I_{2m}, \; \varepsilon > 0, \tag{34}$$

has the unique positive solution

$$P_\xi = \begin{bmatrix}P_{\xi 1} & \overline{P}\\ \overline{P}^{\mathrm{T}} & P_{\xi 2}\end{bmatrix} \in \mathcal{R}^{2m\times 2m}, \tag{35}$$
$$\overline{P} = H_1, \;\; P_{\xi 2} = H_2, \;\; P_{\xi 1} = \varepsilon(H_1 H_2^{-1} + H_1^{-1}H_2) + H_1 H_2^{-1} H_1.$$

It is clear P_ξ of (35) is a positive matrix on $\varepsilon > 0$ from Schur complement (see e.g. (Iwasaki, 1997)) because $P_{\xi 2} = H_2 > 0$, $P_{\xi 1} - \overline{P}P_{\xi 2}^{-1}\overline{P}^{\mathrm{T}} = \varepsilon(H_1 H_2^{-1} + H_1^{-1}H_2) > 0$.

Furthermore since A_η of (24) is asymptotic stable matrix from Assumption 3, there exists an unique solution $P_\eta \in \mathcal{R}^{(n-2m)\times(n-2m)} > 0$ satisfying

$$P_\eta A_\eta + A_\eta^{\mathrm{T}} P_\eta = -I_{n-2m}. \tag{36}$$

Now, by using P_ξ of (35) and P_η of (36), we consider the following Lyapunov function candidate:

$$V(\xi(t), \eta(t), \psi_I(t), \Psi_{P1}(t), \Psi_{P2}(t), \Psi_{D1}(t), \Psi_{D2}(t))$$
$$= \begin{bmatrix}\xi(t)\\ \eta(t)\end{bmatrix}^{\mathrm{T}}\begin{bmatrix}P_\xi & 0\\ 0 & P_\eta\end{bmatrix}\begin{bmatrix}\xi(t)\\ \eta(t)\end{bmatrix} + \psi_I(t)^{\mathrm{T}}\gamma_I^{-1}(CAB)^{-1}\psi_I(t)$$
$$+ \mathrm{Tr}\big[\Psi_{P1}(t)^{\mathrm{T}}\Gamma_{P1}^{-1}(CAB)^{-1}\Psi_{P1}(t)\big] + \mathrm{Tr}\big[\Psi_{D1}(t)^{\mathrm{T}}\Gamma_{D1}^{-1}(CAB)^{-1}\Psi_{D1}(t)\big]$$
$$+ \mathrm{Tr}\big[\Psi_{P2}(t)^{\mathrm{T}}\Gamma_{P2}^{-1}(CAB)^{-1}\Psi_{P2}(t)\big] + \mathrm{Tr}\big[\Psi_{D2}(t)^{\mathrm{T}}\Gamma_{D2}^{-1}(CAB)^{-1}\Psi_{D2}(t)\big] \tag{37}$$

where $\Gamma_{P1}, \Gamma_{D1}, \Gamma_{P2}, \Gamma_{D2} \in \mathcal{R}^{m\times m}$ are arbitrary positive definite matrices, γ_I is positive scalar. $\mathrm{Tr}[\cdot]$ denotes trace of a square matrix. Here put $V(t) := V(\xi(t), \eta(t), \psi_I(t), \Psi_{P1}(t), \Psi_{P2}(t), \Psi_{D1}(t), \Psi_{D2}(t))$ for simplicity. The derivative of (37) along the trajectories of the error system (30) $\sim$ (32d) can be calculated as

$$
\begin{aligned}
\dot{V}(t) &= 2\begin{bmatrix}\dot{\xi}(t)\\ \dot{\eta}(t)\end{bmatrix}^{\mathrm{T}}\begin{bmatrix}P_\xi & 0\\ 0 & P_\eta\end{bmatrix}\begin{bmatrix}\xi(t)\\ \eta(t)\end{bmatrix} + 2\psi_I(t)^{\mathrm{T}}\gamma_I^{-1}(CAB)^{-1}\dot{\psi}_I(t)\\
&\quad + 2\mathrm{Tr}\big[\Psi_{P1}(t)^{\mathrm{T}}\Gamma_{P1}^{-1}(CAB)^{-1}\dot{\Psi}_{P1}(t)\big] + 2\mathrm{Tr}\big[\Psi_{D1}(t)^{\mathrm{T}}\Gamma_{D1}^{-1}(CAB)^{-1}\dot{\Psi}_{D1}(t)\big]\\
&\quad + 2\mathrm{Tr}\big[\Psi_{P2}(t)^{\mathrm{T}}\Gamma_{P2}^{-1}(CAB)^{-1}\dot{\Psi}_{P2}(t)\big] + 2\mathrm{Tr}\big[\Psi_{D2}(t)^{\mathrm{T}}\Gamma_{D2}^{-1}(CAB)^{-1}\dot{\Psi}_{D2}(t)\big]\\
&= \begin{bmatrix}\xi(t)\\ \eta(t)\end{bmatrix}^{\mathrm{T}}\begin{bmatrix}P_\xi(A_\xi - B_\xi K_\xi) + (A_\xi - B_\xi K_\xi)^{\mathrm{T}}P_\xi & P_\xi\overline{Q}_1 + \overline{Q}_2^{\mathrm{T}}P_\eta\\ (P_\xi\overline{Q}_1 + \overline{Q}_2^{\mathrm{T}}P_\eta)^{\mathrm{T}} & P_\eta A_\eta + A_\eta^{\mathrm{T}}P_\eta\end{bmatrix}\begin{bmatrix}\xi(t)\\ \eta(t)\end{bmatrix}\\
&\quad + 2\psi_I(t)^{\mathrm{T}}\Big[-B_\xi^{\mathrm{T}}P_\xi\xi(t) + \gamma_I^{-1}(CAB)^{-1}\dot{\psi}_I(t)\Big]\\
&\quad + 2\mathrm{Tr}\Big[\Psi_{P1}(t)^{\mathrm{T}}\Big(-B_\xi^{\mathrm{T}}P_\xi\xi(t)\xi_1(t)^{\mathrm{T}} + \Gamma_{P1}^{-1}(CAB)^{-1}\dot{\Psi}_{P1}(t)\Big)\Big]\\
&\quad + 2\mathrm{Tr}\Big[\Psi_{D1}(t)^{\mathrm{T}}\Big(-B_\xi^{\mathrm{T}}P_\xi\xi(t)\xi_2(t)^{\mathrm{T}} + \Gamma_{D1}^{-1}(CAB)^{-1}\dot{\Psi}_{D1}(t)\Big)\Big]\\
&\quad + 2\mathrm{Tr}\Big[\Psi_{P2}(t)^{\mathrm{T}}\Big(-B_\xi^{\mathrm{T}}P_\xi\xi(t)y_M(t)^{\mathrm{T}} + \Gamma_{P2}^{-1}(CAB)^{-1}\dot{\Psi}_{P2}(t)\Big)\Big]\\
&\quad + 2\mathrm{Tr}\Big[\Psi_{D2}(t)^{\mathrm{T}}\Big(-B_\xi^{\mathrm{T}}P_\xi\xi(t)\dot{y}_M(t)^{\mathrm{T}} + \Gamma_{D2}^{-1}(CAB)^{-1}\dot{\Psi}_{D2}(t)\Big)\Big]\\
&= \begin{bmatrix}\xi(t)\\ \eta(t)\end{bmatrix}^{\mathrm{T}}\begin{bmatrix}P_\xi(A_\xi - B_\xi K_\xi) + (A_\xi - B_\xi K_\xi)^{\mathrm{T}}P_\xi & P_\xi\overline{Q}_1 + \overline{Q}_2^{\mathrm{T}}P_\eta\\ (P_\xi\overline{Q}_1 + \overline{Q}_2^{\mathrm{T}}P_\eta)^{\mathrm{T}} & P_\eta A_\eta + A_\eta^{\mathrm{T}}P_\eta\end{bmatrix}\begin{bmatrix}\xi(t)\\ \eta(t)\end{bmatrix}\\
&\quad + 2\psi_I(t)^{\mathrm{T}}\Big[-B_\xi^{\mathrm{T}}P_\xi\xi(t) + \gamma_I^{-1}(K_{I0}\xi_1(t) + K_{P0}\xi_2(t))\Big]\\
&\quad + 2\mathrm{Tr}\Big[\Psi_{P1}(t)^{\mathrm{T}}\Big(-B_\xi^{\mathrm{T}}P_\xi\xi(t)\xi_1(t)^{\mathrm{T}} + \Gamma_{P1}^{-1}\dot{K}_{P1}(t)\Big)\Big]\\
&\quad + 2\mathrm{Tr}\Big[\Psi_{D1}(t)^{\mathrm{T}}\Big(-B_\xi^{\mathrm{T}}P_\xi\xi(t)\xi_2(t)^{\mathrm{T}} + \Gamma_{D1}^{-1}\dot{K}_{D1}(t)\Big)\Big]\\
&\quad + 2\mathrm{Tr}\Big[\Psi_{P2}(t)^{\mathrm{T}}\Big(-B_\xi^{\mathrm{T}}P_\xi\xi(t)y_M(t)^{\mathrm{T}} + \Gamma_{P2}^{-1}\dot{K}_{P2}(t)\Big)\Big]\\
&\quad + 2\mathrm{Tr}\Big[\Psi_{D2}(t)^{\mathrm{T}}\Big(-B_\xi^{\mathrm{T}}P_\xi\xi(t)\dot{y}_M(t)^{\mathrm{T}} + \Gamma_{D2}^{-1}\dot{K}_{D2}(t)\Big)\Big].
\end{aligned}
\tag{38}
$$

Therefore from $\xi(t) = [\xi_1^{\mathrm{T}}, \xi_2^{\mathrm{T}}]^{\mathrm{T}} = [e_y^{\mathrm{T}}, \dot{e}_y^{\mathrm{T}}]^{\mathrm{T}}$ and $B_\xi^{\mathrm{T}}P_\xi = [H_1\ H_2]$, giving the constant gain matrices K_{I0}, K_{P0} as (18), (20) and the adaptive tuning laws of $K_{Pi}(t)$, $K_{Di}(t)$, $i = 1,2$ as (19a) $\sim$ (19d), (20) , we can get (38) be

$$
\dot{V}(t) = \begin{bmatrix}\xi(t)\\ \eta(t)\end{bmatrix}^{\mathrm{T}}\begin{bmatrix}P_\xi(A_\xi - B_\xi K_\xi) + (A_\xi - B_\xi K_\xi)^{\mathrm{T}}P_\xi & P_\xi\overline{Q}_1 + \overline{Q}_2^{\mathrm{T}}P_\eta\\ (P_\xi\overline{Q}_1 + \overline{Q}_2^{\mathrm{T}}P_\eta)^{\mathrm{T}} & P_\eta A_\eta + A_\eta^{\mathrm{T}}P_\eta\end{bmatrix}\begin{bmatrix}\xi(t)\\ \eta(t)\end{bmatrix}. \tag{39}
$$

Here the symmetric matrix of (39) can be expressed as

$$
\begin{bmatrix}-2\varepsilon I_{2m} & P_\xi\overline{Q}_1 + \overline{Q}_2^{\mathrm{T}}P_\eta\\ (P_\xi\overline{Q}_1 + \overline{Q}_2^{\mathrm{T}}P_\eta)^{\mathrm{T}} & -I_{n-2m}\end{bmatrix} \tag{40}
$$

from (36), (34). Using Schur complement, we have the following necessary and sufficient conditions such that (40) is negative definite:

$$- 2\varepsilon I_{2m} < 0, \tag{41}$$

$$- I_{n-2m} + (P_\xi \overline{Q}_1 + \overline{Q}_2^{\mathrm{T}} P_\eta)^{\mathrm{T}} \frac{1}{2\varepsilon} (P_\xi \overline{Q}_1 + \overline{Q}_2^{\mathrm{T}} P_\eta) < 0 \tag{42}$$

where

$$P_\xi \overline{Q}_1 = \begin{bmatrix} \overline{P} \\ P_{\xi 2} \end{bmatrix} Q_{23} = \begin{bmatrix} H_1 \\ H_2 \end{bmatrix} Q_{23} \tag{43}$$

from (27), (35). Obviously, the first inequality (41) is hold. The second inequality (42) is also achieved under large $\varepsilon > 0$ (because $\overline{Q}_2^{\mathrm{T}} P_\eta$ and $P_\xi \overline{Q}_1$ are independent of ε). At this time, (40) becomes negative definite matrix and (39) is

$$\dot{V}(t) = \begin{bmatrix} \xi(t) \\ \eta(t) \end{bmatrix}^{\mathrm{T}} \begin{bmatrix} -2\varepsilon I_{2m} & P_\xi \overline{Q}_1 + \overline{Q}_2^{\mathrm{T}} P_\eta \\ (P_\xi \overline{Q}_1 + \overline{Q}_2^{\mathrm{T}} P_\eta)^{\mathrm{T}} & -I_{n-2m} \end{bmatrix} \begin{bmatrix} \xi(t) \\ \eta(t) \end{bmatrix} \leq 0. \tag{44}$$

Hence, giving the constant gain matrices K_{I0},K_{P0} as (18), (20) and the adaptive law of $K_{Pi}(t)$, $K_{Di}(t)$, $i = 1, 2$ as (19a) $\sim$ (19d), (20), we have shown that there exists the Lyapunov function which derivative is (44). Therefore, all variables in $V(\cdot)$ is bounded, that is $\xi(t), \eta(t), \psi_I(t)$, $\Psi_{P1}(t)$, $\Psi_{P2}(t)$, $\Psi_{D1}(t)$, $\Psi_{D2}(t) \in \mathcal{L}_\infty$. Furthermore, $\dot{\xi}(t), \dot{\eta}(t)$ are bounded from (30) and $\xi(t), \eta(t) \in \mathcal{L}_2$ from (44). Accordingly, since $\xi(t), \eta(t) \in \mathcal{L}_2 \cap \mathcal{L}_\infty$, $\dot{\xi}(t), \dot{\eta}(t) \in \mathcal{L}_\infty$, the origin of the error system $(\xi, \eta) = (0, 0)$, namely $e_x = 0$ is asymptotically stable from Barbalat's lemma, and $K_{Pi}(t)$, $K_{Di}(t)$, $i = 1, 2$ are bounded from $\Psi_{P1}(t)$, $\Psi_{P2}(t)$, $\Psi_{D1}(t)$, $\Psi_{D2}(t) \in \mathcal{L}_\infty$. □

Remark 1: In proposed method, it is important how to select $H_1, H_2, h_{ij} > 0$ which always guarantee the asymptotic stability because they also affect the transient response. Especially, taking large h_{ij} causes the large over shoot of inputs at first time range because of the proportional gain matrix K_{P0} with h_{ij}. So it seems to be appropriate to adjust h_{ij} from small values slowly such that better response is gotten although it is difficult to show concrete guide because system's parameters are unknown. But it is also one of the characteristic in our proposed method that the designer can adjust transient response manually under guaranteeing stability.

4.2 Case B

Corollary 1: Suppose Assumption 3 and Assumption 4(b). Give the constant gain matrices K_{I0}, K_{P0} as (18) and the adaptive tuning law of the adjustable gain matrices $K_{Pi}(t), K_{Di}(t), i = 1, 2$ as (19a) $\sim$ (19d) where $H_1 = \mathrm{diag}\{h_{1j}, \cdots, h_{1m}\}$, $H_2 = 0, h_{1j} > 0$, $j = 1, \cdots, m$, then (17) is asymptotically stable and the adjustable gain matrices are bounded. Here Γ_{P1}, Γ_{P2}, $\Gamma_{D1}, \Gamma_{D2} \in \mathcal{R}^{m \times m}$ are arbitrary positive definite matrices and γ_I is arbitrary positive scalar.

Proof : After transforming the error system (17) into the normal form (see e.g. (Isidori, 1995)) based on Assumption 4(b), do the procedure like Theorem 1, it can be proved more easily than Theorem 1. □

5. Simulations

Example 1

Consider the missile control system (Bar-Kana & Kaufman, 1985):

$$\dot{x}(t) = \begin{bmatrix} 3.23 & 12.5 & -476 & 0 & 228 & 0 & 0 & 0 \\ -12.5 & -3.23 & 0 & 476.0 & 0 & -228 & 0 & 0 \\ 0.39 & 0 & -1.93 & -10 & -415 & 0 & 0 & 0 \\ 0 & -0.39 & 10 & -1.93 & 0 & -415 & 0 & 0 \\ 0 & 0 & 0 & 0 & 0 & 0 & 75 & 0 \\ 0 & 0 & 0 & 0 & 0 & 0 & 0 & -75 \\ 0 & 0 & 22.4 & 0 & -300 & 0 & -150 & 0 \\ 0 & 0 & 0 & -22.4 & 0 & 300 & 0 & -150 \end{bmatrix} x(t) + \begin{bmatrix} 0 & 0 \\ 0 & 0 \\ 0 & 0 \\ 0 & 0 \\ 0 & 0 \\ 0 & 0 \\ -1 & 0 \\ 0 & -1 \end{bmatrix} u(t) + d_i.$$

$$y(t) = \begin{bmatrix} -2.99 & 0 & -1.19 & 1.5375 & -27.64 & 0 & 0 & 0 \\ 0 & -2.99 & 1.5375 & 1.19 & 0 & 27.64 & 0 & 0 \end{bmatrix} x(t) + d_o.$$

Let the reference system be

$$\dot{x}_M(t) = \begin{bmatrix} 0 & q_{M1} & 0 & 0 \\ -q_{M1} & 0 & 0 & 0 \\ 0 & 0 & 0 & q_{M2} \\ 0 & 0 & -q_{M2} & 0 \end{bmatrix} x_M(t), \; y_M(t) = \begin{bmatrix} 0 & q_{M3} & 0 & 0 \\ 0 & 0 & q_{M4} & 0 \end{bmatrix} x_M(t).$$

which means $y_M(t) = \begin{bmatrix} q_{M3} \cos q_{M1}t & q_{M4} \sin q_{M2}t \end{bmatrix}^{\mathrm{T}}$ at $x_M(0) = \begin{bmatrix} 0\,1\,0\,1 \end{bmatrix}^{\mathrm{T}}$.
Set disturbances d_i, d_o and parameters of the reference system q_M as follows:

$$q_{M1} = 1,\; q_{M2} = 2.0,\; q_{M3} = 0.5,\; q_{M4} = 1,\; d_i = \begin{bmatrix} 0\,0\,0\,0\,0\,0\,1\,2 \end{bmatrix}^{\mathrm{T}},\; d_o = \begin{bmatrix} 0.5 & -1 \end{bmatrix}^{\mathrm{T}}.$$

Select arbitrary H_1, H_2 as $H_1 = \begin{bmatrix} 0.5 & 0 \\ 0 & 0.5 \end{bmatrix}, H_2 = \begin{bmatrix} 0.5 & 0 \\ 0 & 0.5 \end{bmatrix}$ based on Remark 1. Set the $\Gamma_{P1} = \Gamma_{P2} = \Gamma_{D1} = \Gamma_{D2} = I_2$ and $\gamma_I = 1$. Put the initial values $x(0) = 0,\; K_{Pi}(0) = K_{Di}(0) = 0,\; i = 1,2$. It is observed from simulation results at Fig. 2 that $K_{P1}(t), K_{P2}(t), K_{D1}(t), K_{D2}(t)$ are on-line adjusted and the asymptotic output tracking is achieved.

Example 2
Consider the following unstable system:

$$\dot{x}(t) = \begin{bmatrix} 1 & 1 & 4 & 3 \\ 1 & 4 & -3 & 1 \\ -1 & 1 & -5 & -1 \\ 1 & 0 & -1 & -1 \end{bmatrix} x(t) + \begin{bmatrix} 1 & 0 \\ 0 & 1 \\ 0 & 0 \\ 0 & 0 \end{bmatrix} u(t) + d_{\mathrm{i}},$$

$$y(t) = \begin{bmatrix} 1 & 0 & 0 & 0 \\ 0 & 1 & 0 & 0 \end{bmatrix} x(t) + d_{\mathrm{o}}.$$

Set the reference system be

$$\dot{x}_M(t) = \begin{bmatrix} 0 & q_{M1} & 0 & 0 \\ -q_{M1} & 0 & 0 & 0 \\ 0 & 0 & 0 & q_{M2} \\ 0 & 0 & -q_{M2} & 0 \end{bmatrix} x_M(t) + \begin{bmatrix} 0 & 0 \\ 0 & 1 \\ -1 & 0 \\ 0 & 0 \end{bmatrix} u_M,$$

$$y_M(t) = \begin{bmatrix} 0 & q_{M3} & 0 & 0 \\ 0 & 0 & q_{M4} & 0 \end{bmatrix} x_M(t),$$

which generates $y_M(t) = \begin{bmatrix} q_{M3} \cos q_{M1}t & q_{M4} \sin q_{M2}t \end{bmatrix}^{\mathrm{T}}$ at $x_M(0) = \begin{bmatrix} 0\,1\,0\,1 \end{bmatrix}^{\mathrm{T}}$ when $u_M = 0$.

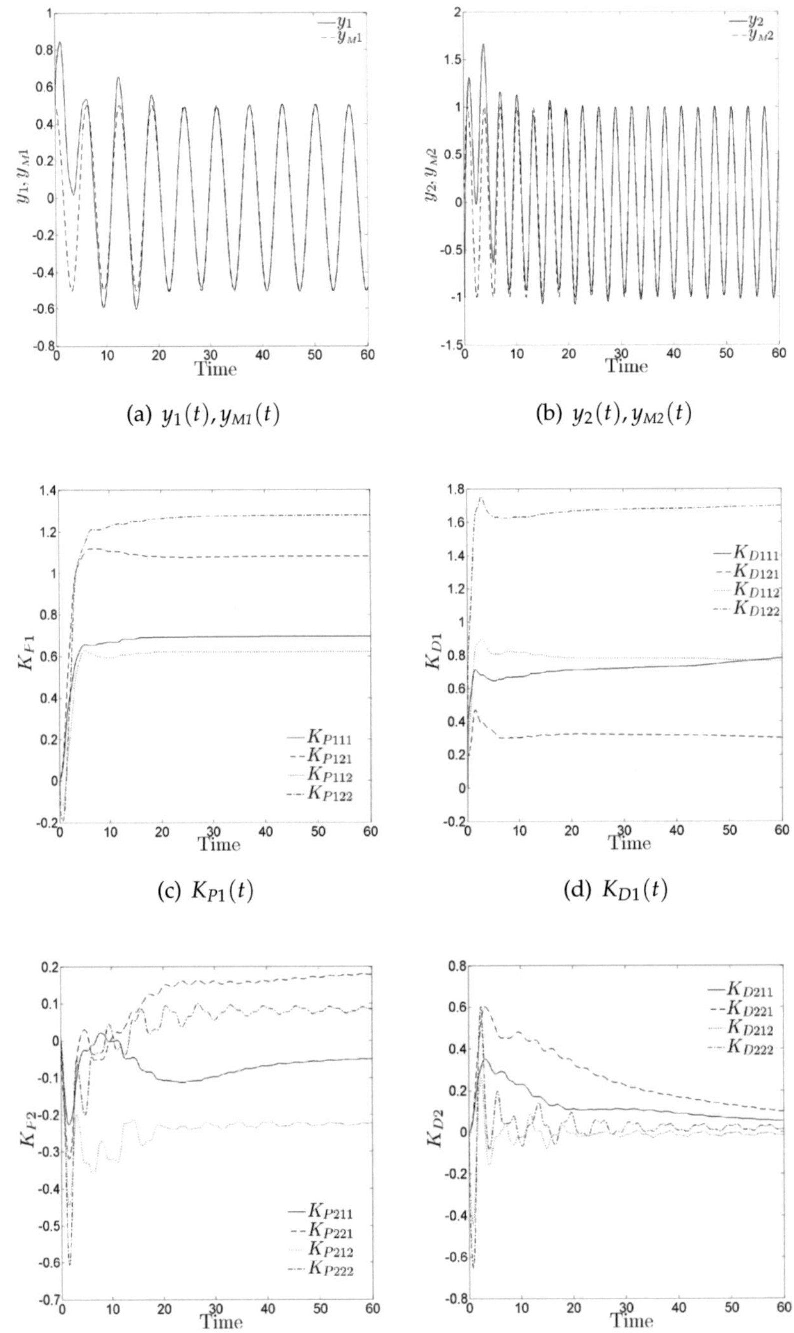

(a) $y_1(t), y_{M1}(t)$ (b) $y_2(t), y_{M2}(t)$

(c) $K_{P1}(t)$ (d) $K_{D1}(t)$

(e) $K_{P2}(t)$ (f) $K_{D2}(t)$

Fig. 2. Simulation Results of Example 1

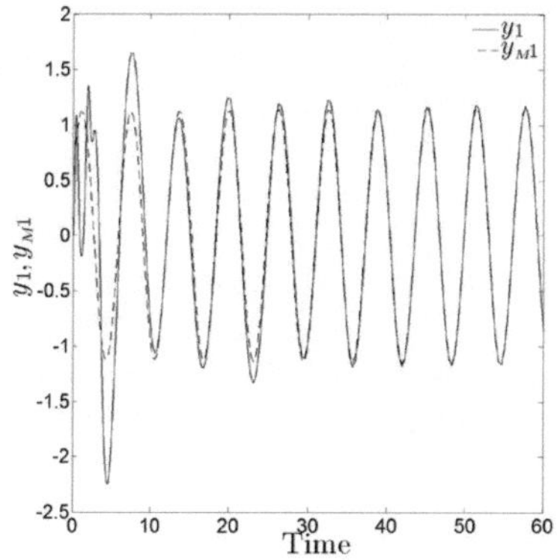

(a) $y_1(t), y_{M1}(t)$

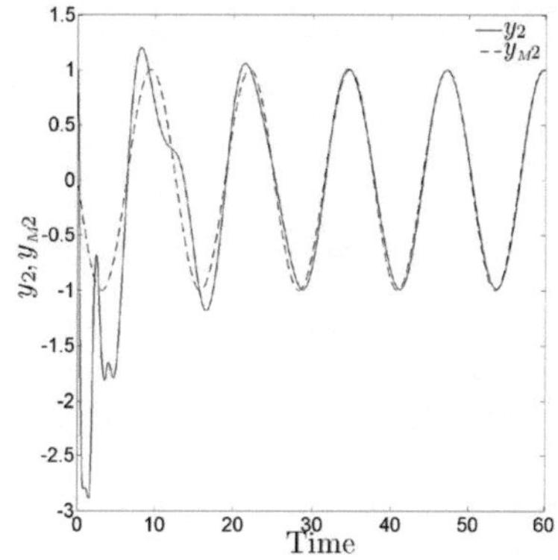

(b) $y_2(t), y_{M2}(t)$

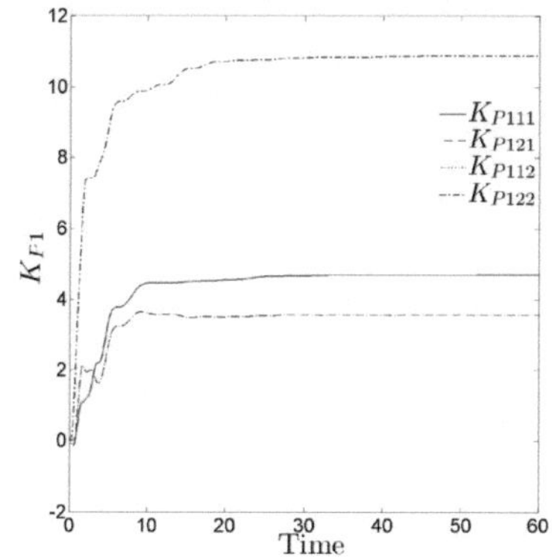

(c) $K_{P1}(t)$

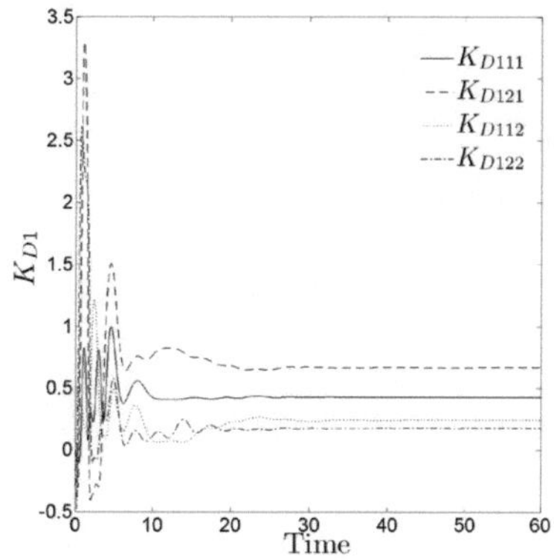

(d) $K_{D1}(t)$

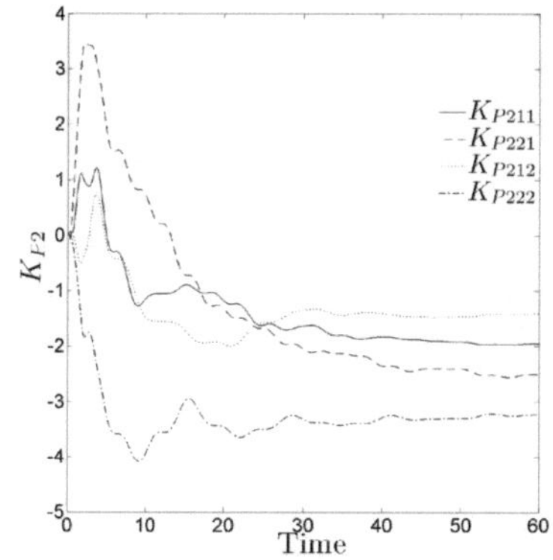

(e) $K_{P2}(t)$

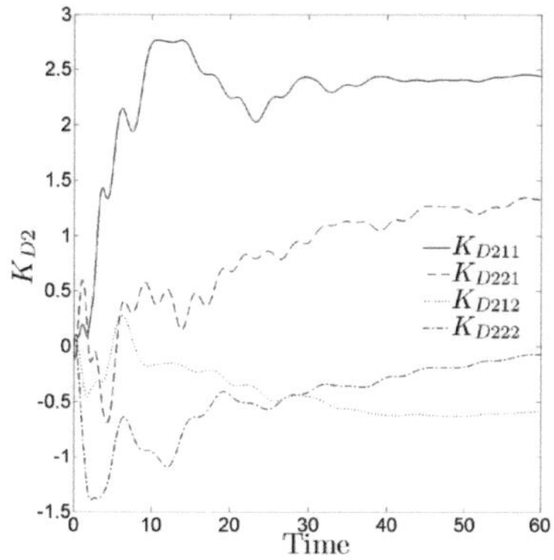

(f) $K_{D2}(t)$

Fig. 3. Simulation Results of Example 2

Disturbances d_i, d_o and parameters of the reference system q_M are set as follows:

$$q_{M1} = 1.0,\ q_{M2} = 0.5,\ q_{M3} = 0.5,\ q_{M4} = 1,\ u_M = \begin{bmatrix} 1\ 2 \end{bmatrix}^{\mathrm{T}},$$
$$d_{\mathrm{i}} = \begin{bmatrix} 1 & -2 & 0 & 0 \end{bmatrix}^{\mathrm{T}},\ d_{\mathrm{o}} = \begin{bmatrix} 0 & 1 \end{bmatrix}^{\mathrm{T}}$$

From Colloraly 1, select arbitrary H_1, H_2 as $H_1 = \begin{bmatrix} 1 & 0 \\ 0 & 1 \end{bmatrix}, H_2 = \begin{bmatrix} 0 & 0 \\ 0 & 0 \end{bmatrix}$.
Set the $\Gamma_{P1} = \Gamma_{P2} = \Gamma_{D1} = \Gamma_{D2} = I_2$ and $\gamma_I = 1$. Put the initial values $x(0) = 0,\ K_{Pi}(0) = K_{Di}(0) = 0,\ i = 1, 2$. We can observe that $K_{P1}(t), K_{P2}(t), K_{D1}(t), K_{D2}(t)$ are on-line adjusted and the asymptotic output tracking is achieved from simulation results at Fig. 3 .

6. Conclusions

We have proposed the new adaptive PID control and its parameter tuning method for the MIMO system. In our method, the asymptotic output tracking can be guaranteed even if the MIMO system is unstable and has unknown system parameters and unknown constant disturbances. The effectiveness of the method is confirmed by numerical simulations. Our future task is extending the controlled system to the nonlinear one.

7. References

Åström, K.J. & Hägglund, T. (1995). *PID Controllers: Theory, Design, and Tuning, 2nd Edition*, ISA, ISBN 978-1-55617-516-9, North Carolina.

Suda, N. (1992). *PID Control*, Asakura Publishing Co., Ltd., ISBN 978-4-254-20966-2, Tokyo.

Ho, W.K.; Lee, T.H.; Xu, W.; Zhou, J.R. & Tay, E.B. (2000). The direct Nyquist array design of PID Controllers. *IEEE Trans. Indust. Elect.*, Vol.47, No.1, 175-185, ISSN 0278-0046

Hara, S.; Iwasaki, T. & Shiokata, D. (2006). Robust PID Control Using Generalized KYP Synthesis. *IEEE Control Systems Magazine*, Vol.26, No.1, 80-91, ISSN 0272-1708

Mattei, M. (2001). Robust multivariable PID control for linear parameter varying systems. *Automatica*, Vol.37, No.12, 1997-2003, ISSN 0005-1098

Saeki, M. (2006). Fixed structure PID controller design for standard H_∞ control problem. *Automatica*, Vol.42, No.1, 93-100, ISSN 0005-1098

Zheng, F.; Wang, Q.G. & Lee, T.H. (2002). On the design of multivariable PID controllers via LMI approach. *Automatica*, Vol.38, No.3, 517-526, ISSN 0005-1098

Gomma, H.W. (2004). Adaptive PID Control Design Based on Generalized Predictive Control (GPC). *Proc. of 2004 IEEE CCA*, 1685-1690, ISBN 0-7803-8633-7, Taipei, Taiwan, Sept. 2004

Yusof, R.; Omatu, S. & Khalid, M. (1994). Self-Tuning PID Control; A Multivariable Derivation and Application. *Automatica*, Vol.30, No.12, 1975-1981, ISSN 0005-1098

Yamamoto, T.; Ishihara, H.; Omatu, S. & Kitamori, T. (1992). A Construction of Multivariable Self-Tuning Controller with Two-Degree-of-Freedom PID Structure. *Transactions of the SICE*, Vol.28, No.4, 484-491, ISSN 0453-4654

Chang, W.D.; Hwang, R.C. & Hsieh, J.G. (2003). A multivariable on-line adaptive PID controllers using auto-tuning neurons. *Engineering Applications of Artificial Intelligence*, Vol.16, No.1, 57-63, ISSN 0952-1976

Hu, J. & Tomizuka, M. (1993). Adaptive asymptotic tracking of repetitive signals – a frequency domain approach. *IEEE Trans. Automatic Control*, Vol.38, No.10, 1752-1578, ISSN 0018-9286

Miyasato, Y. (1998). A design method of universal adaptive servo controller, *Proc. the 37th IEEE CDC*, 2294-2299, ISBN 0-7803-4394-8, Tampa, Florida, Dec. 1998

Ortega, R. & Kelly, R. (1985). On Stability-Instability Conditions of a Simple Robust Adaptive Servo Controller, *Proc. the 24th IEEE CDC*, 502 - 507, Fort Lauderdale, Florida, Dec. 1985

Chang, M. & Davison, E.J. (1995). The Adaptive Servomechanism Problem for MIMO Systems, *Proc. 34th IEEE CDC*, 1738-1743, ISBN 0-7803-2685-7, New Orleans, LA, Dec. 1995

Dang, H & Owens, D.H. (2006). MIMO multi-periodic control system: universal adaptive control schemes. *Int. J. Adaptive Control and Signal Processing*, Vol.20, No.9, 409-429, ISSN 0890-6327

Johansson, R. (1987). Parametric models of linear multivariable systems for adaptive control, *IEEE Trans. Automatic Control*, Vol.32, No.4, 303-313, ISSN 0018-9286

Miyamoto, M. (1999). Design of Fixed-Structure H_∞ Controllers Based on Coprime Factorization and LMI Optimization. *Journal of ISCIE: Systems, Control and Information*, Vol.42, No.2, 80-86, ISSN 0916-1600

Kaufman, H.; Bar-Kana, I. & Sobel, S. (1994). *Direct Adaptive Control Algorithms – Theory and Applications, 2nd edition*, Springer-Verlag, ISBN 0-387-94884-8, New York.

Isidori, A. (1995). *Nonlinear Control Systems, 3rd edition*, Springer-Verlag, ISBN 978-3-540-19916-8, New York.

Iwasaki, T. (1997). *LMI and Control*, SHOKODO Co.,Ltd., ISBN 978-4-7856-9053-3, Tokyo.

Bar-Kana, I. & Kaufman, H. (1985). Global Stability and Performance of a Simplified Adaptive Algorithm. *International Journal of Control*, Vol.42, No.6, 1491-1505, ISSN 0020-7179

Kodama, S. & Suda, N. (1995). *Matrix Theory for System Control*, CORONA PUBLISHING CO.,LTD., ISBN 978-4-339-08330-9, Tokyo.

A. (Proof)

(7), (8) are rewritten as

$$\begin{bmatrix} \dot{x}^*(t) \\ y_M(t) \end{bmatrix} = \begin{bmatrix} A & B \\ C & 0 \end{bmatrix} \begin{bmatrix} x^*(t) \\ u^*(t) \end{bmatrix} + \begin{bmatrix} d_i \\ d_o \end{bmatrix}. \tag{45}$$

Now we prove that the above equation is hold under Assumption 1 by substituting (9), (10). First, we calculate the right side of (45). Since (9), (10) are expressed as

$$\begin{bmatrix} x^*(t) \\ u^*(t) \end{bmatrix} = \begin{bmatrix} T_{11} & T_{12} \\ T_{21} & T_{22} \end{bmatrix} \begin{bmatrix} x_M(t) \\ u_M \end{bmatrix} - \begin{bmatrix} M_{11} & M_{12} \\ M_{21} & M_{22} \end{bmatrix} \begin{bmatrix} d_i \\ d_o \end{bmatrix}, \tag{46}$$

substitute (46) into the right side of (45) to get

$$\text{The right side of (45)} = \begin{bmatrix} A & B \\ C & 0 \end{bmatrix} \begin{bmatrix} T_{11} & T_{12} \\ T_{21} & T_{22} \end{bmatrix} \begin{bmatrix} x_M(t) \\ u_M \end{bmatrix} \tag{47}$$

by using the relation

$$\begin{bmatrix} M_{11} & M_{12} \\ M_{21} & M_{22} \end{bmatrix} = \begin{bmatrix} A & B \\ C & 0 \end{bmatrix}^{-1} \tag{48}$$

from Assumption 1.

Then we calculate the left side of (45). Substituting $\dot{x}^*(t) = T_{11}\dot{x}_M(t)$ which is the time derivative of (9) and using the relation of (3), (4), we can get

$$\text{The left side of (45)} = \begin{bmatrix} T_{11}A_M & T_{11}B_M \\ C_M & 0 \end{bmatrix} \begin{bmatrix} x_M(t) \\ u_M \end{bmatrix}. \tag{49}$$

Therefore from (47), (49), the equation obtained from substituting (9), (10) into (45) is

$$\begin{bmatrix} T_{11}A_M & T_{11}B_M \\ C_M & 0 \end{bmatrix} \begin{bmatrix} x_M(t) \\ u_M \end{bmatrix} = \begin{bmatrix} A & B \\ C & 0 \end{bmatrix} \begin{bmatrix} T_{11} & T_{12} \\ T_{21} & T_{22} \end{bmatrix} \begin{bmatrix} x_M(t) \\ u_M \end{bmatrix}. \tag{50}$$

This equation is always hold for all $x_M(t)$ and u_M if

$$\begin{bmatrix} T_{11}A_M & T_{11}B_M \\ C_M & 0 \end{bmatrix} = \begin{bmatrix} A & B \\ C & 0 \end{bmatrix} \begin{bmatrix} T_{11} & T_{12} \\ T_{21} & T_{22} \end{bmatrix}$$

is hold. This is the matrix linear equation with variables T_{11}.

Now we will show that this matrix equation is solvable. Multiplying both left side of above equation by the nonsingular matrix (48), we have

$$\begin{bmatrix} M_{11} & M_{12} \\ M_{21} & M_{22} \end{bmatrix} \begin{bmatrix} T_{11}A_M & T_{11}B_M \\ C_M & 0 \end{bmatrix} = \begin{bmatrix} T_{11} & T_{12} \\ T_{21} & T_{22} \end{bmatrix}.$$

Obviously, T_{11} is the solution to the linear matrix equation

$$T_{11} = M_{11}T_{11}A_M + M_{12}C_M, \tag{51}$$

and there exists unique solution T_{11} under Assumption 1 (see (Kodama & Suda, 1995)). Therefore rests of T_{ij} exist uniquely as

$$\begin{aligned} T_{12} &= M_{11}T_{11}B_M, \ T_{22} = M_{21}T_{11}B_M, \\ T_{21} &= M_{21}T_{11}A_M + M_{22}C_M. \end{aligned}$$

We have proved that (9), (10) satisfy the relation (7), (8) for all d_o, d_i, u_M under Assumption 1. □

B. (Proof)

Using (4), we can calculate (15) as

$$(T_{21} - S_1C_M - S_2C_MA_M)x_M(t) = 0.$$

This equation is always hold for all $x_M(t)$ if

$$T_{21} - S_1C_M - S_2C_MA_M = 0$$

is satisfied, that is if

$$[S_1 \ S_2] \begin{bmatrix} C_M \\ C_MA_M \end{bmatrix} = T_{21}$$

is solvable on S_1, S_2. In fact this equation is solvable from Assumption 2 (see (Kodama & Suda, 1995)), so there exist S_1, S_2 satisfying (15). □

Pre-compensation for a Hybrid Fuzzy PID Control of a Proportional Hydraulic System

Pornjit Pratumsuwan and Chaiyapon Thongchaisuratkrul
King Mongkut's University of Technology North Bangkok
Thailand

1. Introduction

Fluid power is a term which was created to include the generation, control, and application of smooth, effective power of pumped or compressed fluids when this power is used to provide force and motion to mechanisms. Fluid power includes hydraulic, which involves liquids, and pneumatic, which involves gases. Hydraulic and pneumatic power offer many advantages over electric motors, especially for systems that require high speed linear travel, moving or holding heavy loads, or very smooth position or pressure control. Compared to other types, hydraulic and pneumatic actuators are smaller and quieter. They also produce less heat and electromagnetic interference (EMI) at the machine than do electric actuators, and in many cases, in particular with high performance hydraulic or pneumatic system, they offer the ability to build machines at considerable savings compared to machines employing purely electrical or mechanical motion (Chuang & Shiu, 2004; Knohl & Unbehauen, 2000).

Hydraulic drives, thanks to their high power intensity, are low in weight and require a minimum of mounting space. They facilitate fast and accurate control of very high energies and forces. The hydraulic actuator (cylinder) represents a cost-effective and simply constructed linear drive. The combination of these advantages opens up a wide range of applications. The increase in automation makes it ever more necessary for pressure, flow rate, and flow direction in hydraulic systems to be controlled by means of an electrical control system. The obvious choice for this is hydraulic proportional valves (or servo valves) as an interface between controller and hydraulic system (Knohl & Unbehauen, 2000).

The hydraulic actuator, usually a cylinder, provides the motion of the load attached to the hydraulic system. A control valve meters the fluid into the actuator as a spool traverses within the valve body. The control valve is either a servo valve or a proportional valve. In hydraulic control applications, proportional valves offer various advantages over servo valves (Eryilmaz &. Wilson, 2006). Proportional valves are much less expensive. They are more suitable for industrial environments because they are less prone to malfunction due to fluid contamination. In addition, since proportional valves do not contain sensitive, precision components, they are easier to handle and service. However, these advantages are offset by their nonlinear response characteristics. Since proportional valves have less precise manufacturing tolerances, they suffer from performance degradation. The larger tolerances on spool geometry result in response nonlinearities, especially in the vicinity of neutral spool position. Proportional valves lack the smooth flow properties of "critical centre" valves, a condition closely approximated by servo valves at the expense of high machining cost. Small changes in spool geometry (in terms of lapping) may have large effects on the hydraulic system dynamics. Especially, a closed-centre spool (overlapped) of proportional valve, which usually provides the motion of the actuator in a proportional hydraulic system (PHS), may result in the steady state error because of its dead zone characteristics in flow gain [(Eryilmaz &. Wilson, 2006). Figure 1 illustrates the characteristics of proportional valve.

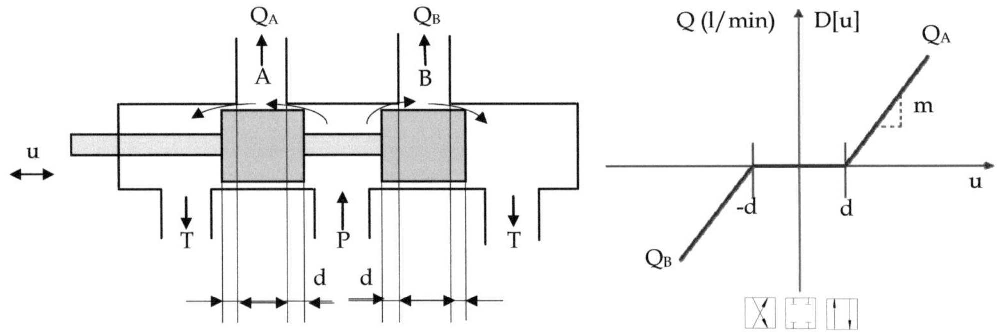

Fig. 1. Characteristics of a closed-centre spool (overlapped) of proportional valve.

Valve lap, or valve overlap, refers to the amount of spool travel from the center position required to start opening between the powered input port and the work (output) port or the tank port. A zero lapped valve is one in which any tiny, differentially small amount of spool shift, either way, starts the opening. However, there is no contact between the OD of the spool and ID of the bore. And even zero lapped valves have some slight amount of overlap. Nonetheless, the zero lapped term persists.
The characteristics of the proportional valve with dead zone *D* (from figure 1) is described by the function

$$Q = D[u] = \begin{cases} m(u-d), & \text{if}\, u \geq d \\ 0, & \text{if} -d \leq u \leq d \\ m(u+d), & \text{if}\, u < -d \end{cases} \tag{1}$$

where d, $m \geq 0$. The parameter $2d$ specifies the width of the deadzone, while m represents the slope of the response outside the dead zone.
The proportional hydraulic system shown in figure 2 is comprised of a double acting cylinder, a 4/3 way proportional valve, and load. The supply pressure P is assumed to be constant, and the control objective is the positioning of the pay load. The proportional valve used in this plant is a low cost, which can be characterized by a relative large and symmetric dead zone. A complete mathematic model of such an electro-hydraulic system, for example, has been given by (Knohl & Unbehauen, 2000). However, these equations are highly complex and difficult to utilize in control design. A more practical model may be obtained through the linearization of the non-linear function.

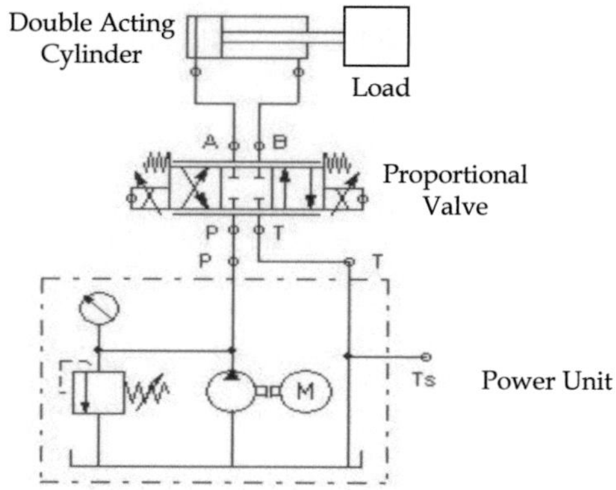

Fig. 2. Schematic diagram of the PHS.

A mathematical model of the plant can be derived from the flow equation of valve, the continuity equation and balance of forces at the piston. The valve flow rate equation is highly non-linear and dependent on the valve displacement from neutral, which is proportional to input voltage u and the pressure drop across the load P_L. A Taylor series linearization leads to

$$Q_L = K_q u - K_c P_L \tag{2}$$

where, K_q = flow gain coefficient and K_c = flow pressure coefficient

The movement of the piston, the change of the oil volume due to compressibility and the leakage oil flow determine the total oil flow Q_L as

$$Q_L = A_P \dot{y} + \frac{V_t}{4\beta_e}\dot{P}_L + f(P_L) \tag{3}$$

where, V_t is the total compressed oil volume, A_P the surface area of the piston, $\dot{y}$ the velocity of the piston, β_e the effective bulk modulus (compressibility) and $f(P_L)$ the non linear influence of the internal and external oil leakage. Here, it is assumed that the rod and the head side areas of the piston are equal. If the leakage function is approximated by a linear relation, equation (3) can be rewritten as

$$\dot{P}_L = \frac{4\beta_e}{V_t}(Q_L - C_{tp}P_L - A_p\dot{y} \tag{4}$$

where C_{tp} is the total leakage coefficient of the piston. The balance of forces at the sliding carriage leads to

$$F_G = A_p P_L = M_t\ddot{y} + B_p\dot{y} + y \tag{5}$$

where F_G is the force generated by the piston, M_t the total mass of the piston and the load, and B_p the damping coefficient of the piston and the load. Neglect the non linear effects of dry and adhesive friction, combining equations (2), (4), and (5) and applying the Laplace transformation to the resulting third order differential equation results in the transfer function

$$\frac{Y(s)}{u(s)} = \frac{b_0}{s\left(a_2 s^2 + a_1 s + a_0\right)} \tag{6}$$

where the abbreviations

$$b_0 = \frac{4\beta_e}{V_t} K_q \tag{7a}$$

$$a_0 = \frac{4\beta_e}{V_t}\left(\frac{B_p}{A_p}(K_C + C_{tp}) + A_p\right) \tag{7b}$$

$$a_1 = \frac{B_p}{A_p} + \frac{4\beta_e M_t}{V_t A_p}(K_C + C_{tp}) \tag{7c}$$

$$a_2 = \frac{M_t}{A_p} \tag{7d}$$

The parameters of the linear model of the hydraulic system given by equation (7), which depend on the constants derived by linearization, are all of varying nature. Variation of the load mass and the damping coefficient can also be interpreted as parameter changes of the linear model. To take the dead zone of the actuating valve into account, a static non linear dead zone that is portrayed in figure 1.

2. Preliminary of Controller Designs

A closed loop system, whither the reference signal is set manually or automatically, can perform control of cylinder position, speed and force. Figure 3 represents typical of an "Automatic Closed Loop" control system. As shown in the figure, the position of a hydraulic cylinder is controlled by a proportional valve. The proportional valve solenoid is receiving driving electrical current from an amplifier card, which is generating the driving current based on a control signal supplied by a controller. The controller responsibility is to continuously compare the reference signal and the actual cylinder position fed back by the position sensor, after consequently generate the adequate control signal.

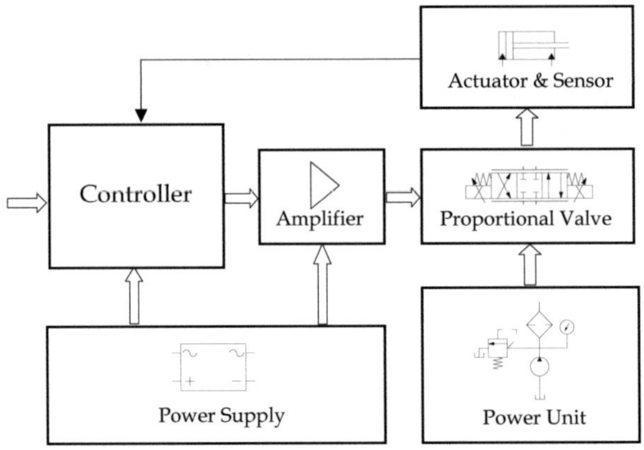

Fig. 3. Conceptual controls of the PHS.

2.1 PID controller

Classical PID controllers are very popular in industries because they can improve both the transient response and steady state error of the system at the same time (Kim *et al.*, 1994). Although great efforts have been devoted to develop PID controller, PID controllers are not robust to the parameter variation to the plants being controlled. Figure 4 shows the use of PID controller controls the PHS.

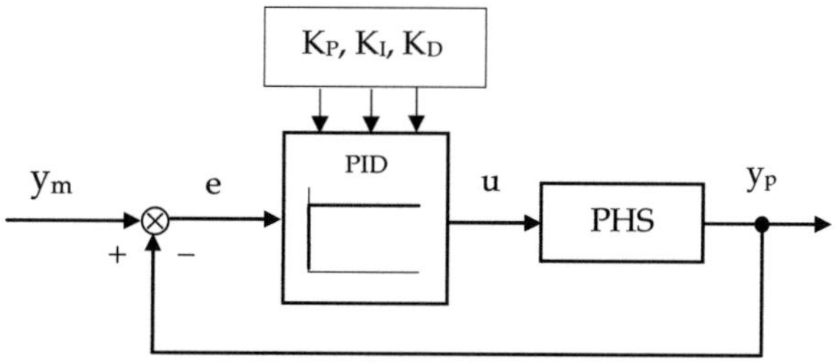

Fig. 4. Block diagram of a PID controller.

The PID control method has been widely used in industry during last several decades because of its simplicity. The implementation of PID control logic, as shown in equation (8), requires finding suitable values for the gain parameters K_P, K_I, and K_D. To tune these parameters, the model is linearized around different equilibrium points,

$$u(k) = K_P e(k) + K_I \sum_{i=0}^{k} e(i) + K_D[e(k) - e(k-1)] \tag{8}$$

where e(k) is the error signal. However, the PID method is not suitable for controlling a system with a large amount of lag, parameter variations, and uncertainty in the model. Thus, PID control logic cannot accurately control position in a hydraulic system.

2.2 Fuzzy logic controller

Fuzzy logic control (FLC) has found many applications in a variety of fields since Prof. Zadeh introduced fuzzy set theory in 1965 (Zadah, 1965). Among the most successful applications of this theory has been the area of FLC initiated by the work of (Mamdani & Assilian, 1975). FLC has the advantage that it does not require an accurate mathematical model of the process. It uses a set of artificial rules in a decision-making table and calculates an output based on the table. Figure 5 shows specific components characteristic of a FLC. Input variables go through the fuzzification interface and are converted to linguistic variables. Then, a database and rule base holding the decision-making logic are used to infer the fuzzy output. Finally, a defuzzification method converts the fuzzy output into a signal to be sent out.

Fuzzy control is robust to the system with variation of system dynamics and the system of model free or the system which precise information is not required. It has been successfully used in the complex ill-defined process with better performance than that of a PID controller. Another important advance of fuzzy controller is a short rise time and small overshoot (Li *et al.*, 2006; Rahbari & Silva, 2000).

FLC has the advantage that it does not require an accurate mathematical model of the process. It uses a set of artificial rules in a decision-making table and calculates an output based on the table (Li *et al.*, 2006; Rahbari & Silva, 2000).

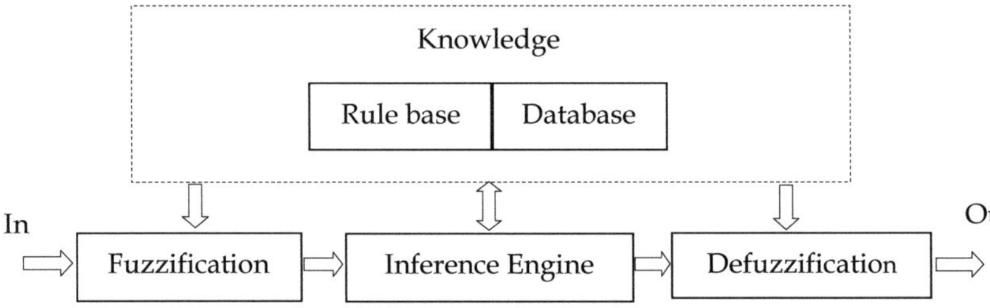

Fig. 5. Structure of FLC.

Figure 6 shows a fuzzy control of the PHS. Input variables go through the fuzzification interface and are converted to linguistic variables. Then, a database and rule base holding the decision-making logic are used to infer the fuzzy output. Finally, a defuzzification method converts the fuzzy output into a signal to be sent out.

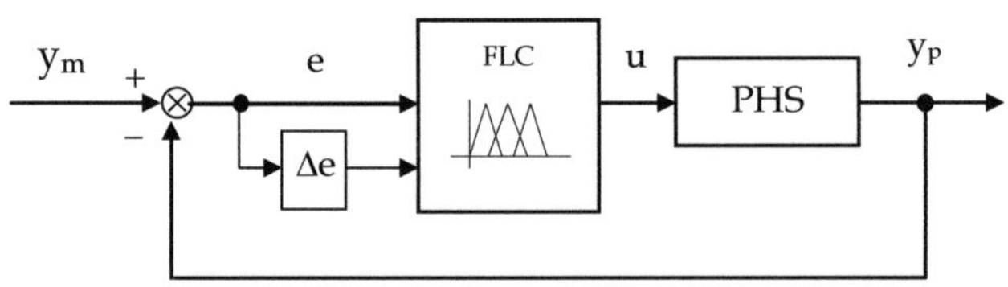

Fig. 6. Block digram of a FLC.

When used in a control system, FLC is robust since it provides a fast rise time and a small amount of overshoot. The control parameters and set of terms that describe each linguistic variable must be determined when designing a FLC. Obviously, the position in the electro-hydraulic is the parameter to be controlled in the system. A two-dimension structure will be used to produce fast calculations. The two input linguistic variables are the error of the position (e) and the error change of the position (Δe). The output is the voltage signal to control the amplifier and proportional valve. Thus, the FLC has two antecedences and one consequence.

The controller structure of a FLC, which the fuzzy states of the inputs and the output, all are chosen to be equal in number and use the same linguistic descriptors: NB = negative big, N = negative, Z = zero, P = positive, and PB = positive big. The fuzzy sets and its memberships function is shown in figure 7. A set of fuzzy rules is shown in the Table 1.

Since the dynamics of each cylinder is not symmetric, due to the difference in the effective area of the rod side and the head side of the piston. The designed fuzzy set of the fuzzy controller accounts for this asymmetry as well (Rahbari & Silva, 2000; Pratumsuwan & Thongchai, 2009).

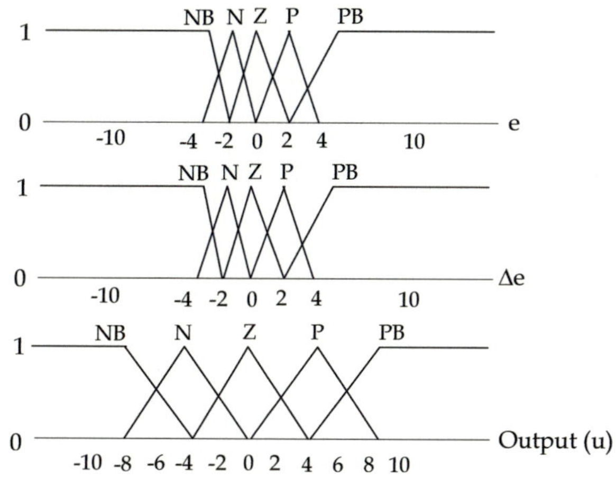

Fig. 7. Fuzzy sets of a FLC.

e \ Δe	NB	N	Z	P	PB
NB	NB	NB	N	N	Z
N	NB	N	N	Z	P
Z	N	N	Z	P	P
P	N	Z	P	P	PB
PB	Z	P	P	PB	PB

Table 1. Fuzzy Rules of a FLC.

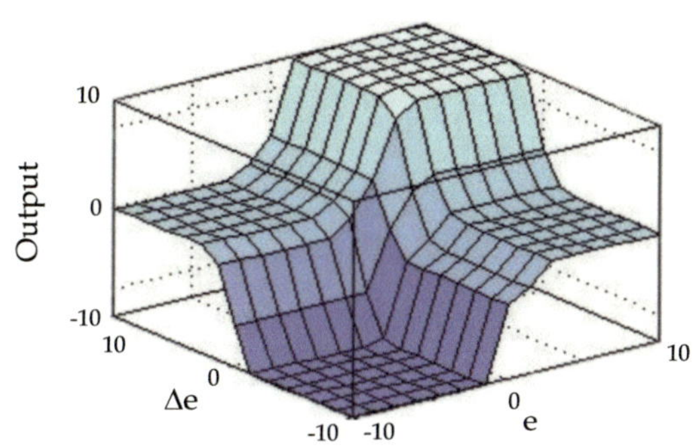

Fig. 8. Input-output mapping of a FLC.

2.3 Hybrid of fuzzy and PID controller

While conventional PID controllers are sensitive to variations in the system parameters, fuzzy controllers do not need precise information about the system variables in order to be effective. However, PID controllers are better able to control and minimize the steady state error of the system. Hence, a hybrid system, as shown in figure 9, was developed to utilize the advantages of both PID controller and fuzzy controller (Parnichkul & Ngaecharoenkul, 2000; Erenoglu *et al.*, 2006; Pratumsuwan *et al.*, 2009;).

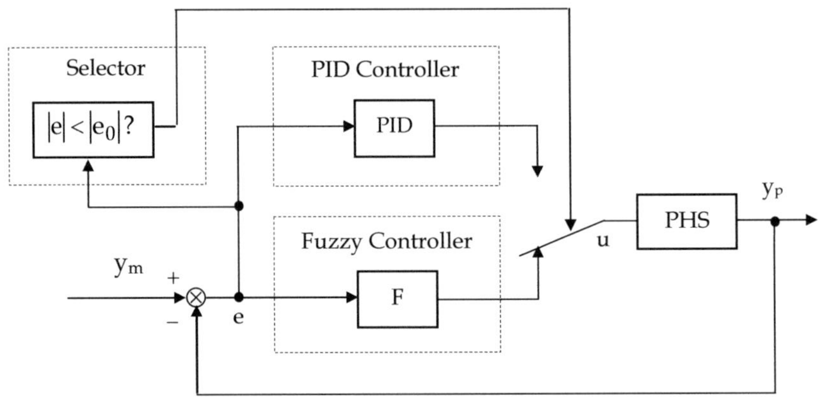

Fig. 9. Block diagram of a hybrid fuzzy PID controller.

Figure 9 shows a switch between the fuzzy controller and the PID controller, where the position of the switch depends on the error between the actual value and set point value. If the error in position reaches a value higher than that of the threshold e_0, the hybrid system applies the fuzzy controller, which has a fast rise time and a small amount of overshoot, to the system in order to correct the position with respect to the set point. When the position is below the threshold e_0 or close to the set point, the hybrid system shifts control to the PID, which has better accuracy near the set position (Parnichkul & Ngaecharoenkul, 2000; Erenoglu *et al.*, 2006; Pratumsuwan *et al.*, 2009;).

2.4 Fuzzy pre-compensated PID controller

Since classical PID controllers are widely used in industrial applications, they exhibit poor performance when applied to the PHS containing unknown nonlinearities, such as dead zones, saturation, and hysteresis. In this section, we will describe a fuzzy logic-based pre-compensation design for PID controllers (Kim *et al.*, 1994)

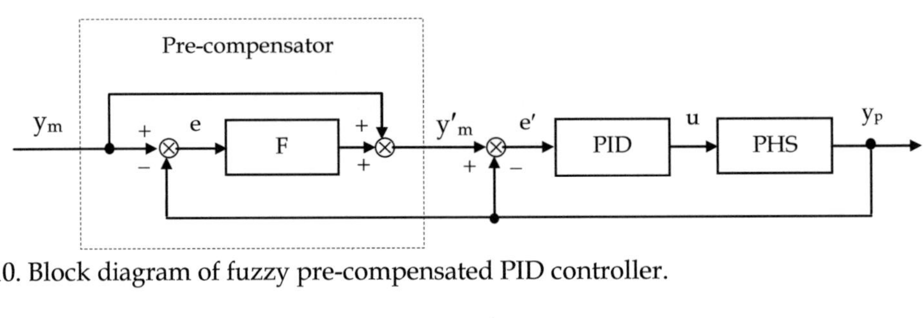

Fig. 10. Block diagram of fuzzy pre-compensated PID controller.

Figure 10 illustrates the basic control structure. The scheme consists of a classical PID control structure together with fuzzy pre-compensator. The fuzzy pre-compensator uses the command input y_m and the PHS output y_p to generate a pre-compensated command signal y'_m, described by the following equations (Kim *et al.*, 1994).

$$e(k) = y_m(k) - y_p(k) \tag{9}$$

$$\Delta e(k) = e(k) - e(k-1) \tag{10}$$

$$\mu(k) = F[e(k), \Delta e(k)] \tag{11}$$

$$y'_m(k) = y_m(k) + \mu(k) \tag{12}$$

In the above, $e(k)$ is the position error between the command input $y_m(k)$ and the PHS output $y_p(k)$ and $\Delta e(k)$ is the change in position error. The term $F[e(k), \Delta e(k)]$ is a nonlinear mapping of $e(k)$ and $\Delta e(k)$ based on fuzzy logic. The term $\mu(k) = [e(k), \Delta e(k)]$ represents a compensation or correction term, so that the compensated command signal $y'_m(k)$ is simply the sum of the external command signal $y_m(k)$ and $\mu(k)$. The correction term is based on the error $e(k)$ and the change of error $\Delta e(k)$. The compensated command signal $y'_m(k)$ is applied to a classical PID scheme, as shown in figure 10.
The equations governing the PID controller are as follows

$$e'(k) = y'_m(k) - y_p(k) \tag{13}$$

$$\Delta e'(k) = e'(k) - e'(k-1) \tag{14}$$

$$u(k) = u(k-1) + K_P \Delta e'(k) + K_I e'(k) + K_D(\Delta e'(k) - \Delta e'(k-1)) \tag{15}$$

The quantity $e'(k)$ is the pre-compensated position error between the pre-compensated command input $y'_m(k)$ and PHS output $y_p(k)$, and $\Delta e'(k)$ is the change in the pre-compensated position error. The control $u(k)$ is applied to the input of the PHS. The purpose of the fuzzy pre-compensator is to modify the command signal to compensate for the overshoots and undershoots present in the output response when the PHS has unknown nonlinearities.
For PID tuning in this paper, we set PID gains with Ziegler-Nichols and trial-error method.
The purpose of the fuzzy pre-compensator is to modify the command signal to compensate for the overshoots and undershoots present in the output response when the PHS has unknown nonlinearities. An expert's experience and knowledge method is used to build a rule base and membership functions (Rahbari & Silva, 2000).
In our description, we think of $e(k)$ and $\Delta e(k)$ as inputs, and $\mu(k)$ as the output. The fuzzy states of the inputs and the output, all are chosen to be equal in number and use the same linguistic descriptors : NB = negative big, NM = negative medium, NS = negative small, ZO = zero, PS = positive small, PM = positive medium, and PB = positive big. (The fuzzy sets is shown in figure 11.) The decision-making output can be obtained using a max-min fuzzy inference where the crisp output is calculated by the center of gravity (COG) method.
Using these fuzzy sets, the fuzzy rules can be designed. The designed rules are presented in table 2. To explain how these rules were obtained, consider for example the rules in table 2. Suppose that the command signal is a constant y_m, the error $e(k)$ is zero, and the change of error $\Delta e(k)$ is a negative number. This mean that the output $y_p(k) = y_m - e(k)$ is increasing, i.e., heading in the direction of an overshoot. To compensate for this, we decrease the command signal. This corresponds to applying a correction term $\mu(k)$ that is negative. Hence, we get the rule "if error is zero and change of error is negative medium, then output a negative big correction term".

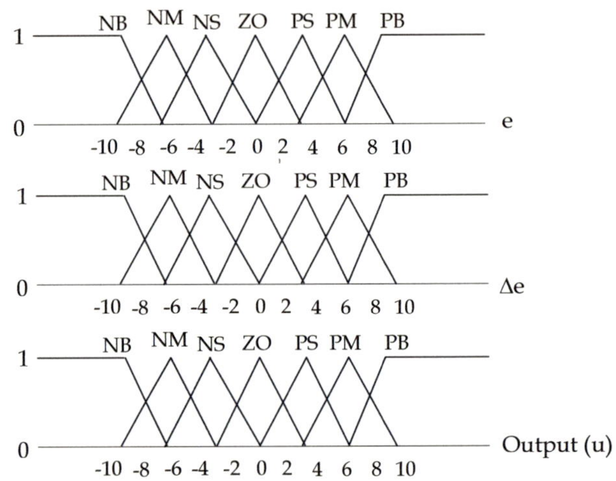

Fig. 11. The fuzzy sets of a pre-compensator.

e \ Δe	NB	NM	NS	ZO	PS	PM	PB
NB		NB	NB	NB	NM		
NM				NM	NM		
NS				NS	PS		PM
ZO	NB	NB	NM	ZO	PS	PM	PM
PS	NB	NB	NM	PS	PM	PB	PB
PM			NM	PM		PB	
PB			PM	PB			

Table 2. Fuzzy rules of a pre-compensator.

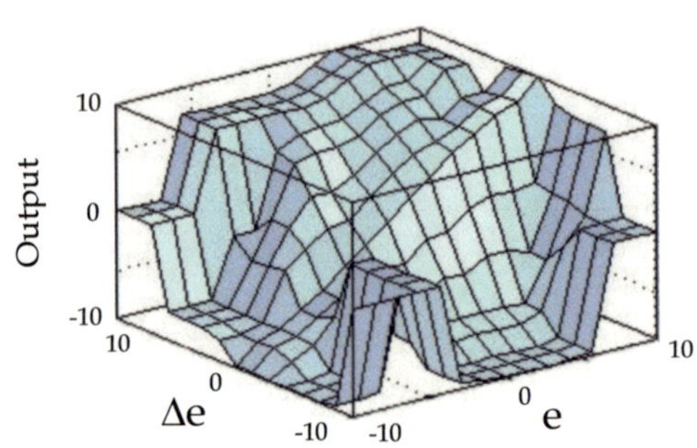

Fig. 12. Input-output mapping of a pre-compensator.

2.5 Two-layered fuzzy logic controller

In this section we describe a fuzzy pre-compensated fuzzy controller or a two-layered fuzzy logic controller (Kim *et al.*, 1994; Pratumsuwan & Thongchai, 2009). The aim is to eliminate the steady state error and improve the performance of the output response for control systems with dead zones.

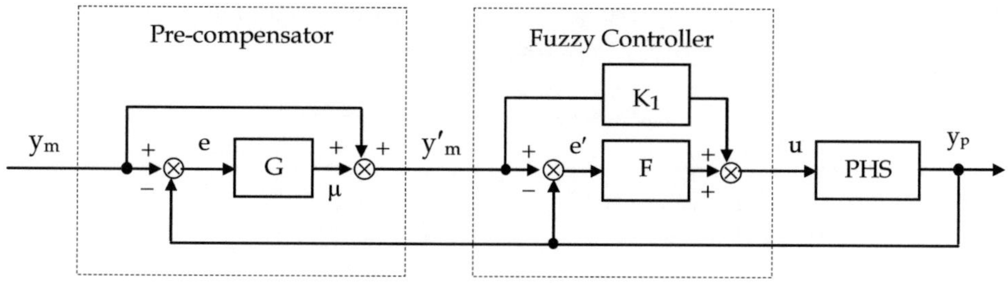

Fig. 13. Block diagram of a two-layered fuzzy logic controller.

We now proceed to describe the proposed controller. First, define the variables $y'_m(k)$ and $e'(k)$ as follows :

$$y'_m(k) = y_m(k) + \mu(k) \tag{16}$$

$$e'(k) = e(k) + \mu(k) \tag{17}$$

where $\mu(k)$ is a compensating term which is generated using a fuzzy logic scheme, which we will describe below. The proposed control scheme is shown in figure 13. The controller consists of two "layers" : a fuzzy pre-compensator, and a usual fuzzy PD controller. Hence we refer to the scheme as a *two-layered fuzzy logic controller*. The error $e(k)$, and change of error $\Delta e(k)$, are inputs to the pre-compensator. The output of the pre-compensator is $\mu(k)$.
The dynamics of overall system is then described by the following equations:

$$e(k) = y_m(k) - y_p(k) \tag{18}$$

$$\Delta e(k) = e(k) - e(k-1) \tag{19}$$

$$\mu(k) = G[e(k), \Delta e(k)] \tag{20}$$

$$y'_m(k) = y_m(k) + \mu(k) \tag{21}$$

$$e'(k) = y'_m(k) - y_p(k) \tag{22}$$

$$\Delta e'(k) = e'(k) - e'(k-1) \tag{23}$$

$$u(k) = K_1 y'_m(k) + F[e'(k), \Delta e'(k)] \tag{24}$$

where K_1 is the feed forward gain.

3. Pre-compensation of a Hybrid Fuzzy PID Controller

Pre-compensation of a hybrid fuzzy PID controller, as shown in figure 14, was developed to combine the advantages of both fuzzy pre-compensated PID controller and fuzzy pre-compensated fuzzy controller, which described in section 2.3, 2.4, and 2.5.
Figure 14 shows a switch between the fuzzy controller and the PID controller, where the position of the switch depends on the error between the actual value and set point value. If the error in PHS reaches a value higher than that of the threshold e_0, the control system applies the fuzzy controller, which has a fast rise time and a small amount of overshoot, to the system in order to correct the position with respect to the set point. When the position is

lower the threshold e_0 or close to the set point, the control system shift switch to the PID controller, which has better accuracy near the set point.

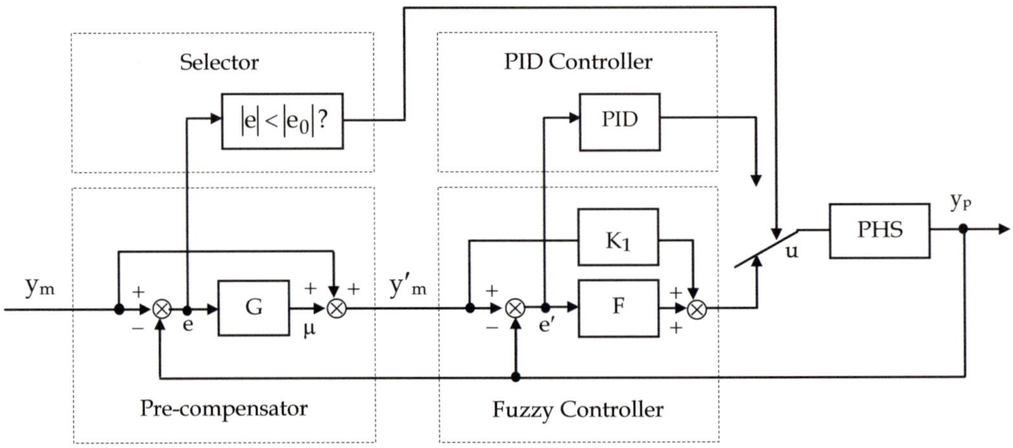

Fig. 14. Block diagram of pre-compensation of a hybrid Fuzzy PID controller.

4. Experimental Description

The specifications of a PHS are depicted in figures. 15, 16 and table 3 respectively. Figure 15 shows a diagram of the tested system. The position control of a PHS procedure is described as follows: upon the intended initial and ending position of the piston (stroke) are given, the computer receives the feedback signal through DAQ card (A/D) from linear potentiometer, realizes various control algorithm and transmits a control signal through DAQ card (D/A) and amplifier card to proportional valve. The spool displacement of proportional valve is proportional to the input signal.

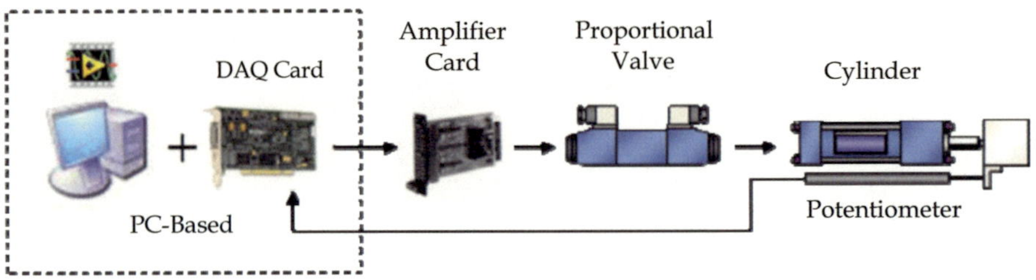

Fig. 15. PC-Based position control of a PHS.

Fig. 16. The experimental setup.

Elements	Descriptions
Cylinder	piston diameter 16 mm, piston rod diameter 10 mm, stroke 200 mm
Proportional valve (4/3 closed-center spool, overlapped)	directly actuated spool valve, grade of filtration 10 μm, nominal flow rate 1.5l/min (at Δp_N = 5 bar/control edge), leakage oil flow < 0.01 l/min (at 60 bar), nominal current 680 mA, resolution < 1 mA, setting time of signal jump 0…100% = 60 ms, repetition accuracy < 1%
Pump (supply pressure)	60 bar
Linear potentiometer	output voltage 0…10V, measuring stroke 200 mm, linearity tolerance 0.5%
Amplifier card	set point values ± 10 VDC, solenoid outputs (PWM signal) 24 V, dither frequency 200 Hz, max current 800 mA,
DAQ Card (NI 6221 PCI)	analog input resolutions 16 bits (input range ±10V), output resolutions 16 bits (output range ±10V), 833 kS/s (6 μs full-scale settling)
Operating systems & Program	Windows XP, and LabVIEW 8.6

Table 3. Specifications of a PHS.

5. The Experimental Results

The control algorithms described in section 2.3, 2.4, and 2.5 were hybridized and applied to the PHS using by LabVIEW, Nation Instruments as the development platform and shown in figure 17.

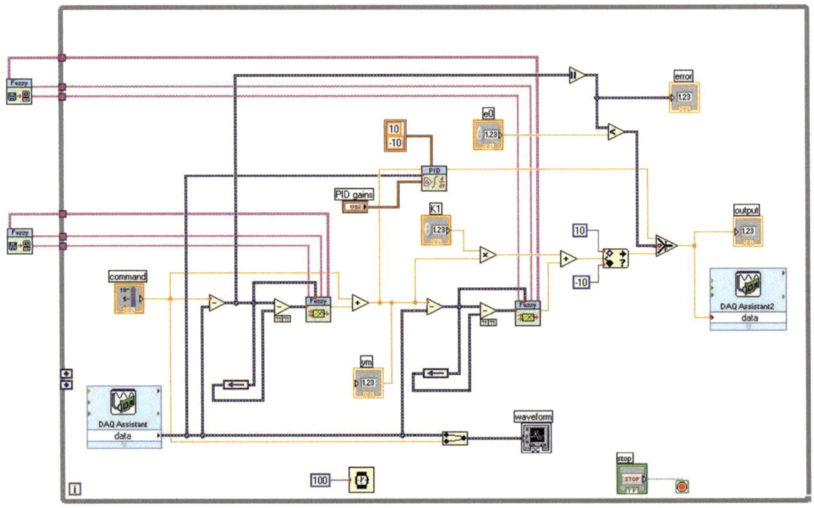

Fig. 17. The control algorithm are used and developed by LabVIEW program.

In our experiments we compare the performance of conventional hybrid fuzzy PID controller to the proposed pre-compensation of a hybrid fuzzy PID controller. A testing of response of the system was performed using a square wave input. The parameter values of the pre-compensation of a hybrid fuzzy PID controller were experimentally determined to be: K_1 = 0.93, K_P = 5.6, e_0 = 0.92. Figures. 18 and 19 shows the output response of a conventional hybrid fuzzy PID system compared to the pre-compensation of a hybrid fuzzy PID system. It is found that the pre-compensation of a hybrid fuzzy PID controller gives the most satisfying results of rise time, overshoot, and steady state error.

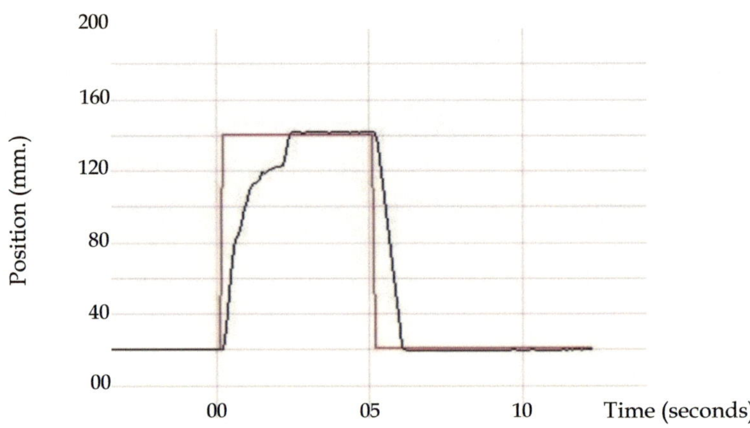

Fig. 18. Output response of conventional (hybrid fuzzy PID) controller.

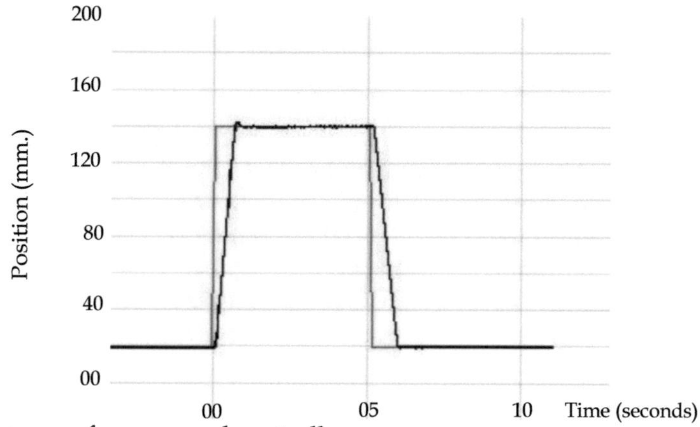

Fig. 19. Output response of a proposed controller.

6. Conclusions

The objective of this study, we proposed the pre-compensation of a hybrid fuzzy PID controller for a PHS with deadzones. The controller consists of a fuzzy pre-compensator followed by fuzzy controller and PID controller. The proposed scheme was tested experimentally and the results have superior transient and steady state performance, compared to a conventional hybrid fuzzy PID controller. An advantage of the present approach is that an existing hybrid fuzzy PID controller can be easily modified into the control structure by adding a fuzzy pre-compensator, without having to retune the internal variables of the existing hybrid fuzzy PID controller.

In this study, an experimental research, so we do not address the problem of analyzing the stability of the control scheme in this paper. This difficult but important problem is a topic of ongoing research.

7. References

Chin-Wen Chuang and Liang-Cheng Shiu.(2004). CPLD based DIVSC of hydraulic position control systems. *Computers and Electrical Engineering*. Vol.30, pp.527-541.

T. Knohl and H. Unbehauen. (2000). Adaptive position control of electrohydraulic servo systems using ANN. *Mechatronics*. Elsevier Science Ltd. vol. 10, pp. 127-143.

Bora Eryilmaz, and Bruce H. Wilson. (2006). Unified modeling and analysis of a proportional valve. *Journal of the Franklin Institute*, pp. 48-68.

L.A. Zadeh. (1965). Fuzzy sets. *Information and Control*. vol.8, pp.338-1588.

E.H. Mandani and S. Assilian. (1975). An experiment in linquistic synthsis with a fuzzy logic control. *Machine Studies*. vol.7, pp. 1-13.

Isin Erenoglu, Ibrahim Eksin, Engin Yesil, and Mujde Guzelkaya. (2006). An Intelligent Hybrid Fuzzy PID Controller," *Proceedings 20th European Conference on Modelling and Simulation* © ECMS. ISBN : 0-9553018-0-7.

Roya Rahbari, and Clarence W. de Silva. (2000). Fuzzy Logic Control of a Hydraulic System. ©IEEE, ISBN: 0-7803-58112-0, pp. 313- 318.

M. Parnichkul and C. Ngaecharoenkul. (2000). Hybrid of Fuzzy and PID in Kinematics of a Pneumatic System. *Proceeding of IEEE Industrial Electronics Society*. IEEE Press, vol. 2 no. 2, pp.1485-1490.

Hao Liu, Jae-Cheon Lee, and Bao-Ren Li. (2007). High Precision Pressure Control of a Pneumatic Chamber using a Hybrid Fuzzy PID Controller. *International Journal of Precision Engineering and Manufacturing*, Vol.8, No3. pp. 8-13.

Jong-Hwan Kim, Kwang-Choon Kim, and Edwin K.P. Chong. (1994). Fuzzy Precompensated PID Controllers. *IEEE Transactions on Control System Technology*, Vol.2, No. 4, December, pp. 406-411.

J.-H. Kim, J.-H. Park, S.-W. Lee, E.K.P. Chong. (1994). A Two-Layered Fuzzy Logic Controller for Systems with Deadzones," *IEEE Transactions on Industrial Electronics*, Vol. 41, No. 2, April. pp. 155-162.

Chang-chun Li, Xiao-dong Liu, Xin Zhou, Xuan Bao, Jing Huang. (2006). Fuzzy Control of Electro-hydraulic Servo Systems Based on Automatic Code Generation. *Proceedings of the Sixth International Conference on Intelligent Systems Design and Applications (ISDA'06).*

Pornjit Pratumsuwan, and Siripun Thongchai. (2009). A Two-Layered Fuzzy Logic Controller for Proportional Hydraulic System. *Proceeding of 4th IEEE International Conference on Industrial Electronic and Applications (ICIEA2009),* pp. 2778-2781.

Pornjit Pratumsuwan, Siripun Thongchai, and Surapun Tansriwong. (2010). A Hybrid of Fuzzy and Proportional-Intrgral-Derivative Controller for Electro-Hydraulic Position Servo System. *Energy Research Journal 1(2),* pp.62-67.

11

A New Approach of the Online Tuning Gain Scheduling Nonlinear PID Controller Using Neural Network

Ho Pham Huy ANH[1] and Nguyen Thanh Nam[2]
[1]*Corresponding author, Ho Chi Minh City University of Technology, Ho Chi Minh City, Viet Nam*
[2]*DCSELAB, Viet Nam National University Ho Chi Minh City (VNU-HCM), Viet Nam*

Abstract

This chapter presents the design, development and implementation of a novel proposed online-tuning Gain Scheduling Dynamic Neural PID (DNN-PID) Controller using neural network suitable for real-time manipulator control applications. The unique feature of the novel DNN-PID controller is that it has highly simple and dynamic self-organizing structure, fast online-tuning speed, good generalization and flexibility in online-updating. The proposed adaptive algorithm focuses on fast and efficiently optimizing Gain Scheduling and PID weighting parameters of Neural MLPNN model used in DNN-PID controller. This approach is employed to implement the DNN-PID controller with a view of controlling the joint angle position of the highly nonlinear pneumatic artificial muscle (PAM) manipulator in real-time through Real-Time Windows Target run in MATLAB SIMULINK® environment. The performance of this novel proposed controller was found to be outperforming in comparison with conventional PID controller. These results can be applied to control other highly nonlinear SISO and MIMO systems.

Keywords: highly nonlinear PAM manipulator, proposed online tuning Gain Scheduling Dynamic Nonlinear PID controller (DNN-PID), real-time joint angle position control, fast online tuning back propagation (BP) algorithm, pneumatic artificial muscle (PAM) actuator.

1. Introduction

The compliant manipulator was used to replace monotonous and dangerous tasks, which has enhanced lots of researchers to develop more and more intelligent controllers for human-friendly industrial manipulators. Due to uncertainties, it is difficult to obtain a precise mathematical model for robot manipulators. Hence conventional control methodologies find it difficult or impossible to handle un-modeled dynamics of a robot manipulator. Furthermore, most of conventional control methods, for example PID controllers, are based on mathematical and statistical procedures for modeling the system and estimation of optimal controller parameters. In practice, such manipulator is often highly non-linear and a mathematical model may be difficult to derive. Thus, as to accommodate system uncertainties and variations, learning methods and adaptive intelligent techniques must be incorporated.

Due to their highly nonlinear nature and time-varying parameters, PAM robot arms present a challenging nonlinear model problem. Approaches to PAM control have included PID control, adaptive control (Lilly, 2003), nonlinear optimal predictive control (Reynolds *et al.*, 2003), variable structure control (Repperger *et al.*, 1998; Medrano-Cerda *et al.*,1995), gain scheduling (Repperger *et al.*,1999), and various soft computing approaches including neural network Kohonen training algorithm control (Hesselroth *et al.*,1994), neural network + nonlinear PID controller (Ahn and Thanh, 2005), and neuro-fuzzy/genetic control (Chan *et al.*, 2003; Lilly *et al.*, 2003). Balasubramanian *et al.*, (2003a) applied the fuzzy model to identify the dynamic characteristics of PAM and later applied the nonlinear fuzzy model to model and to control of the PAM system. Lilly (2003) presented a direct continuous-time adaptive control technique and applied it to control joint angle in a single-joint arm. Tsagarakis *et al.* (2000) developed an improved model for PAM. Hesselroth *et al.* (1994) presented a neural network that controlled a five-link robot using back propagation to learn the correct control over a period of time. Repperger *et al.* (1999) applied a gain scheduling model-based controller to a single vertically hanging PAM. Chan *et al.*, (2003) and Lilly *et al.*, (2003) introduced a fuzzy P+ID controller and an evolutionary fuzzy controller, respectively, for the PAM system. The novel feature is a new method of identifying fuzzy models from experimental data using evolutionary techniques. Unfortunately, these fuzzy models are clumsy and have only been tested in simulation studies. (Ahn and Anh, 2006) applied a modified genetic algorithm (MGA) for optimizing the parameters of a linear ARX model of the PAM manipulator which can be modified online with an adaptive self-tuning control algorithm, and then (Ahn and Anh, 2007b) successfully applied recurrent neural networks (RNN) for optimizing the parameters of neural NARX model of the PAM robot arm. Recently, we (Ahn and Anh, 2009) successfully applied the modified genetic algorithm (MGA) for optimizing the parameters of the NARX fuzzy model of the PAM robot arm.

Although these control systems were partially successful in obtaining smooth actuator motion in response to input signals, the manipulator must be controlled slowly in order to get stable and accurate position control. Furthermore the external inertia load was also assumed to be constant or slowly varying. It is because PAM manipulators are multivariable non-linear coupled systems and frequently subjected to structured and/or unstructured uncertainties even in a well-structured setting for industrial use or human-friendly applications as well.

To overcome these drawbacks, the proposed online tuning DNN-PID algorithm in this chapter is a newly developed algorithm that has the following good features such as highly simple and dynamic self-organizing structure, fast learning speed, good generalization and flexibility in learning. The proposed online tuning DNN-PID controller is employed to compensate for environmental variations such as payload mass and time-varying parameters during the operation process. By virtue of on-line training by back propagation (BP) learning algorithm and then auto-tuned gain scheduling K and PID weighting values K_p, K_i and K_d, it learns well the nonlinear robot arm dynamics and simultaneously makes control decisions to both of joints of the robot arm. In effect, it offers an exciting on-line estimation scheme.

This chapter composes of the section 1 for introducing related works in PAM robot arm control. The section 2 presents procedure of design an online tuning gain scheduling DNN-

PID controller for the 2-axes PAM robot arm. The section 3 presents and analyses experiment studies and results. Finally, the conclusion belongs to the section 4.

2. Control System

2.1. Experimental apparatus

The PAM manipulator used in this paper is a two-axis, closed-loop activated with 2 antagonistic PAM pairs which are pneumatic driven controlled through 2 proportional valves. Each of the 2-axes provides a different motion and contributes to 1 degree of freedom of the PAM manipulator (Fig. 1). In this paper, the 1st joint of the PAM manipulator is fixed and proposed online tuning Gain Scheduling neural DNN-PID control algorithm is applied to control the joint angle position of the 2nd joint of the PAM manipulator. A general configuration of the investigated 2-axes PAM manipulator shown through the schematic diagram of the 2-axes PAM robot arm and the experimental apparatus presented in Fig.1 and Fig.2, respectively.

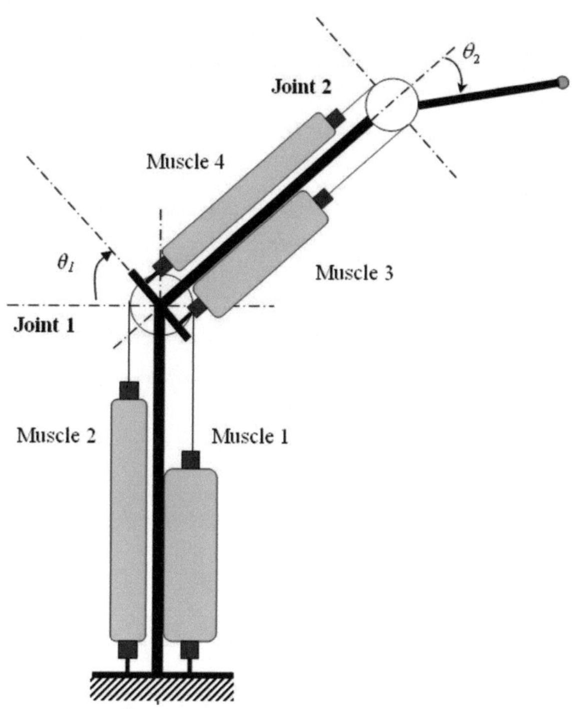

Fig. 1. Working principle of the 2-axes PAM robot arm.

The experiment system is illustrated in Fig.2. The air pressure proportional valve manufactured by FESTO Corporation is used. The angle encoder sensor is used to measure the output angle of the joint. The entire system is a closed loop system through computer. First, initial control voltage value $u_0(t)$=5[V] is sent to proportional valve as to inflate the artificial muscles with air pressure at P_0 (initial pressure) to render the joint initial status. Second, by changing the control output $u(t)$ from the D/A converter, we could set the air pressures of the two artificial muscles at $(P_0 + \Delta P)$ and $(P_0 - \Delta P)$, respectively. As a result, the joint is forced to rotate for a certain angle. Then we can measure the joint angle rotation through the rotary encoder and the counter board and send it back to PC to have a closed loop control system.

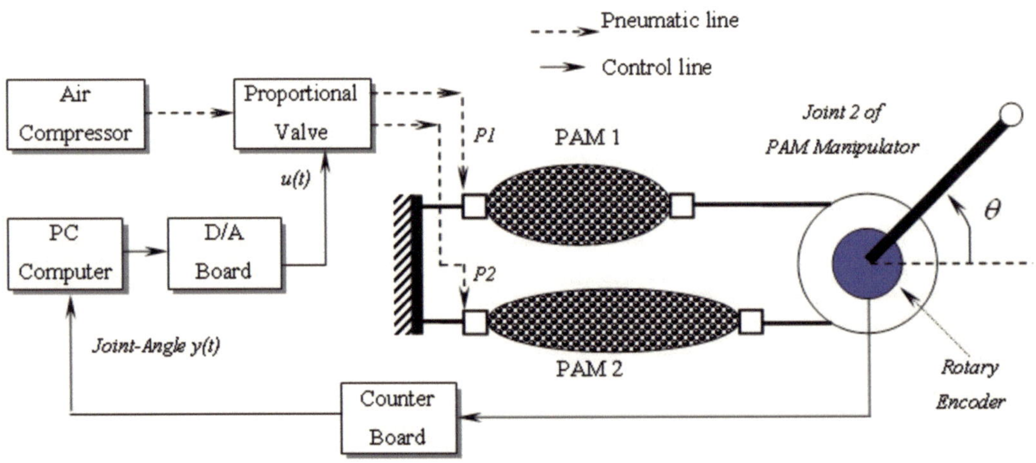

Fig. 2. Experimental Set-up Configuration of the PAM robot arm.

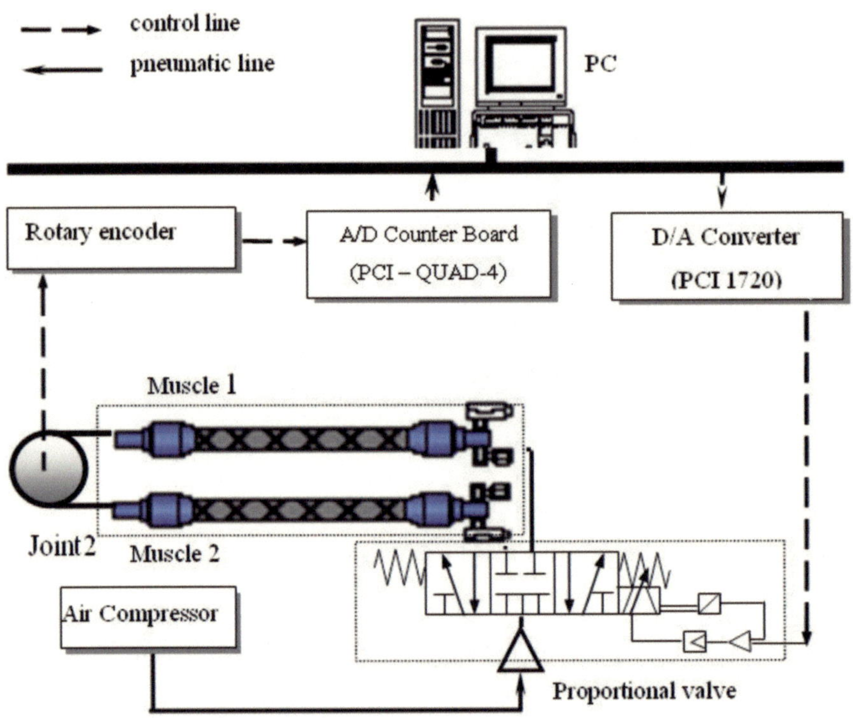

Fig. 3. Schematic diagram of the experimental apparatus.

The experimental apparatus is shown in Fig.3. The hardware includes an IBM compatible PC (Pentium 1.7 GHz) which sends the control voltage signal *u(t)* to control the proportional valve (FESTO, MPYE-5-1/8HF-710B), through a D/A board (ADVANTECH, PCI 1720 card) which change digital signal from PC to analog voltage *u(t)*. The rotating torque is generated by the pneumatic pressure difference supplied from air-compressor between the antagonistic artificial muscles. Consequently, the 2nd joint of PAM manipulator will be rotated. The joint angle, θ [deg], is detected by a rotary encoder (METRONIX, H40-8-3600ZO) with a resolution of 0.1[deg] and fed back to the computer through an 32-bit counter board (COMPUTING MEASUREMENT, PCI QUAD-4 card) which changes digital pulse signals to joint angle value *y(t)*. The external inertia load could be changed from 0.5[kg] to 2[kg], which is a 400 (%) change with respect to the minimum inertia load condition. The experiments are conducted under the pressure of 4[bar] and all control software is coded in MATLAB-SIMULINK with C-mex S-function.

Table 1 presents the configuration of the hardware set-up installed from Fig.2 and Fig.3 as to control of the 2nd joint of the PAM manipulator using the novel proposed online tuning Gain Scheduling DNN-PID control algorithm.

No.	Name	Model name	Company
1	Proportional valve	MPYE-5-1/8HF-710 B	FESTO
2	Pneumatic artificial muscle (x2)	MAS-10-N-220-AA-MCFK	FESTO
3	D/A board	PCI 1720	ADVANTECH
4	A/D board	PCI QUAD-4	COMPUTING MEASUREMENT
5	Rotary encoder	H40-8-3600ZO	METRONIX

Table 1. Lists of the experimental hardware set-up.

2.2. Controller design

The structure of the newly proposed online tuning Gain Scheduling DNN-PID control algorithm using neural network is shown in Fig. 4. This control algorithm is a new one and has the characteristics such as simple structure and little computation time, compared with the previous neural network controller using auto-tuning method (Ahn K.K., Thanh T.D.C., 2005). This system with the set point filter and controller using neural network can solve the problems, which were mentioned in the introduction and is also useful for the PAM manipulator with nonlinearity properties.

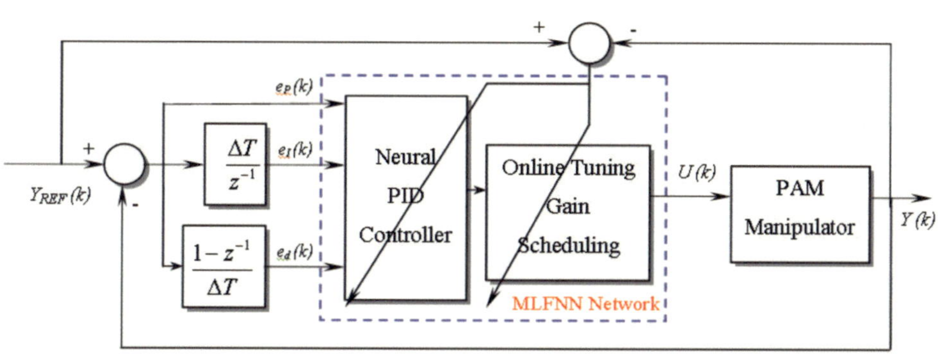

Fig. 4. Block diagram of proposed online tuning gain scheduling DNN-PID position control system.

The block diagram of proposed online tuning Gain Scheduling DNN-PID control based on Multi-Layer Feed-Forward Neural Network (MLFNN) composed of three layers is shown in Figure 5.

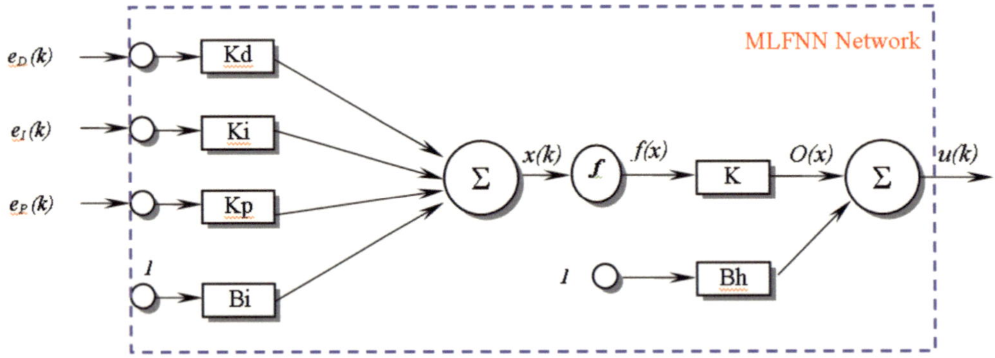

Fig. 5. Structure of MLFNN network system used in proposed online tuning DNN-PID controller.

The structure of the newly proposed online tuning Gain Scheduling DNN-PID control algorithm using Multi-Layer Feed-forward Neural Network (MLFNN) is shown in Fig.5. This control algorithm is a new one and has the characteristics such as simple structure, little computation time and more robust control, compared with the previous neural network controller using auto-tuning method (Ahn K.K., Thanh T.D.C., 2005).
From Figures 4 and 5, a control input u applied to the 2nd joints of the 2-axes PAM manipulator can be obtained from the following equation.

$$u = K\, f(x) + B_h \tag{1}$$

with x is input of Hyperbolic Tangent function $f(.)$ which is presented in Equation (2), K and B_h are the bias weighting values of input layer and hidden layer respectively. The Hyperbolic Tangent function $f(.)$ has a nonlinear relationship as explained in the following equation.

$$f(x) = \frac{\left(1 - e^{-x}\right)}{\left(1 + e^{-x}\right)} \tag{2}$$

The block diagram of proposed online tuning Gain Scheduling DNN-PID control based on Multi-Layer Feed-Forward Neural Network (MLFNN) composed of three layers is shown in Figure 5. In this figure, K, K_p, K_i and K_d, are scheduling, proportional, integral and derivative gain while e_p, e_i and e_d are system error between desired set-point output and output of joint of the PAM manipulator, integral of the system error and the difference of the system error, respectively.
MLFNN network is trained online by the fast learning back propagation (FLBP) algorithm as to minimize the system error between desired set-point output and output of joint of the PAM manipulator.

From Figure 5, the input signal of the Hyperbolic Tangent function *f(.)* becomes

$$\begin{aligned} x(k) &= K_p(k)e_p(k) + K_i(k)e_i(k) + K_d(k)e_d(k) + B_i(k) \\ O(k) &= f(x(k)) \\ u(k) &= K(k)O(k) + B_h(k) \end{aligned} \tag{3}$$

with

$$\begin{aligned} e_p(k) &= y_{REF}(k) - y(k) \\ e_i(k) &= e_p(k).\Delta T \\ e_d(k) &= \frac{e_p(k)(1 - z^{-1})}{\Delta T} \end{aligned} \tag{4}$$

ΔT is the sampling time, *z* is the operator of Z-Transform, *k* is the discrete sequence, $y_{REF}(k)$ and $y(k)$ are the desired set-point output and output of joint of the PAM manipulator. Furthermore, B_i, K_p, K_i and K_d are weighting values of Input layer and B_h *and* K are weighting values of Hidden layer. These weighting values will be tuned online by fast learning back propagation (FLBP) algorithm.
As to online tuning the gain scheduling *K* and PID parameters K_p, K_i and K_d, the gradient descent method used in FLBP learning algorithm using the following equations were applied.

$$\begin{aligned} K(k+1) &= K(k) - \eta \frac{\partial E(k)}{\partial K} \\ K_p(k+1) &= K_p(k) - \eta_p \frac{\partial E(k)}{\partial K_p} \\ K_i(k+1) &= K_i(k) - \eta_i \frac{\partial E(k)}{\partial K_i} \\ K_d(k+1) &= K_d(k) - \eta_d \frac{\partial E(k)}{\partial K_d} \end{aligned} \tag{5}$$

and the Bias weighting values $B_i(k)$ and $B_h(k)$ are updated as follows:

$$\begin{aligned} B_i(k+1) &= B_i(k) - \eta_{Bi} \frac{\partial E(k)}{\partial B_i} \\ B_h(k+1) &= B_h(k) - \eta_{Bh} \frac{\partial E(k)}{\partial B_h} \end{aligned} \tag{6}$$

where η, η_p, η_i, η_d, η_{Bi} and η_{Bh} are learning rate values determining the convergence speed of updated weighting values; E(k) is the error defined by the gradient descent method as follows

$$E(k)=\frac{1}{2}\left(y_{REF}(k)-y(k)\right)^2 \tag{7}$$

Apply the chain rule with equation 5 and 6, it leads to

$$\frac{\partial E(k)}{\partial K}=\frac{\partial E(k)}{\partial y}\frac{\partial y(k)}{\partial u}\frac{\partial u(k)}{\partial K}$$

$$\frac{\partial E(k)}{\partial K_p}=\frac{\partial E(k)}{\partial y}\frac{\partial y(k)}{\partial u}\frac{\partial u(k)}{\partial O}\frac{\partial O(k)}{\partial x}\frac{\partial x(k)}{\partial K_p} \tag{8}$$

$$\frac{\partial E(k)}{\partial K_i}=\frac{\partial E(k)}{\partial y}\frac{\partial y(k)}{\partial u}\frac{\partial u(k)}{\partial O}\frac{\partial O(k)}{\partial x}\frac{\partial x(k)}{\partial K_i}$$

$$\frac{\partial E(k)}{\partial K_d}=\frac{\partial E(k)}{\partial y}\frac{\partial y(k)}{\partial u}\frac{\partial u(k)}{\partial O}\frac{\partial O(k)}{\partial x}\frac{\partial x(k)}{\partial K_d}$$

And

$$\frac{\partial E(k)}{\partial B_i}=\frac{\partial E(k)}{\partial y}\frac{\partial y(k)}{\partial u}\frac{\partial u(k)}{\partial O}\frac{\partial O(k)}{\partial x}\frac{\partial x(k)}{\partial B_i}$$

$$\frac{\partial E(k)}{\partial B_h}=\frac{\partial E(k)}{\partial y}\frac{\partial y(k)}{\partial u}\frac{\partial u(k)}{\partial B_h} \tag{9}$$

From equations 1, 3 and 6, the following equations can be derived

$$\begin{aligned}
&\frac{\partial E(k)}{\partial y}=-\left(y_{REF}(k)-y(k)\right)=-e_p(k)\\
&\frac{\partial y(k)}{\partial u}\approx\frac{\Delta y}{\Delta u}=\frac{\left(y(k)-y(k-1)\right)}{\left(u(k)-u(k-1)\right)}=\Delta\\
&\frac{\partial u(k)}{\partial O}=K\\
&\frac{\partial O(k)}{\partial x}=f'\left(x(k)\right)\\
&\frac{\partial x(k)}{\partial B_i}=1;\frac{\partial x(k)}{\partial K_p}=e_p(k);\frac{\partial x(k)}{\partial K_i}=e_i(k);\frac{\partial x(k)}{\partial K_d}=e_d(k)\\
&\frac{\partial u(k)}{\partial K}=O(k);\frac{\partial u(k)}{\partial B_h}=1
\end{aligned} \tag{10}$$

From equations 8, 9 and 10, the following resulting equations can be derived

$$\frac{\partial E(k)}{\partial K}=\frac{\partial E(k)}{\partial y}\frac{\partial y(k)}{\partial u}\frac{\partial u(k)}{\partial K}=-e_p(k)\Delta O(k)$$

$$\frac{\partial E(k)}{\partial K_p}=\frac{\partial E(k)}{\partial y}\frac{\partial y(k)}{\partial u}\frac{\partial u(k)}{\partial O}\frac{\partial O(k)}{\partial x}\frac{\partial x(k)}{\partial K_p}=-e_p(k)\Delta Kf'(x(k))e_p(k)=-\Delta Kf'(x)e_p^2(k) \quad (11)$$

$$\frac{\partial E(k)}{\partial K_i}=\frac{\partial E(k)}{\partial y}\frac{\partial y(k)}{\partial u}\frac{\partial u(k)}{\partial O}\frac{\partial O(k)}{\partial x}\frac{\partial x(k)}{\partial K_i}=-e_p(k)\Delta Kf'(x(k))e_i(k)=-\Delta Kf'(x)e_p(k)e_i(k)$$

$$\frac{\partial E(k)}{\partial K_d}=\frac{\partial E(k)}{\partial y}\frac{\partial y(k)}{\partial u}\frac{\partial u(k)}{\partial O}\frac{\partial O(k)}{\partial x}\frac{\partial x(k)}{\partial K_d}=-e_p(k)\Delta Kf'(x(k))e_d(k)=-\Delta Kf'(x)e_p(k)e_d(k)$$

and

$$\frac{\partial E(k)}{\partial B_i}=\frac{\partial E(k)}{\partial y}\frac{\partial y(k)}{\partial u}\frac{\partial u(k)}{\partial O}\frac{\partial O(k)}{\partial x}\frac{\partial x(k)}{\partial B_i}=-e_p(k)\Delta Kf'(x(k))1=-\Delta Kf'(x)e_p(k)$$

$$\frac{\partial E(k)}{\partial B_h}=\frac{\partial E(k)}{\partial y}\frac{\partial y(k)}{\partial u}\frac{\partial u(k)}{\partial B_h}=-e_p(k)\Delta 1=-\Delta e_p(k) \quad (12)$$

and with

$$f'(x)=2.\frac{e^{-x}}{\left(1+e^{-x}\right)^2} \quad (13)$$

From equation 5 and 6, the final equations for online tuning gain scheduling K and PID parameters K_p, K_i and K_d are expressed as follows:

$$\begin{aligned}
K(k+1) &= K(k)+\eta .e_p(k)\Delta .O(k)\\
K_p(k+1) &= K_p(k)+\eta_p .e_p^2(k)\Delta K.\frac{2e^{-x}}{\left(1+e^{-x}\right)^2}\\
K_i(k+1) &= K_i(k)+\eta_i .e_p(k)e_i(k)\Delta K\,\frac{2e^{-x}}{\left(1+e^{-x}\right)^2}\\
K_d(k+1) &= K_d(k)+\eta_d .e_p(k)e_i(k)\Delta K\,\frac{2e^{-x}}{\left(1+e^{-x}\right)^2}
\end{aligned} \quad (14)$$

and the Bias weighting values $B_i(k)$ and $B_h(k)$ are updated as follows:

$$B_i(k+1) = B_i(k) + \eta_{Bi}.e_p(k)\Delta K \frac{2e^{-x}}{\left(1+e^{-x}\right)^2}$$
$$B_h(k+1) = B_h(k) + \eta_{Bh} e_p(k)\Delta \tag{15}$$

3. Experimental Results

The performance of proposed online tuning gain scheduling DNN-PID control scheme is verified on joint angle position control of the 2nd joint of the 2-axes PAM robot arm. Fig.3 and Fig.4 describes the working diagram of this control scheme.

Fig.6 presents the experiment SIMULINK diagram of proposed online tuning DNN-PID control algorithm run in Real-time Windows Target with DYNAMIC_NEURAL_PID being subsystems written in C then compiled and run in real-time C-mex file. Three initial PID parameters K_P, K_I, K_D and gain scheduling G value are chosen by trial and error method and determined as G=0.8; K_P=0.09; K_I=0.089 and K_D=0.07.

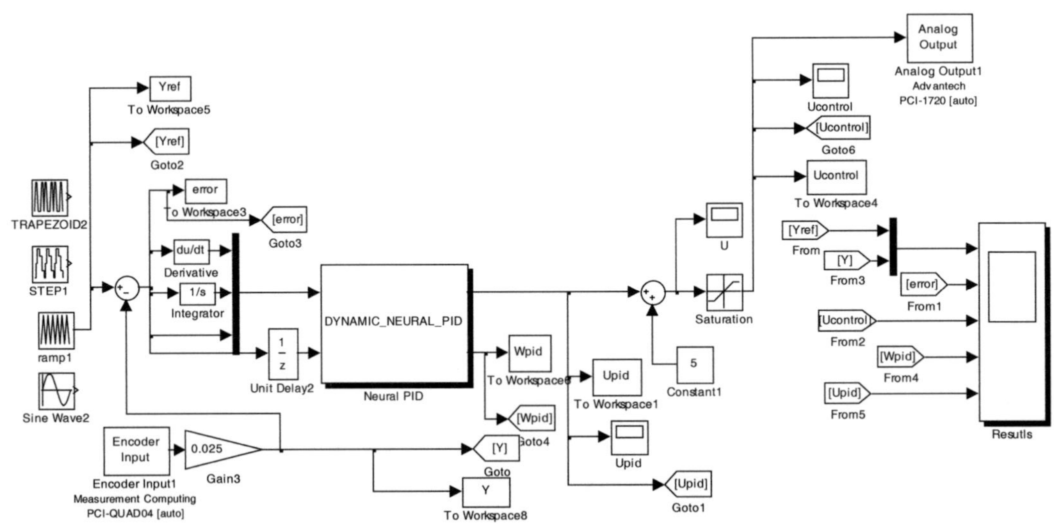

Fig. 6. Experiment SIMULINK Model of PAM robot arm control using online tuning DNN-PID control.

The experiment SIMULINK diagram of the 2nd joint of the 2-axes PAM robot arm position control using conventional PID controller in order to compare as to demonstrate the superiority of proposed control system. Three PID parameters K_P, K_I, K_D and gain scheduling K value of conventional PID controller are chosen by trial and error method.

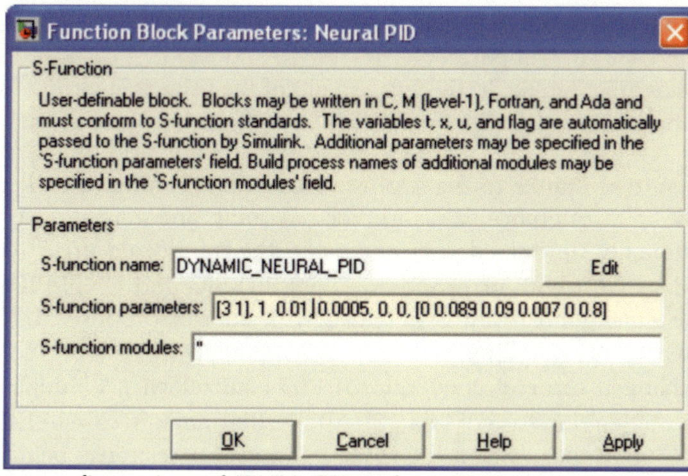

Fig. 7. Parameter configuration of DNN_PID subsystem used in proposed online tuning DNN-PID control

Fig. 7 shows that the parameter configuration of DYNAMIC_NEURAL_PID subsystem composes of seven parameters. The 1st vector parameter contains number of inputs and outputs of neural DYNAMIC_NEURAL_PID subsystem; the 2nd relates to the number of neurons of Hidden Layer used; the 3rd declares the step size used in real-time operation of PAM system; the 4th declares the learning rate value used in real-time operation of PAM manipulator; the 5th parameter contains logic value as to choose the sigmoid function (1) or the hyperbolic tangent function (0); the 6th parameter contains logic value as to choose the linear function (0) or the sigmoid/hyperbolic tangent function (1) of Output layer; and the 7th vector parameter contains the initial K, K_P, K_I, K_D weighting values and two initial bias weighting values B_i and B_h.

The final purpose of the PAM manipulator is to be used as an elbow and wrist rehabilitation robot device. Thus, the experiments were carried out with respect to 3 different waveforms as reference input (Triangular, Trapezoidal and Sinusoidal reference) with 2 different end-point Payloads (Load 0.5[kg] and Load 2[kg]) as to demonstrate the performance of novel proposed online tuning DNN-PID controller. Furthermore, the comparisons of control performance between the conventional PID and two different methods of the proposed online tuning DNN-PID controller were performed.

These two novel proposed methods compose of proposed online tuning DNN-PID-SIG and proposed online tuning DNN-PID-HYP. The 1st method possesses the activation function of hidden layer of DYNAMIC_NEURAL_PID subsystem being Sigmoid function and the 2nd method corresponds to the Hyperbolic Tangent function respectively.

The initial gain scheduling value G and PID controller parameters K_p, K_i and K_d were set to be $G = 0.8$, $Kp = 0.089$, $Ki = 0.09$, $Kd = 0.07$. These parameters of PID controller were obtained by trial-and-error through experiment. Forwardly, the two initial bias weighting values B_i and B_h are chosen equal 0.

First, the experiments were carried out to verify the effectiveness of the proposed online tuning DNN-PID controller using neural network with triangular reference input. Fig.8a shows the experimental results between the conventional PID controller and the proposed nonlinear DNN-PID controller in 2 cases of Load 0.5[kg] and Load 2[kg] respectively. The online updating of each control parameter (*G, Kp, Ki* and *Kd*) in 2 cases of Load 0.5[kg] and Load 2[kg] was shown in Fig. 8b. In the experiment of the proposed online tuning DNN-PID controller, the initial values of *G, Kp, Ki* and *Kd* are set to be the same as that of conventional PID controller.

These figures show that thanks to the sophisticated online tuning of *G, Kp, Ki* and *Kd,* the error between desired reference y_{REF} and actual joint angle response y of the PAM manipulator continually optimized. Consequently, the minimized error decreases only in the range $\pm$0.5[deg] with both of proposed DNN-PID-SIG and DNN-PID-HYP in case of Load 0.5[kg]. The same good result is obtained with both of proposed DNN-PID-SIG and DNN-PID-HYP in case of Load 2[kg]. These results are really impressive in comparison with the bad and unchanged error of conventional PID controller ($\pm$1.5[deg] in case of Load 0.5[kg] and up to $\pm$2[deg] in case of Load 2[kg]). Furthermore, in case of Load 2[kg], Figure 8a shows that PID controller caused the PAM manipulator response being oscillatory and unstable. Otherwise, proposed online tuning DNN-PID controller continues to assert robust control to keep PAM manipulator response stable and accurate tracking.

In comparison between proposed DNN-PID-SIG and DNN-PID-HYP, both of proposed control algorithms obtain the excellent robustness and accuracy as well and thus are considered the performance equivalent. However in initial stage, proposed DNN-PID-SIG possesses significant overshoot which may cause unstable to PAM manipulator in its initial operation.

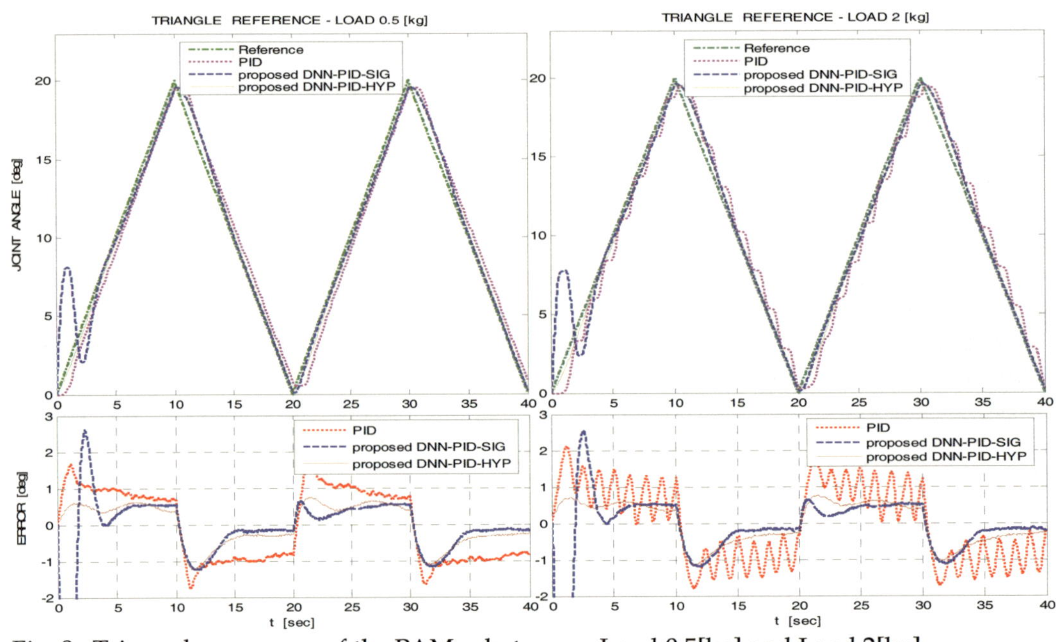

Fig. 8a.Triangular response of the PAM robot arm – Load 0.5[kg] and Load 2[kg].

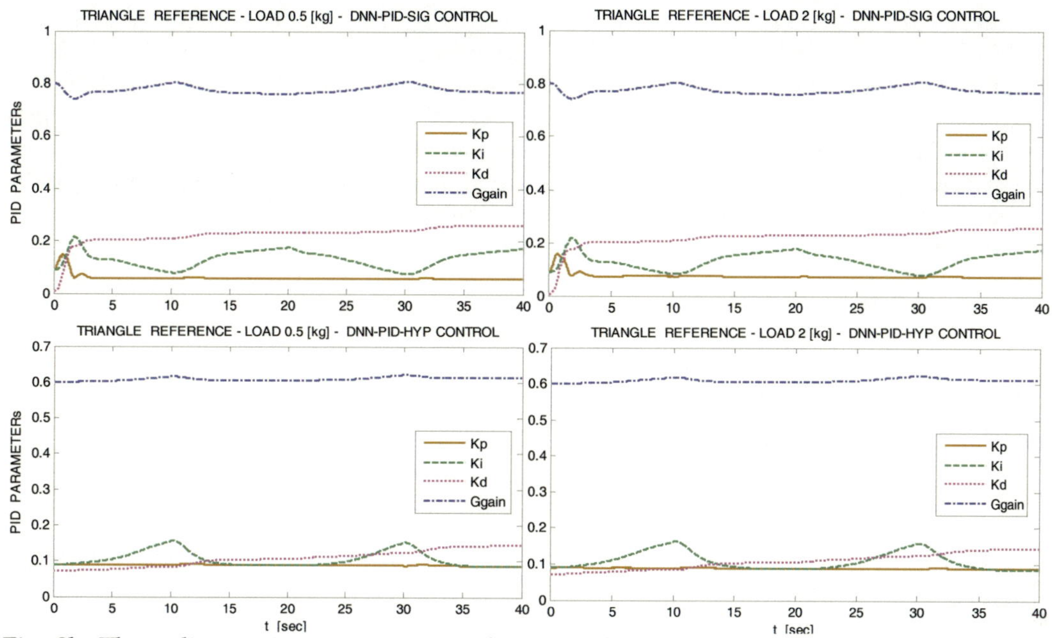

Fig. 8b. The online tuning convergence of proposed DNN-PID controller parameters with triangular reference.

Figure 8c shows the resulted shape of control voltage *U* applied to the joint of PAM manipulator, which is generated by the proposed online tuning DNN-PID controller as to assure the performance and the accuracy of the PAM manipulator response. This figure shows that PID controller generates an oscillatory and unstable control voltage in case of Load 2[kg]. On the contrary, proposed online tuning DNN-PID controller continues to robustly control with refined control voltage as to keep PAM manipulator response stable and accurate tracking.

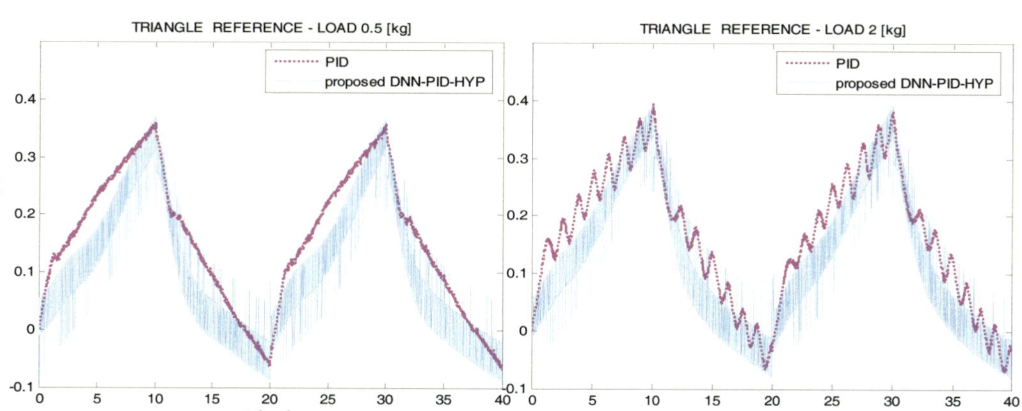

Fig. 8c.The voltage control applied to the PAM robot arm with triangular reference.

Forwardly, the experiments were carried out to verify the effectiveness of the proposed DNN-PID controller using neural network with trapezoidal reference input. Fig.9a shows the experimental results in comparison between the conventional PID controller and the two proposed nonlinear DNN-PID-SIG and DNN-PID-HYP controllers in 2 cases of Load 0.5[kg] and Load 2[kg] respectively. The online updating of each control parameter (*G, Kp, Ki* and *Kd*) in 2 cases of Load 0.5[kg] and Load 2[kg] was shown in Fig. 9b. In the experiment of the proposed online tuning DNN-PID controller, the initial values of *G, Kp, Ki* and *Kd* are set to be the same as that of conventional PID controller.

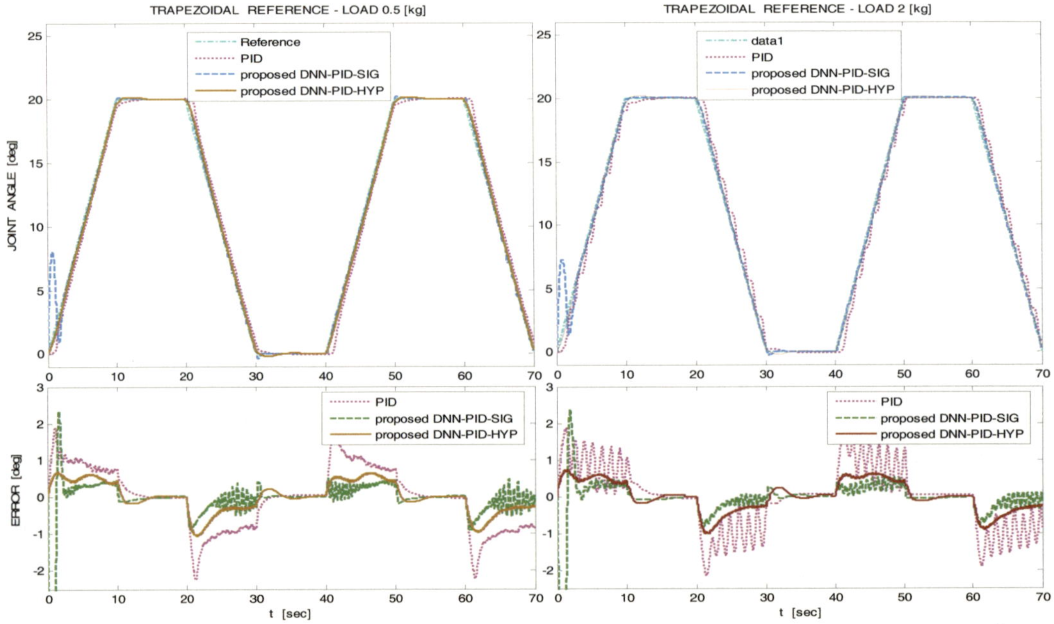

Fig. 9a.Trapezoidal response of both joints of the 2-axes PAM robot arm - Load 0.5[kg] and Load 2[kg].

These figures show that thanks to the refined online tuning of *G, Kp, Ki* and *Kd,* the error between desired reference y_{REF} and actual joint angle response y of the PAM manipulator continually optimized. Consequently, the minimized error decreases only in the range $\pm$ 0.5[deg] with both of proposed DNN-PID-SIG and DNN-PID-HYP in case of Load 0.5[kg]. The same good result is also obtained with both of proposed DNN-PID-SIG and DNN-PID-HYP in case of Load 2[kg]. These results are really superioir in comparison with the passive and unchanged error of conventional PID controller ($\pm$ 2[deg] in case of Load 0.5[kg] and up to $\pm$ 2.2[deg] in case of Load 2[kg]). Furthermore, in case of Load 2[kg], Figure 9a shows that PID controller caused the PAM manipulator response being oscillatory and unstable. On the contrary, proposed online tuning DNN-PID controller continues to assert robust control to keep PAM manipulator response stable and accurate tracking.

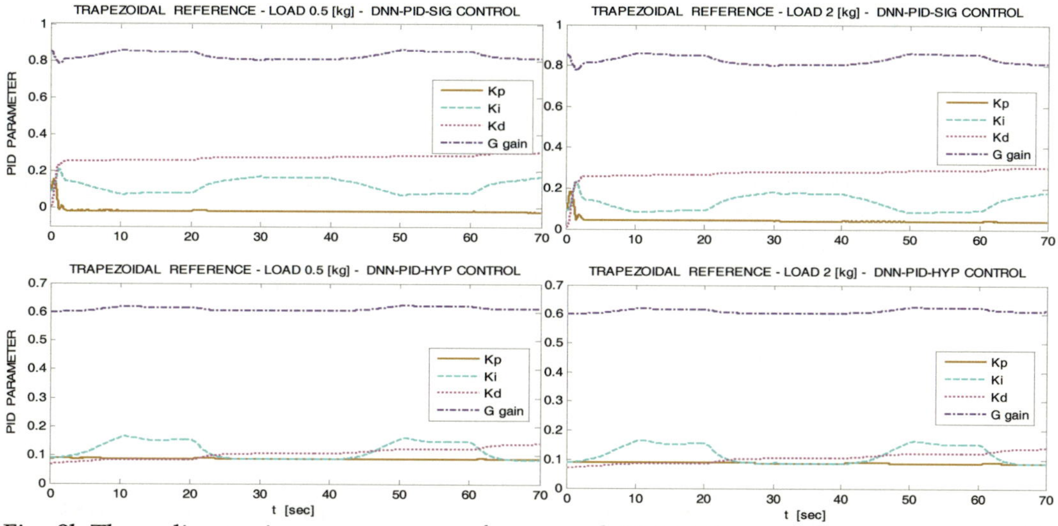

Fig. 9b.The online tuning convergence of proposed DNN-PID controller parameters with triangular reference.

Figure 9c presents the refined shape of control voltage *U* applied to the joint of PAM manipulator, which is generated by the proposed online tuning DNN-PID controller as to assure the performance and the accuracy of the PAM manipulator response. This figure proves that PID controller generates an oscillatory and unstable control voltage in case of Load 2[kg]. On the contrary, proposed online tuning DNN-PID controller continues to robustly control with refined control voltage as to keep PAM manipulator response stable and accurate tracking.

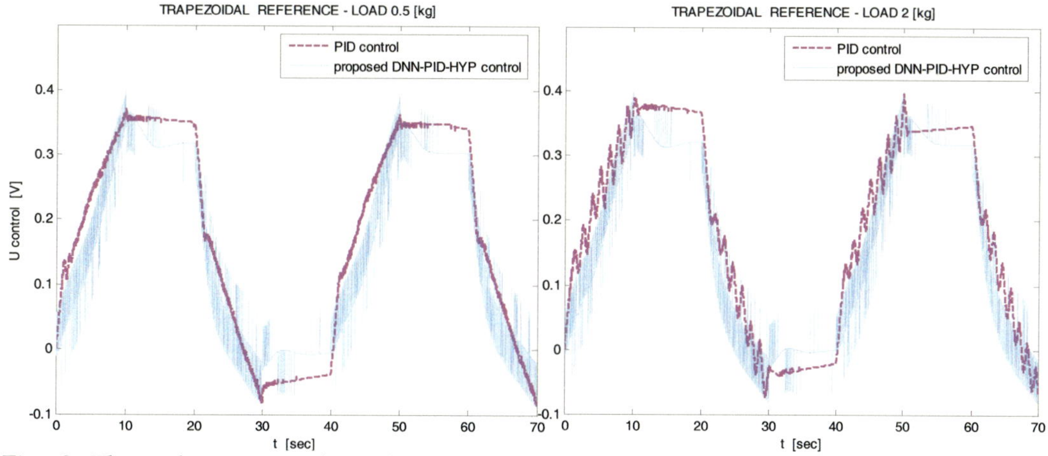

Fig. 9c.The voltage control applied to both joints of the 2-axes PAM robot arm with triangular reference.

Next, the experiments were carried out to verify the effectiveness of the proposed DNN-PID controller using neural network with sinusoidal reference 0.05[Hz]. Fig.10a shows the experimental results in comparison between the conventional PID controller and the two proposed DNN-PID-SIG and DNN-PID-HYP controllers in 2 cases of Load 0.5[kg] and Load 2[kg] respectively. The online tuning of each control parameter (*G, Kp, Ki* and *Kd*) in 2 cases of Load 0.5[kg] and Load 2[kg] was shown in Fig. 10b.

These figures show that thanks to the refined online tuning of *G, Kp, Ki* and *Kd,* the error between desired reference y_{REF} and actual joint angle response y of the PAM manipulator continually optimized. Consequently, the minimized error decreases excellently in the range $\pm$ 1[deg] with proposed DNN-PID-HYP and in the range $\pm$ 1.5[deg] with proposed DNN-PID-SIG in case of Load 0.5[kg]. The same good result is also obtained with proposed DNN-PID-SIG and DNN-PID-HYP in case of Load 2[kg]. These results are really superior in comparison with the passive and unchanged error of conventional PID controller ($\pm$ 3[deg] in case of Load 0.5[kg] and up to $\pm$ 4[deg] in case of Load 2[kg]). Furthermore, in case of Load 2[kg], Figure 10a shows that PID controller caused the PAM manipulator response oscillatory and unstable. Otherwise, proposed online tuning DNN-PID controller continues to keep robust control as to maintain PAM manipulator response stable and accurate tracking.

In comparison between proposed DNN-PID-SIG and DNN-PID-HYP, proposed DNN-PID-HYP obtains the excellent robustness and accuracy in comparison with proposed DNN-PID-SIG and thus the proposed DNN-PID-HYP controller is considered to possess the best performance. Furthermore, in initial stage, proposed DNN-PID-SIG possesses again significant overshoot which may cause unstable to PAM manipulator in its initial operation.

Figure 10c depicts the refined control voltage *U* applied to the joint of PAM manipulator, which is generated by the proposed online tuning DNN-PID controller as to assure the performance and the accuracy of the PAM manipulator response.

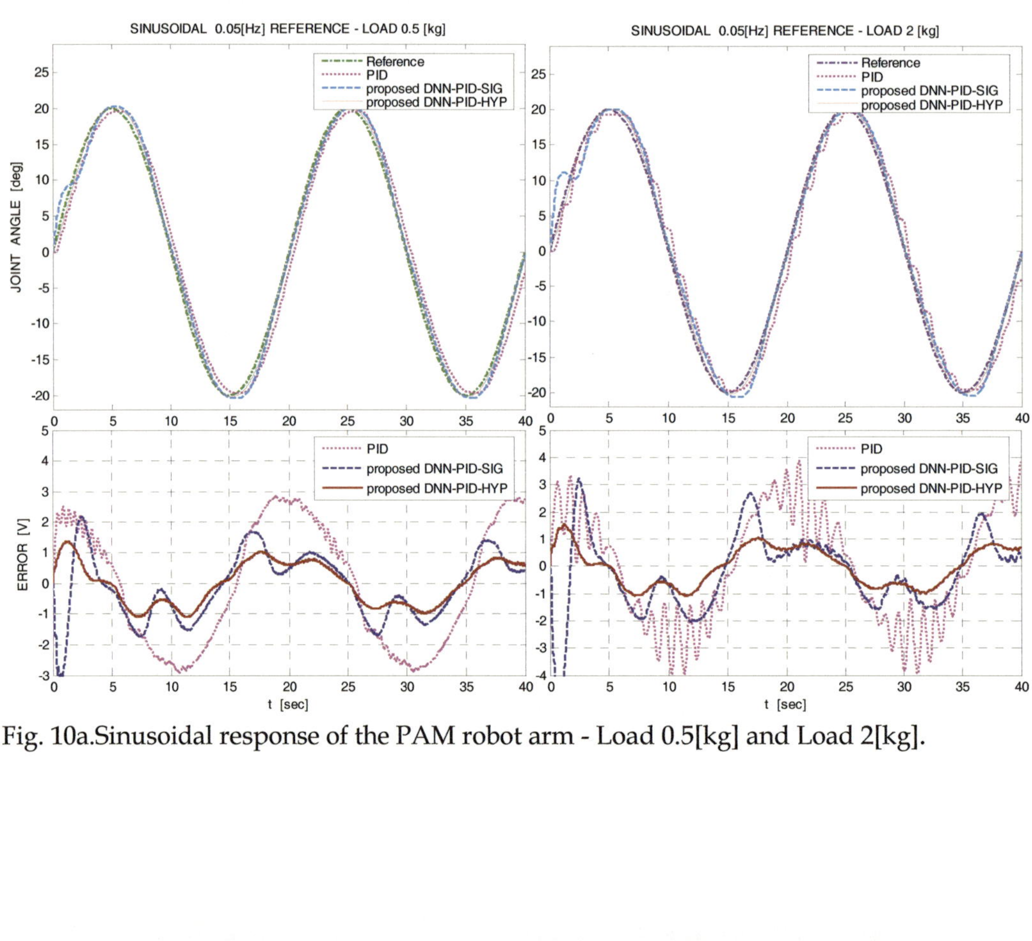

Fig. 10a.Sinusoidal response of the PAM robot arm - Load 0.5[kg] and Load 2[kg].

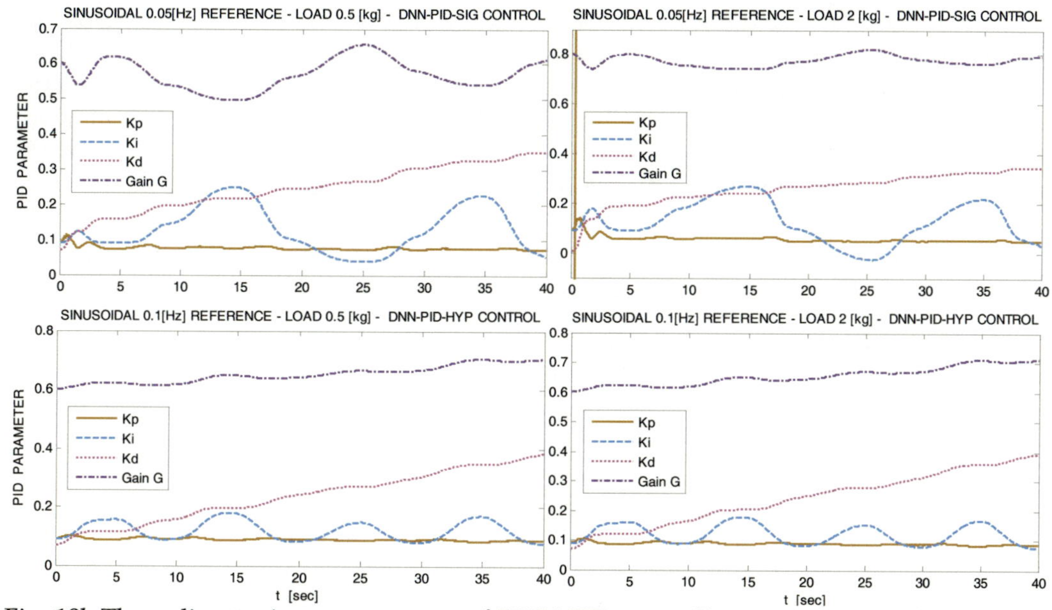

Fig. 10b.The online tuning convergence of DNN-PID controller parameters with sinusoidal reference.

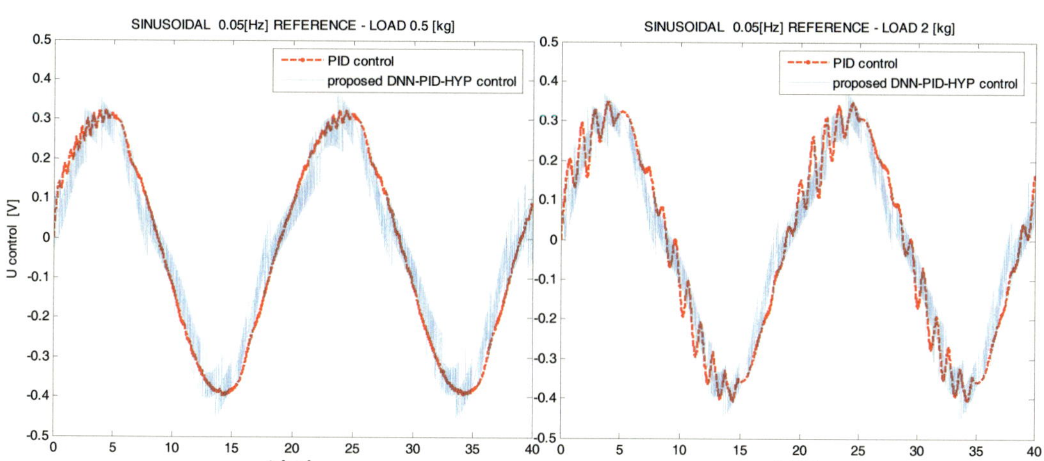

Fig. 10c.The voltage control applied to the 2nd joint of the 2-axes PAM robot arm with sinusoidal reference.

Finally, the experiments were carried out with critical sinusoidal reference input 0.2[Hz]. Fig.11a shows the experimental results in comparison between the two proposed DNN-PID-SIG and DNN-PID-HYP controllers in 2 cases of Load 0.5[kg] and Load 2[kg] respectively. The online tuning of each control parameter (*G*, *Kp*, *Ki* and *Kd*) in 2 cases of Load 0.5[kg] and Load 2[kg] was shown in Fig. 11b. It's important to note that PID controller is impossible to apply with critical sinusoidal reference input 0.2[Hz] because it caused uncontrollable and unstable as well to the operation of PAM manipulator.

These figures show that thanks to the refined online tuning of *G, Kp, Ki* and *Kd,* the error between desired reference y_{REF} and actual joint angle response y of the PAM manipulator continually optimized. Consequently, the minimized error decreases spectacularly in the range $\pm$ 1[deg] with proposed DNN-PID-HYP in case of Load 2[kg] and in the range $\pm$ 1.5[deg] with proposed DNN-PID-SIG in case of Load 0.5[kg]. In critical sinusoidal reference input 0.2[Hz], proposed online tuning DNN-PID controller continues to keep robust control as to maintain PAM manipulator response stable and accurate tracking.

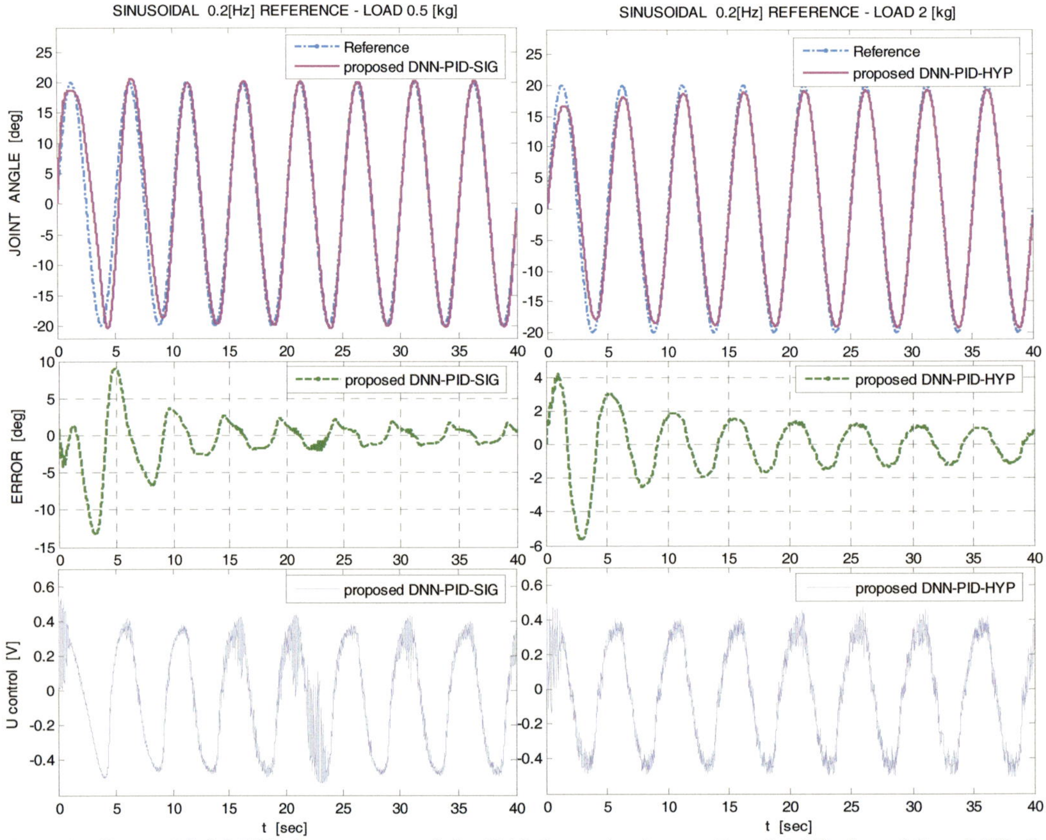

Fig. 11a Sinusoidal 0.2[Hz] response of the PAM manipulator - Load 0.5[kg] and Load 2[kg].

In comparison between proposed DNN-PID-SIG and DNN-PID-HYP, in this case of critical sinusoidal reference input 0.2[Hz], proposed DNN-PID-HYP once more obtains the excellent robustness and accuracy in comparison with proposed DNN-PID-SIG and thus the proposed DNN-PID-HYP controller is considered to possess the best performance between them. Furthermore, in initial stage, proposed DNN-PID-SIG possesses again significant overshoot which may cause unstable to PAM manipulator in its initial operation.

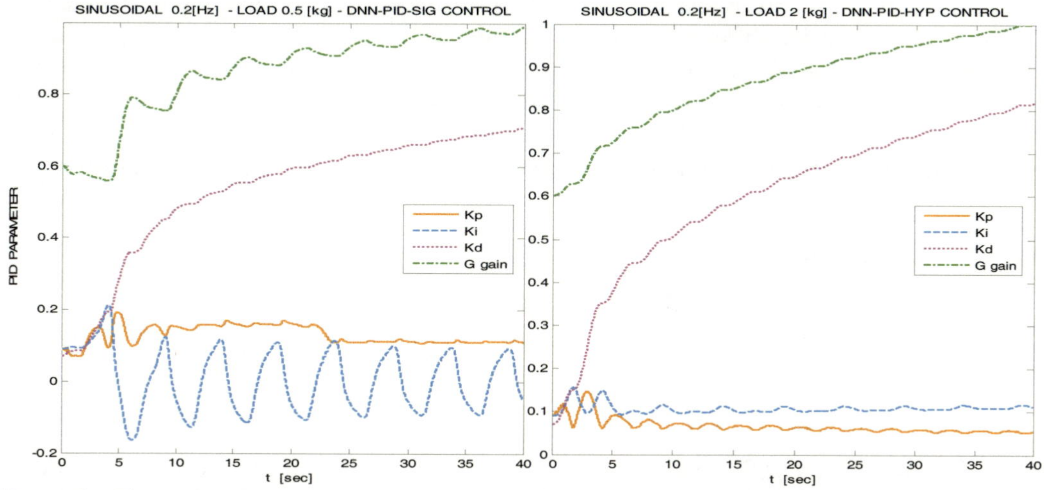

Fig. 11b. The online tuning convergence of proposed DNN-PID controller parameters in case of sinusoidal reference.

In summary, novel DNN-PID controller using neural network was investigated in this paper. It has shown that the proposed method had a good control performance for the highly nonlinear system, such as the PAM manipulator. The controller had an adaptive control capability and the control parameters were optimized via the back propagation algorithm. The controller designed by this method does not need any training procedure in advance, but it uses only the input and output of the plant for the adaptation of proposed control parameters and can tune these parameters online iteratively. From the experiments of the position control of the PAM manipulator, it was verified that the proposed control algorithm presented in this paper was online control with simple structure and had better dynamic property, strong robustness and it was suitable for the control of various plants, including linear and nonlinear process, compared to the conventional PID controller. In comparison between 2 proposed DNN-PID-SIG and DNN-PID-HYP control algorithms, based on experiment results, proposed DNN-PID-HYP control obtains the excellent robustness and accuracy in comparison with proposed DNN-PID-SIG and thus the proposed DNN-PID-HYP controller is considered to possess the better performance than the proposed DNN-PID-SIG one.

4. Conclusions

An innovative online tuning gain scheduling neural DNN-PID Controller suitable for real-time human-friendly industrial applications has been designed, developed and implemented for position control the joint angle of the experimental PAM manipulator in this paper. Experiment results show that the proposed online tuning Gain Scheduling DNN-PID controller is able to learn the nonlinear and dynamic characteristics of the PAM manipulator quickly and thus reduce the tracking error to nearly zero in its operation. The performance of the online tuning gain scheduling DNN-PID controller was found to be very good and robust in the presence of external disturbances. Furthermore, with this proposed online tuning DNN-PID control algorithm, gain scheduling value G and PID parameters K_p,

K_i and K_d can be modified in real time and actual trajectories can be monitored as well. This facilitates testing under different input conditions and ensures future applications of the PAM manipulator as a rehabilitation device for stroke patients. It determines confidently that the proposed online tuning Gain Scheduling DNN-PID controller not only proves its superb performance in control the highly nonlinear PAM manipulator but also would be very efficient in control of other real-time industrial and human-friendly applications.

Acknowledgements

This research was supported by the DCSELAB, Viet Nam National University Ho Chi Minh City (VNU-HCM), Viet Nam.

5. References

Ahn K.K., Anh H.P.H., 2006. System modeling and identification of the two-link pneumatic artificial muscle (PAM) manipulator optimized with genetic algorithm. In: Proceedings of the 2006 IEEE-ICASE Int. Conf., Busan, Korea, pp. 356–61.

Ahn K.K., Anh H.P.H., 2007b. A new approach of modeling and identification of the pneumatic artificial muscle (PAM) manipulator based on recurrent neural network. In Proc. IMechE, Part I: Journal of Systems and Control Engineering, 2007, 221(I8), 1101-1122.

Ahn K.K., Anh H.P.H., 2009. Identification of the pneumatic artificial muscle manipulators by MGA-based nonlinear NARX fuzzy model. In MECHATRONICS, 2009, Volume 19, Issue 1, pp. 106-133.

Ahn K.K., Thanh T.D.C., 2005. Nonlinear PID control to improve the control performance of PAM manipulators using neural network. In KSME, Int., Jour., 19(1):pp.106~15.

Balasubramanian K, Rattan K.S., 2003a. Fuzzy logic control of a pneumatic muscle system using a linearizing control scheme. In: Proceedings of Int. Conf., North American Fuzzy Information Processing Society, pp. 432-6.

Chan S.W., Lilly J., Berlin J.E., May 2003. Fuzzy PD+I learning control for a pneumatic muscle. In: Proceedings of IEEE Int. Conf. Fuzzy Systems, St. Louis, MO, pp. 278–83.

Hesselroth T, Sarkar K, Van der Smagt P, Schulten K., 1994. Neural network control of a pneumatic robot arm. IEEE Trans. System Man Cybernetics 24(1): pp.28–38.

Lilly J., Sep. 2003.Adaptive tracking for pneumatic muscle actuators in bicep and tricep configurations. IEEE Trans. Neural Syst. Rehabil. Eng. 11(3):pp.333–9.

Lilly J.H., Chang X., Sep.2003. Tracking control of a pneumatic muscle by an evolutionary fuzzy controller. In IEEE Intell. Automat. Soft Comput., 9(3):pp. 227–44.

Medrano-Cerda G.A., Bowler C.J., Caldwell D.G., Aug. 1995. Adaptive position control of antagonistic pneumatic muscle actuators. In: Proceedings of IEEE Int. Conf. Intelligent Robots and Systems, Pittsburgh, PA, pp. 378–83.

Nelles O., Nonlinear system identification, Springer, 2000.

Repperger D.W., Johnson K.R., Phillips C.A., 1998. VSC position tracking system involving a large scale pneumatic muscle actuator. In: Proceedings of IEEE Conf. Decision Control, Tampa, FL, Dec. pp. 4302–7.

Repperger D.W., Phillips C.A., Krier M., Aug. 1999. Controller design involving gain scheduling for a large scale pneumatic muscle actuator," In: Proceedings of IEEE Conf. Control Applications, Kohala Coast, HI, pp. 285–90.

Reynolds D.B., Repperger D.W., Phillips C.A., Bandry G., 2003. Dynamic characteristics of pneumatic muscle. In IEEE Ann. Biomed. Eng., 31(3):pp.310–7.

Tsagarakis N, Darwin G.C., 2000. Improved modeling and assessment of pneumatic muscle actuators. In: Proceedings of IEEE Int. Conf. Robotics and Automation, San Francisco, CA, pp. 3641–6.